Bruno Apolloni, Ashish Ghosh, Ferda Alpaslan, Lakhmi C. Jain,
Srikanta Patnaik (Eds.)

Machine Learning and Robot Perception

Studies in Computational Intelligence, Volume 7

Editor-in-chief
Prof. Janusz Kacprzyk
Systems Research Institute
Polish Academy of Sciences
ul. Newelska 6
01-447 Warsaw
Poland
E-mail: kacprzyk@ibspan.waw.pl

Further volumes of this series
can be found on our homepage:
springeronline.com

Vol. 1. Tetsuya Hoya
Artificial Mind System – Kernel Memory Approach, 2005
ISBN 3-540-26072-2

Vol. 2. Saman K. Halgamuge, Lipo Wang (Eds.)
Computational Intelligence for Modelling and Prediction, 2005
ISBN 3-540-26071-4

Vol. 3. Bożena Kostek
Perception-Based Data Processing in Acoustics, 2005
ISBN 3-540-25729-2

Vol. 4. Saman Halgamuge, Lipo Wang (Eds.)
Classification and Clustering for Knowledge Discovery, 2005
ISBN 3-540-26073-0

Vol. 5. Da Ruan, Guoqing Chen, Etienne E. Kerre, Geert Wets (Eds.)
Intelligent Data Mining, 2005
ISBN 3-540-26256-3

Vol. 6. Tsau Young Lin, Setsuo Ohsuga, Churn-Jung Liau, Xiaohua Hu, Shusaku Tsumoto (Eds.)
Foundations of Data Mining and Knowledge Discovery, 2005
ISBN 3-540-26257-1

Vol. 7. Bruno Apolloni, Ashish Ghosh, Ferda Alpaslan, Lakhmi C. Jain, Srikanta Patnaik (Eds.)
Machine Learning and Robot Perception, 2005
ISBN 3-540-26549-X

Bruno Apolloni
Ashish Ghosh
Ferda Alpaslan
Lakhmi C. Jain
Srikanta Patnaik
(Eds.)

Machine Learning and Robot Perception

Springer

Professor Bruno Apolloni
Department of Information Science
University of Milan
Via Comelico 39/41
20135 Milan
Italy
E-mail: apolloni@dsi.unimi.it

Professor Lakhmi C. Jain
School of Electrical & Info Engineering
University of South Australia
Knowledge-Based Intelligent Engineering
Mawson Lakes Campus
5095 Adelaide, SA
Australia
E-mail: lakhmi.jain@unisa.edu.au

Professor Ashish Ghosh
Machine Intelligence Unit
Indian Statistical Institute
203 Barrackpore Trunk Road
Kolkata 700108
India
E-mail: ash@isical.ac.in

Professor Srikanta Patnaik
Department of Information
and Communication Technology
F. M. University
Vyasa Vihar
Balasore-756019
Orissa, India
E-mail: patnaik_srikanta@yahoo.co.in

Professor Ferda Alpaslan
Faculty of Engineering
Department of Computer Engineering
Middle East Technical University - METU
06531 Ankara
Turkey
E-mail: alpaslan@ceng.metu.edu.tr

Library of Congress Control Number: 2005929885

ISSN print edition: 1860-949X
ISSN electronic edition: 1860-9503
ISBN 978-3-642-06586-6   ISBN 978-3-540-32409-6 (eBook)

Springer is a part of Springer Science+Business Media
springeronline.com

Softcover reprint of the hardcover 1st edition 2005

Typesetting: by the authors and TechBooks using a Springer LaTeX macro package

Printed on acid-free paper   SPIN: 11504634   89/TechBooks   5 4 3 2 1 0

Preface

This book presents some of the most recent research results in the area of machine learning and robot perception. The book contains eight chapters.

The first chapter describes a general-purpose deformable model based object detection system in which evolutionary algorithms are used for both object search and object learning. Although the proposed system can handle 3D objects, some particularizations have been made to reduce computational time for real applications. The system is tested using real indoor and outdoor images. Field experiments have proven the robustness of the system for illumination conditions and perspective deformation of objects. The natural application environments of the system are predicted to be useful for big public and industrial buildings (factories, stores), and outdoor environments with well-defined landmarks such as streets and roads.

Fabrication of space-variant sensor and implementation of vision algorithms on space-variant images is a challenging issue as the spatial neighbourhood connectivity is complex. The lack of shape invariance under translation also complicates image understanding. The retino-cortical mapping models as well as the state-of-the-art of the space-variant sensors are reviewed to provide a better understanding of foveated vision systems in Chapter 2. It is argued that almost all the low level vision problems (i.e., shape from shading, optical flow, stereo disparity, corner detection, surface interpolation etc.) in the deterministic framework can be addressed using the techniques discussed in this chapter. The vision system must be able to determine where to point its high-resolution fovea. A proper mechanism is expected to enhance image understanding by strategically directing fovea to points which are most likely to yield important information.

In Chapter 3 a discrete wavelet based model identification method has been proposed in order to solve the online learning problem. The method minimizes the least square residual parameter estimation in noisy environments. It offers significant advantages over the classical least square estimation methods as it does not need prior statistical knowledge of measurement of noises. This claim is supported by the experimental results on estimating the mass and length of a nonholonomic cart having a wide range of applications in complex and dynamic environments.

Chapter 4 proposes a reinforcement learning algorithm which allows a mobile robot to learn simple skills. The neural network architecture works with continuous input and output spaces, has a good resistance to forget previously learned actions and learns quickly. Nodes of the input layer are allocated dynamically. The proposed reinforcement learning algorithm has been tested on an autonomous mobile robot in order to learn simple skills showing good results. Finally the learnt simple skills are combined to successfully perform more complex skills called *visual approaching* and go to *goal avoiding obstacles*.

In Chapter 5 the authors present a simple but efficient approach to object tracking combining active contour framework and the optical-flow based motion estimation. Both curve evolution and polygon evolution models are utilized to carry out the tracking. No prior shape model assumptions on targets are made. They also did not make any assumption like static camera as is widely employed by other object tracking methods. A motion detection step can also be added to this framework for detecting the presence of multiple moving targets in the scene.

Chapter 6 presents the state-of-the-art for constructing geometrically and photometrically correct 3D models of real-world objects using range and intensity images. Various surface properties that cause difficulties in range data acquisition include specular surfaces, highly absorptive surfaces, translucent surfaces and transparent surfaces. A recently developed new range imaging method takes into account of the effects of mutual reflections, thus providing a way to construct accurate 3D models. The demand for constructing 3D models of various objects has been steadily growing and we can naturally predict that it will continue to grow in the future.

Systems that visually track human motion fall into three basic categories: analysis-synthesis, recursive systems, and statistical methods including particle filtering and Bayesian networks. Each of these methods has its uses. In Chapter 7 the authors describe a computer vision system called DYNA that employs a three-dimensional, physics-based model of the human body and a completely recursive architecture with no bottom-up processes. The system is complex but it illustrates how careful modeling can improve robustness and open the door to very subtle analysis of human motion. Not all interface systems require this level of subtlety, but the key elements of the DYNA architecture can be tuned to the application. Every level of processing in the DYNA framework takes advantage of the constraints implied by the embodiment of the observed human. Higher level processes take advantage of these constraints explicitly while lower level processes gain the advantage of the distilled body knowledge in the form of predicted probability densities.

Chapter 8 advocates the concept of user modelling which involves dialogue strategies. The proposed method allows dialogue strategies to be determined by maximizing mutual expectations of the pay-off matrix. The authors validated the proposed method using iterative prisoner's dilemma problem that is usually used for modelling social relationships based on reciprocal altruism. Their results suggest that in principle the proposed dialogue strategy should be implemented to achieve maximum mutual expectation and uncertainty reduction regarding pay-offs for others.

We are grateful to the authors and the reviewers for their valuable contributions. We appreciate the assistance of Feng-Hsing Wang during the evolution phase of this book.

June 2005

Bruno Apolloni
Ashish Ghosh
Ferda Alpaslan
Lakhmi C. Jain
Srikanta Patnaik

Table of Contents

1 Learning Visual Landmarks for Mobile Robot Topological Navigation

Mario Mata[1], Jose Maria Armingol[2], Arturo de la Escalera[2]

1. Computer Architecture and Automation Department, Universidad Europea de Madrid, 28670 Villaviciosa de Odon, Madrid, Spain. mmata@uem.es
2. Systems Engineering and Automation Department. Universidad Carlos III de Madrid, 28911 Leganés, Madrid, Spain. {armingol,escalera}@ing.uc3m.es

1.1 Introduction

Relevant progress has been done, within the Robotics field, in mechanical systems, actuators, control and planning. This fact, allows a wide application of industrial robots, where manipulator arms, Cartesian robots, etc., widely outcomes human capacity. However, the achievement of a robust and reliable autonomous mobile robot, with ability to evolve and accomplish general tasks in unconstrained environments, is still far from accomplishment. This is due, mainly, because autonomous mobile robots suffer the limitations of nowadays perception systems. A robot has to perceive its environment in order to interact (move, find and manipulate objects, etc.) with it. Perception allows making an internal representation (model) of the environment, which has to be used for moving, avoiding collision, finding its position and its way to the target, and finding objects to manipulate them. Without a sufficient environment perception, the robot simply can't make any secure displacement or interaction, even with extremely efficient motion or planning systems. The more unstructured an environment is, the most dependent the robot is on its sensorial system. The success of industrial robotics relies on rigidly controlled and planned environments, and a total control over robot's position in every moment. But as the environment structure degree decreases, robot capacity gets limited.

Some kind of model environment has to be used to incorporate perceptions and taking control decisions. Historically, most mobile robots are based on a geometrical environment representation for navigation tasks. This facilitates path planning and reduces dependency on sensorial system, but forces to continuously monitor robot's exact position, and needs precise

M. Mata et al.: *Learning Visual Landmarks for Mobile Robot Topological Navigation*, Studies in Computational Intelligence (SCI) **7**, 1–55 (2005)
www.springerlink.com

environment modeling. The navigation problem is solved with odometry-relocalization, or with an external absolute localization system, but only in highly structured environments. Nevertheless, the human beings use a topological environment representation to achieve their amazing autonomous capacity. Here, environment is sparsely modeled by a series of identifiable objects or places and the spatial relations between them. Resultant models are suitable to be learned, instead of hard-coded. This is well suited for open and dynamic environments, but has a greater dependency on the perception system. Computer Vision is the most powerful and flexible sensor family available at the present moment. The combination of topological environment modeling and vision is the most promising selection for future autonomous robots. This implies the need for developing visual perception systems able to learn from the environment.

Following these issues, a new learning visual perception system for robots is presented in this chapter based on a generic landmark detection and recognition system. Here, a landmark is a localized physical feature that the robot can sense and use to estimate its own position in relation to some kind of "map" that contains the landmark's relative position and/or other mark characterization. It is able to learn and use nearly any kind of landmark on structured and unstructured environments. It uses deformable models as the basic representation of landmarks, and genetic algorithms to search them in the model space. Deformable models have been studied in image analysis through the last decade, and are used for detection and recognition of flexible or rigid templates under diverse viewing conditions. Instead of receiving the model definition from the user, our system extracts, and learns, the information from the objects automatically. Both 2D and 3D models have been studied, although only 2D models have been tested on a mobile robot. One of the major contributions of this work is that the visual system is able to work with any 2D (or nearly 2D) landmark. This system is not specifically developed for only one object. In the experiments carried out, several different landmarks have been learnt. Two of these have been tested in a mobile robot navigation application, employing the same searching algorithm: an artificial landmark (green circles placed on the walls) and a natural landmarks (office's nameplates attached at the entrance of each room), shown in Fig. 1.1.a and Fig. 1.1.b. All of them have been automatically learnt by the system, with very little human intervention (only several training images, with the landmarks to learn marked, must be provided).

The deformable model carries the landmark information inside it, so this information is adapted to the model's deformation and can be used to evaluate the model fitness. This is achieved using a genetic algorithm, where each individual represents a deformed model. The population then explores thc image during its evolution. The genetic search algorithm is able to handle landmark's perspective deformation problems. The second relevant aspect is the system capacity for reading text or icons inside landmarks designed for human use, such as those shown in Fig. 1.2, so the system can be used to find and read signs, panels and icons in both indoor and outdoor environments. This allows the robot to make high-level decisions, and results in a higher degree of integration of mobile robotics in everyday life. Various experimental results in real environments have been done, showing the effectiveness and capacity of landmark learning, detection and reading system. These experiments are high-level topological navigation tasks. Room identification from inside, without any initialization, is achieved through its landmark signature. Room search along a corridor is done by reading the content of room nameplates placed around for human use; this allows the robot to take high-level decisions, and results in a higher integration degree of mobile robotics in real life. Finally, although the presented system is being tested for mobile robot topological navigation, it is general enough for its direct use in a wide range of applications, such as geometric navigation, inspection and surveillance systems, etc.

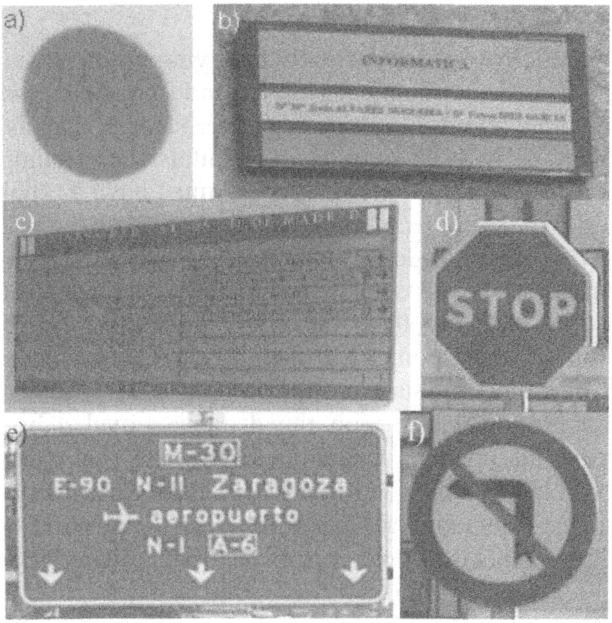

Fig. 1.1. Some of the landmarks learned

The structure of this chapter is, following this introduction, a brief state of the art concerning actual work on mobile robot navigation. Then an overview about deformable models, and how they are used in the core of the landmark learning and recognition system, is described. It is followed by introducing how to learn new landmark's parameters; after that, the landmark detection system structure is presented. Once the system is described, its application to a mobile robot and several experimental results are presented, and also a practical study of the system's limitations. The chapter concludes with the relevant conclusions and future work.

Fig. 1.2. Landmarks with iconic information used for topological navigation

1.2 State of the Art

Autonomous mobile robots are currently receiving an increasing attention as well in the scientific community as in the industry. Mobile robots have many potential applications in routine or dangerous task such as operations in a nuclear plant, delivery of supplies in hospitals and cleaning of offices and houses [30]. A mobile autonomous robot must have a reliable navigation system for avoiding objects in its path and recognizing important objects of the environment to identify places in order to understand the surrounding environment. A prerequisite for geometric navigation of a mobile robot is a position-finding method. Odometry is the most used localization method for mobile robot geometrical navigation. The problem is that the accumulation of small measure errors will cause large position errors, which increase proportionally with the distance traveled by the robot. Wheel slippage and unequal wheel diameters are the most important source of error [11]. As a mobile robot moves through its environment, its actual position and orientation always differ from the position and orientation that it is commanded to hold. Errors accumulate and the localization uncertainty increases over time.

An alternative approach is topological navigation. It allows overcoming some of the classical problems of geometric navigation in mobile robots, such as simultaneously reducing the uncertainty of localization and of perception of the environment [42]. On the other hand, topological navigation is heavily dependent on a powerful perception system to identify elements of the environment. Chosen elements for recognition, or landmarks, should be simple enough to allow an easy identification from different view angles and distances.

Visual recognition is the problem of determining the identity and position of a physical element from an image projection of it. This problem is difficult in practical real-life situations because of uncontrolled illumination, distances and view angles to the landmarks. Machine learning techniques are being applied with remarkable success to several problems of computer vision and perception [45]. Most of these applications have been fairly simple in nature and still can not handle real-time requirements [8, 31, 37]. The difficulty with scaling up to complex tasks is that inductive learning methods require a very large number of training patterns in order to generalize correctly from high density sensor information (such as video cameras). However, recent results in mobile robot learning have demonstrated that robots can learn simple objects to identify from very little initial knowledge in restricted environments [9, 21, 23, 33].

There are two major approaches in the use of landmarks for topological navigation in related literature. One approach uses as landmarks regions of the environment that can be recognized later, although they are not a single object. Colin and Crowley [12] have developed a visual recognition technique in which objects are represented by families of surfaces in a local appearance space. In [4] a spatial navigation system based on visual templates is presented; templates are created by selecting a number of high-contrast features in the image and storing them together with their relative spatial location. Argamon [2] describes a place recognition method for mobile robots based on image signature matching. Thompson and Zelinsky [47] present a method for representing places using a set of visual landmarks from a panoramic sensor, allowing an accurate local positioning. [19] has developed a vision based system for topological navigation in open environments. This system represents selected places by local 360° views of the surrounding scenes. The second approach uses objects of the environment as landmarks, with perception algorithms designed specifically for each object. In [10] a system for topologically localizing a mobile robot using color histogram matching of omni directional images is presented. In [44], images are encoded as a set of visual features. Potential landmarks are detected using an attention mechanism implemented as a measure of uniqueness. [6] describes a series of motor and perceptual behaviors used for indoor navigation of mobile robots; walls, doors and corridors are used as landmarks. In [27] an indoor navigation system is proposed, including the teaching of its environment; the localization of the vehicle is done by detecting fluorescent tubes with a camera. However, there are still few practical implementations of perceptual systems for topological navigation.

1.3 Deformable Models

Much work has been done in visual-based general object detection systems in the last decades, with encouraging results, but only a few systems have been used in practice, within uncontrolled real-world scenes. Furthermore, most of the systems are based on hand-made object representations and searching rules which difficult system adaptability. There is a need for general and practical object detection systems that can be adapted to different applications quick and easily. This need for practical systems inexorably leads to some restrictions, usually opposed to generality requirements:

1. *Computation time* cannot exceed usability limits. Although the proposed system is general enough for handling general 3D objects, time restrictions obligates to particularize for planar objects, or single faces of 3D objects. However, the system is designed for, and can be easily extended to, 3D object detection if desired.

2. *Flexibility* and *generality* points toward general systems which can learn and use new objects with minimal human intervention.

3. *Robustness* is encouraged by the learning ability. No learning can take place without a certain evaluation of its performance.

The proposed particularized system maintains enough generality to cope with the detection of nearly any planar object in cluttered, uncontrolled real images, in useful times, by only software means. It uses a simple but effective representation objects by means of deformable models, and is easily adaptable to detect new objects by training from images, with mini mal human intervention (only marking the object to learn in the training images).

1.3.1 Related Work

Deformable models have been intensively studied in image analysis through the last decade [13, 55], and are used for detection and recognition of flexible or rigid models under various viewing conditions [7]. They have been applied for querying a database given the object shape, color and texture [54]; motion-based segmentation of deformable structures undergoing nonrigid movements through shape and optical flow [24]; for Intelligent Vehicles, they have been used to detect road signs [7, 17], vehicles [56] and road borders [25]; after the work of [55], they are commonly used for human face detection and tracking [20, 28]; recognizing characters and lineal symbols in handwritten line drawings [49, 50, 52]; in medical imagery they have been used for the segmentation of deep brain nuclei in 3D MRI [39], cell segmentation [29] or human melanoma cancer cells in confocal microscopy imaging [41].

As noted in [14], a global shaped model based image segmentation scheme consists of the following blocks:

1. The initial model, M, a model with a fix area, located in the center of the image.

2. The deformable model M(Z). This model is obtained from de previous one through the deformation parameters, Z. They can be position, horizontal and vertical scale, rotation and additional deformation parameter.

3. The likelihood probability density function, p(I|Z), that means the probability of the deformation set Z occurs in the image I.

4. A search algorithm to find the maximum of the posterior probability p(Z|I).

In a latter stage, if the detected object contains symbolic information – like text or icons-, it is interpreted using an empirically selected neural network-based classifier.

Potential fields of application are mobile robotic (landmarks in navigation tasks), industrial robotic (object detection and handling), driving assistance systems (traffic signs, road informative panels, vehicle detection) and industrial tasks (object detection and inspection, tag reading).

Various works on human cognition points that humans use view-point based object representations rather than object-centered ones [15, 46]. This is the focus used in some approaches to object detection and representation issues, like appearance and combination of views [22, 43, 51]. Model-views of objects are a simple but rich way of representing objects, but it has a major drawback in the sense of object aspect changes with perspective and illumination.

In the proposed system, illumination changes are handled using an adequate color representation system, while perspective-related aspect changes are coped with the use of deformable models.

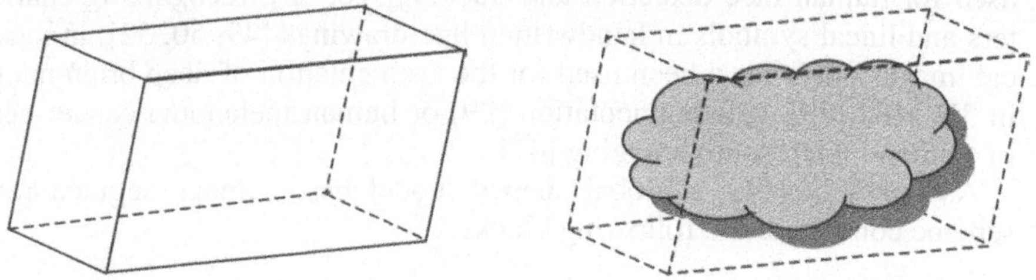

Fig. 1.3. (**a**) Basic deformable model, and (**b**) object-specific added detail

1.3.2 Deformable Model

The proposed deformable model is a very basic geometrical figure, a 3D parallelepiped whose only mission is bounding or enclosing the considered object, independently of its type or shape (Fig. 1.3.a). The geometrical parameters of the deformable model must follow the object aspect changes with perspective. Then, some kind of detail (object-specific data) has to be added over the basic deformable model in order to distinguish one object from another and from the background (Fig. 1.3.b). The only restriction here is that this detail will have to be used in a way that allows following model's deformations. So each object is represented by a set of specific details, which can be "glued" to a general deformable model. The object search is then translated to a search for the deformable model parameters that makes the details to match with the background.

For a practical 2D case, the deformable model needs 6 degrees of freedom (d.o.f.) to follow object translations and rotations, and some perspective deformations, as shown in Fig. 1.4. Object translation in the image is covered by the (X, Y) d.o.f. of Fig. 1.4.a, representing the pixel coordinates of the reference point for the model (the upper left corner). Object scaling (distance from the camera) is handled with the pair (ΔX, ΔY), as shown in Fig. 1.4.b. The parameter α from Fig. 1.4.c manages object rotation. Finally, object skew due to affine perspective deformation is only considered over the vertical axis, heavily predominant in real images; the general skew along the vertical axis can be decomposed as the combination of the basic deformations illustrated in Fig. 1.4.d and Fig. 1.4.e. In practice, only the component in Fig. 1.4.e, measured by the d.o.f. SkY, is frequent; the deformation in Fig. 1.4.d is only dominant for relatively large and narrow objects and when they are at the same vertical level of the optical axis. These simplifications of the perspective distortions could be easily avoided, but they provide a reduction of the number of degrees of freedom considered, saving computation time with little impact on real scenes, as will be shown later.

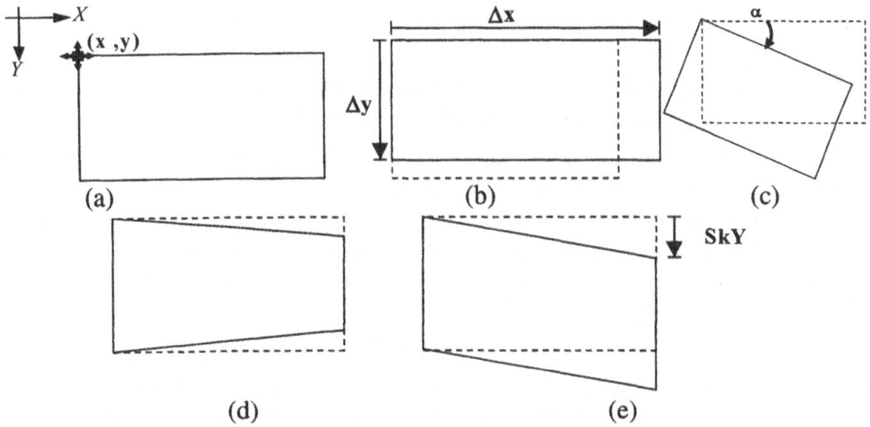

Fig. 1.4. 2D degrees of freedom for the basic deformable model. (**a**) traslation, (**b**)scaling, (**c**)rotation, (**d**)–(**e**) skew by perspective deformation

These six degrees of freedom are valid for planar objects. When considering 3D objects, more degrees of freedom must be added. In the proposed approach, only two new ones are needed, the pair (X', Y') with the pixel coordinates of the frontal side of the 3D deformable model (Fig. 1.5.a), covering object displacements over the plane parallel to the image and rotations over the vertical axis. Rotations that are not covered by αX', Y' can be handled without adding any other d.o.f., simply by allowing the ΔY and

ΔY parameters to be negative. The effect of a negative value of ΔX is shown in Fig. 1.5.b, while a negative ΔY is shown in Fig. 1.5.c.

Of course this set of 8 d.o.f. does not cover precisely all possible perspective deformations of an object, but they allow to approximate them enough to recognize a generic object if adequate added details are used, and provides a reduction of the parameter search space.

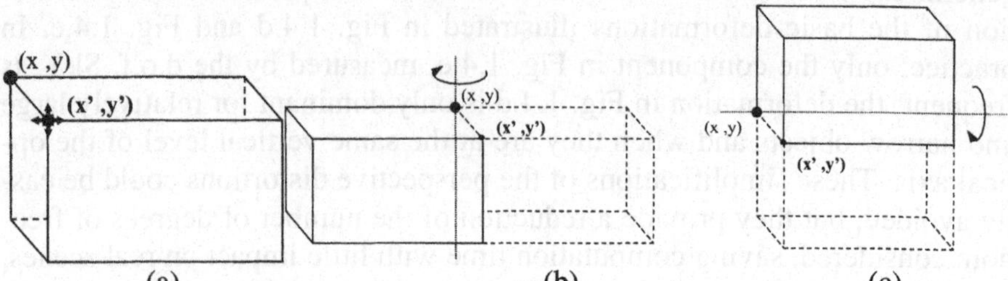

Fig. 1.5. 3D-extension degrees of freedom for the basic deformable model

The detection of the object is now a search process in the model's parameter space, comparing the detail added to the model with the background in the place and with the size and deformation that the parameters determine. Two reasons make this a complex search: the high dimensionality of the search space (8 d.o.f.), and the comparative function between the added detail and the background. This comparative (or cost) function is not a priori predefined, and it can be very complex and not necessarily a parametrical function.

Genetic evolutionary algorithms have shown to be a very useful tool for these kinds of search processes [26, 53]. If the deformable model's geometric parameters are encoded into the genome of the individuals from a genetic algorithm, each individual become a deformable model trying to match the desired object through the image. The fitness function for each individual is the perfect place for doing the matching between the model's added detail and the background (the cost function). A classical genetic algorithm (GA) is used to make the search in model parameter space, with standard binary coding, roulette selection, standard mutation and single-point crossover. Single individual elitism is used to ensure not to loose the best individual. No optimization of the GA code, or evaluation of other GA variants, has been done yet, it is one of the pending tasks to do, so the search still can be speeded up considerably. One consideration has been taken into account for achieving small enough computation times to make the system of practical use: a proper GA initialization is used to speed up the convergence. If the initialization is good enough, GA convergence is extremely quick, as will be shown.

1.3.3 General Scheme

There is a large collection of 2D pattern search techniques in the literature [40]. In this application, a classical technique is used: normalized correlation with an image of the pattern to be found (usually named model). The advantages and drawbacks of this technique are well known. Its strongest drawback is its high sensitivity to pattern aspect changes (mainly size and perspective), which makes this method unpractical in most cases. A two step modified method is proposed for overcoming this problem. First, in a segmentation stage, relevant regions in the image are highlighted; then the regions found (if any) are used for initializing the genetic pattern search process. The main problems when trying to detect objects that humans use as landmarks is perspective deformation and illumination. Object aspect changes in the image with distance and angle of view, and under different illumination conditions. Deformable models are used to handle perspective deformations, and HSL color space and real image-based training cope with illumination.

As an overview, objects are represented as a basic deformable model that encloses the object, plus some specific detail ("glued" to the basic model) to distinguish and recognize objects. Eight degrees of freedom are considered for the deformable model to follow with sufficient approximation all object aspect changes with relative position, distance and perspective. These model parameters are encoded as the genome of individuals from a genetic algorithm's population. Object search is a search in the model parameter space, for the set of parameters that best matches the object-specific detail to the image in the location they determine. This comparison between model's added detail and the background is then the fitness function for the GA individuals (deformed models). The only restriction to the fitness function is that deformed models that better matches the desired object in the image should have associated higher fitness values.

Before starting the GA search, it is a good idea to properly initialize the algorithm, in order to decrease the usually long convergence times of evolutionary algorithms; the way used to select the regions of interest (ROI) can be nearly anything. And once the algorithm has finished, if the object has been found in the image, some useful information must be extracted from it. This working line leads to a three stage structure for the object detection system: initialization, object search, and information extraction, as shown in Fig. 1.6. In order to speed up landmark detection, a three-stage algorithm is used. First, regions of interest (ROI) are extracted. Then, extracted ROI are used to initialize a genetic algorithm (GA) for the landmark search through the image. Each individual of this GA encodes a deformable model. The fitness of an individual is a measure of the matching between the deformed model it encodes and the landmark searched for. Finally, if a landmark is found, symbols are extracted and identified with a classical backpropagation neural network.

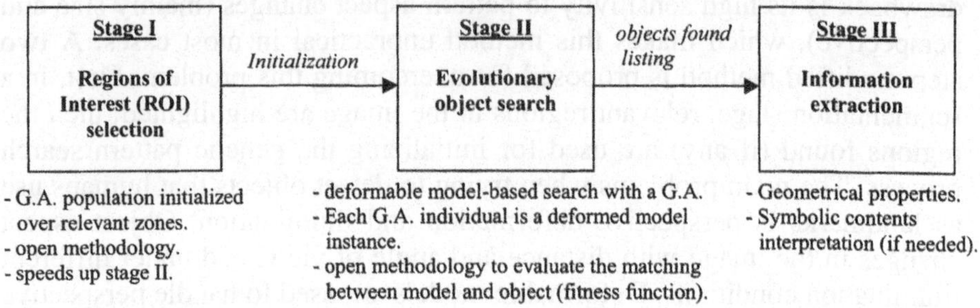

Fig. 1.6. Three stage structure of the proposed object detection system

1.4 Learning Recognition Parameters for New Objects

For the learning process, the human teacher must provide several training images, where the object or landmark to learn is bounded by rectangular boxes (this boxes will be referred to as target boxes in the rest of the chapter). There are no a priori restrictions for the training set provided. However, the wider the conditions this set of images covers (illumination, background, perspective distortions, etc), the best results the learned parameters will achieve on real situations.

As the recognition process, it can be sequentially divided in two steps: candidate hypotheses generation (through color ROI segmentation) and hypotheses verification or rejection (with the genetic search). Consequently, the learning process for a new landmark is also divided in two stages. In the first step, thresholding levels for HSL segmentation are found. The second step is dedicated to determine the location of the correlation pattern-windows inside an individual.

1.4.1 Parameters for ROI Extraction

Any method to segment regions of the image with good probabilities of belonging to the selected object can be used here. After several trials, the use of object's color information to generate regions of interest was decided. Color vision is a powerful tool to handle illumination related aspect changes of the objects in the image. After evaluating different color spaces (RGB, normalized rgb, CIE(L*a*b*) and HSL) , HSL space (Hue, Saturation and Luminance) has been selected as the system base space (Fig. 1.7).

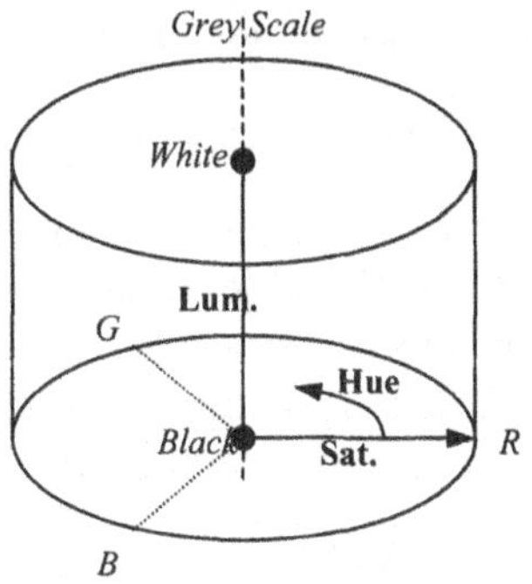

Fig. 1.7. Cylindrical interpretation of HSL space

According with [38], HSL system presents some interesting properties:

1. Hue is closely related to human color sensation as specifies the "perceptual" color property of the considered pixel. Many objects have colors selected to be easily distinguishable by humans, especially those suited to carry symbolic information inside. Furthermore, this component is heavily independent of illumination and shadows.

2. Saturation indicates the "purity" or dominance of one color as it indicates how much of the particular color does the pixel have. Another meaning is how far from gray scale is the pixel because the gray scale, from black to white, has saturation equal to zero (it has the same amount of all colors). This component is somehow insensible to moderate changes of illumination.

3. Luminance takes into account all illumination information of the scene; the L component is the black-and-white version of the scene as it measures the amount of light which has arrived at each pixel.

On the other hand, Hue presents some drawbacks. First, it is an angular component, so the values 0 and 256 are exactly the same (circular continuity); this must be taken into account when segmenting a color interval. Second, Hue is not defined for low or null values of saturation; in these situations, the pixels are achromatic, and Hue can take erratic values. The first issue is easy to overcome segmenting in two steps, but the second one requires a more complex treatment. In this work, the 255 value for Hue is reserved and labeled as achromatic. Hue component is rescaled to 0-254, and pixels having low saturation are set to the achromatic value. For the rest of the processes, when a pixel is set as achromatic, only L component is used for it. Let any HLS color image, sized Xd x Yd pixels, be I(x,y):

$$I(x,y)=(H(x,y),L(x,y),S(x,y)),\quad x\in[0,Xd),y\in[0,Yd) \tag{1}$$

A simple and effective way to generate object-dependant ROI is to select a representative color for the object, and segment image regions having this color. In order to handle intra-class color variations in objects, as well as luminance effects, a representative color interval is learned by the system for each class of objects to detect, defined by

$$CI=(H\pm\Delta H,S\pm\Delta S,L\pm\Delta L) \tag{2}$$

The color segmentation is made in H, S and L components of the image I(x,y) separately, and combining them with an AND logical operation, leading to binary image B(x, y):

$$B(x,y)=\begin{cases} & \{(H-\Delta H)\leq H(x,y)\leq(H+\Delta H)\} \\ 1 & \text{AND}\,\{(L-\Delta L)\leq L(x,y)\leq(L+\Delta L)\} \\ & \text{AND}\,\{(S-\Delta S)\leq S(x,y)\leq(S+\Delta S)\} \\ 0 & \textit{otherwise} \end{cases} \tag{3}$$

Segmentation is done by thresholding in a corrected HLS space followed by some morphologic transformations. In the first training step, the system has to learn the best threshold values for the segmentation of the landmark. Upper and lower thresholds for Hue, Saturation and Luminance components are estimated. This six values (G=5) are made to compose the genome of the individuals of a new GA, used for searching through the training image color space: $[C_0]$=H, $[C_1]$=ΔH, $[C_2]$=L, $[C_3]$=ΔL, $[C_4]$=S, $[C_5]$=ΔS.

The fitness function for this GA must encourage the segmented regions, generated by each individual, to match the target boxes defined in the N_T training images. Each training image $I_T^n(x,y)$, $n\in[0,N_T)$, will contain t_n target boxes A_j^n, $j\in[0,t_n]$. On the other hand, segmented regions outside the target boxes are not desirable. The ideal segmentation result should be a binary black image with the target boxes corresponding zones in white, $B_T^n(x,y)$:

$$B_T^n(x,y)=\begin{cases}1, (x,y)\in\forall A_j^n, j=0,\ldots,t_n-1\\ 0, otherwise\end{cases} \quad (4)$$

This "ideal image" can be matched with the binary image resulting from an individual's genome $[C]_i$, ($B_i^n(x, y, [C]_i)$, calculated with equation (3) with the thresholds carried in $[C]_i$), using a pixel-level XOR logical function. Pixels that survive this operation are misclassified, since they have been included in segmented regions while they do not have to (or the other way around). The number of white pixels after the XOR pass is then a useful measure of the segmentation error for the considered training image. The total segmentation error for one individual is obtained by repeating this operation for all the training image set, and accumulating the misclassified pixels in each image:

$$E([C]_i)=\frac{1}{N_T}\sum_{n=0}^{N_T-1}\left[\sum_{x=0}^{Xd-1}\sum_{y=0}^{Yd-1}\left(B_T^n(x,y)\text{XOR}\,B^n\left(x,y,[C]_i^k\right)\right)\right] \quad (5)$$

The fitness function is then chosen as an inverse function of this total error.

Fig. 1.8. Regions used for LDH calculation

Before the learning search starts, a coarse initialization of the GA is done for decreasing search time. A set of initial threshold H, L and S values are obtained from any of the training images using local histograms. Two sets of histograms for H, L and S are computed from the inner and outer regions adjacent to the target box boundaries (Fig. 1.8). Inner histograms contain information from the object, the background and noise.

Outer histograms contain information from the background, other objects and noise. For each component, the outer histogram is subtracted from the corresponding inner histogram, with negative resulting values forced to zero. The resulting Local Difference Histogram (LDH) will contain only information belonging to the desired object and not present in the outer regions. Initialization values are taken from a peak search over the LDH.

This way several values for H, L and S thresholds are estimated, and their possible combinations generate a good part of the GA's initial population. The rest of the population is randomly generated. This initialization speeds up considerably the training process; training time is of the order of five seconds per training image. Fig. 1.9 shows learned segmentations for different objects.

(a) (b) (c)

Fig. 1.9. Learned segmentation examples: (**a**) pedestrian crossing traffic sign, (**b**) highway informative panel, c) room informative panel

The color interval learning stage makes unnecessary color camera calibration, since thresholds are selected using images captured by the same camera. However, if a new camera needs to be used with an object database learned with a different camera, it is enough to make a coarse color adjustment by any approximate method.

1.4.2 Parameters for Evaluation of the Fitness Function

In order to accomplish practical processing times, one new particularization has been made to the system. Many of everyday objects are planar, or its third dimension is small compared to the other, and many of 3D objects has nearly planar faces that can be considered as a separate planar objects.

Furthermore, certain objects are always seen with the same orientation: objects attached to walls or beams, lying on the floor on or a table, and so on. With these restrictions in mind, it is only necessary to consider five of the eight d.o.f. previously proposed: X, Y, ΔX, ΔY, SkY. This reduction of the deformable model parameter search space increases significantly computation time.

This simplification reduces the applicability of the system to planar objects or faces of 3D objects, but this is not a loose of generality, only a time-reduction operation: issues for implementing the full 3D system will be given along this text. However, many interesting objects for various applications can be managed in despite of the simplification, especially all kind of informative panels.

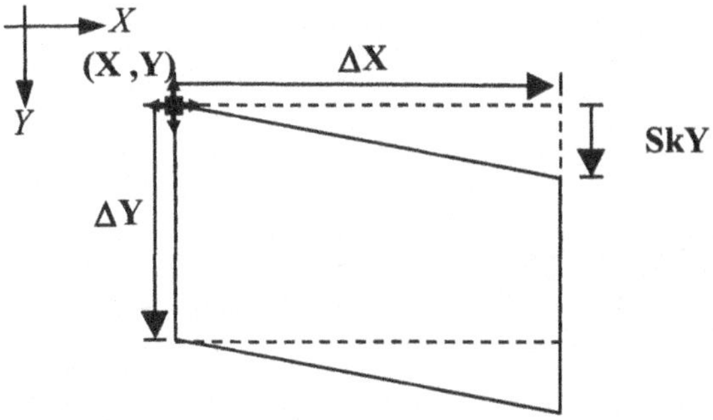

Fig. 1.10. Planar deformable model

The 2D reduced deformable model is shown in Fig. 1.10. Its five parameters are binary coded into any GA individual's genome: the individual's Cartesian coordinates (X, Y) in the image, its horizontal and vertical size in pixels (ΔX, ΔY) and a measure of its vertical perspective distortion (SkY), as shown in equation (4) for the ith individual, with G=5 d.o.f. and q=10 bits per variable (for covering 640 pixels). The variations of these parameters make the deformable model to rover by the image searching for the selected object.

$$[C]_i = \left(\underbrace{\left(b_{11}^i, b_{12}^i, \ldots, b_{1q}^i\right)}_{X_i}; \underbrace{\left(b_{21}^i, b_{22}^i, \ldots, b_{2q}^i\right)}_{Y_i}; \ldots; \underbrace{\left(b_{G1}^i, b_{G2}^i, \ldots, b_{Gq}^i\right)}_{SkY_i} \right) \quad (6)$$

For these d.o.f., a point (x_0, y_0) in model reference frame (no skew, sized ΔX_0, ΔY_0), will have (x, y) coordinates in image coordinate system for a deformed model:

$$\begin{pmatrix} x \\ y \end{pmatrix} = \begin{pmatrix} \frac{\Delta X}{\Delta X_0} & 0 \\ \frac{\Delta X . SkY}{\Delta X_0^2} & \frac{\Delta Y}{\Delta Y_0} \end{pmatrix} \begin{pmatrix} x_0 \\ y_0 \end{pmatrix} + \begin{pmatrix} X \\ Y \end{pmatrix} \quad (7)$$

A fitness function is needed that compares the object-specific detail over the deformed model with the image background. Again nearly any method can be used to do that.

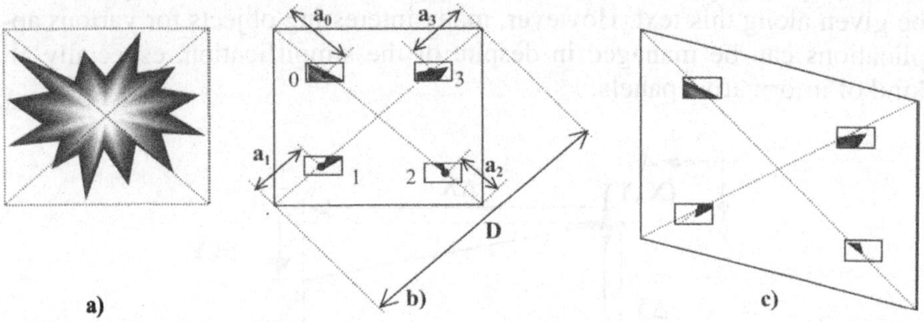

Fig. 1.11. Selected object-specific detail set. (**a**) object to be learned, (**b**) possible locations for the patter-windows, (**c**) memorized pattern-windows following model deformation

Some global detail sets were evaluated: grayscale and color distribution function, and average textureness, but they proved unable to make a precise matching and were excessively attracted by incorrect image zones. Some local detail sets were then evaluated: vertical line detection and corner detection. They proved the opposite effect: several very precise matchings were found, but after a very low convergence speed: it was difficult to get the model exactly aligned over the object, and fitness was low if so. The finally selected detail set is composed of four small size "pattern-windows" that are located at certain learned positions along the model diagonals, as shown in Fig. 1.11.b. These pattern-windows have a size between 10 and 20 pixels, and are memorized by the system during the learning of a new object, at learned distances a_i (i=0,…,3). The relative distances d_i from the corners of the model to the pattern-windows,

$$d_i = a_i / D \quad (8)$$

are memorized together with its corresponding pattern-windows. These relative distances are kept constant during base model deformations in the search stage, so that the position of the pattern-windows follows them, as shown in Fig. 1.11.c, as equation (7) indicates. The pattern-windows will be learned by the system in positions with distinctive local information, such as internal or external borders of the object.

Normalized correlation over the L component (equation 9) is used for comparing the pattern-windows, $M_k(x,y)$, with the image background, L(x,y), in the positions fixed by each individual parameters, for providing an evaluation of the fitness function.

$$r_k(x,y)=\frac{\sum_i\sum_j\left(L(x+i,y+j)-\overline{L}\right).\left(M_k(i,j)-\overline{M_k}\right)}{\sqrt{\sum_i\sum_j\left(L(x+i,y+j)-\overline{L}\right)^2.\sum_i\sum_j\left(M_k(i,j)-\overline{M_k}\right)^2}};\qquad(9)$$

$$\rho_k(x,y)=\max(r_k(x,y),0)^2$$

Normalized correlation makes fitness estimation robust to illumination changes, and provides means to combine local and semi-global range for the pattern-windows. First, correlation is maximal exactly in the point where a pattern-window is over the corresponding detail of the object in the image, as needed for achieving a precise alignment between model and object. Second, the correlation falls down as the pattern-window goes far from the exact position, but it keeps a medium value in a small neighborhood of it; this gives a moderate fitness score to individuals located near an object but not exactly over it, making the GA converge faster.

Furthermore, a small biasing is introduced during fitness evaluation that speeds up convergence. The normalized correlation for each window is evaluated not only in the pixel indicated by the individual's parameters, but also in a small (around 7 pixels) neighborhood of this central pixel, with nearly the same time cost. The fitness score is then calculated and the individual parameters are slightly modified so the individual pattern-windows approach the higher correlation points in the evaluated neighborhood. This modification is limited to five pixels, so it has little effect on individuals far from interesting zones, but allows very quick final convergence by promoting a good match to a perfect alignment, instead of waiting for a lucky crossover or mutation to do this.

The fitness function $F([C]_i)$ used is then a function of the normalized correlation of each pattern-window $\rho_k([C]_i)$, $(0<\rho_v<1)$, placed over the image points established by $[C]_i$ using equation (7). It has been empirically tested, leading to the function in equation (10):

$$E([C]_i)=\frac{\dfrac{3-\rho_0([C]_i)\rho_2([C]_i)-\rho_1([C]_i)\rho_3([C]_i)}{3}}{\dfrac{-\rho_0([C]_i)\rho_1([C]_i)\rho_2([C]_i)\rho_3([C]_i)}{3}}\qquad(10a)$$

$$F([C]_i)=\frac{1}{0.1+E([C]_i)} \tag{10b}$$

The error term E in equation (10a) is a measure of how different from the object is the deformed model. It includes a global term with the product of the correlation of the four pattern-windows, and two terms with the product of correlations of pattern-windows in the same diagonal. These last terms forces the deformed models to match the full extent of the object, and avoids matching only a part of it. Note that these terms can have low values, but will never be zero in practice, because correlation never reaches this value. Finally, the fitness score in equation (10b) is a bounded inverse function of the error.

Fig. 1.12. Individual fitness evaluation process

The whole fitness evaluation process for an individual is illustrated in Fig. 1.12. First, the deformed model (individual) position and deformation is established by its parameters (Fig. 1.12.a) where the white dot indicates the reference point. Then, the corresponding positions of the pattern-windows are calculated with the individual deformation and the stored d_i values, in Fig. 1.12.b, marked with dots; finally, normalized correlation of the pattern-windows are calculated in a small neighborhood of its positions, the individual is slightly biased, and fitness is calculated with equation (10).

Normalized correlation with memorized patterns is not able to handle any geometric aspect change. So, how can it work here? The reason for this is the limited size of the pattern-windows. They only capture information of a small zone of the object. Aspect changes affect mainly the overall appearance of the object, but its effect over small details is much reduced. This allows to use the same pattern-windows under a wide range of object size and skew (and some rotation also), without a critical reduction of their correlation. In the presented application, only one set of pattern-windows is used for each object. The extension to consider more degrees of freedom (2D rotation d 3D) is based on the use of various sets of pattern-windows for the same object. The set to use during the correlation is directly decided by the considered deformed model parameters. Each of the sets will cover a certain range of the model parameters. As a conclusion, the second training step deals with the location of the four correlation-windows (object-specific detail) over the deformable model's diagonals, the adimensional values $d_0, \ldots, d_3$ described before. A GA is used to find these four values, which will compose each individual's genome.

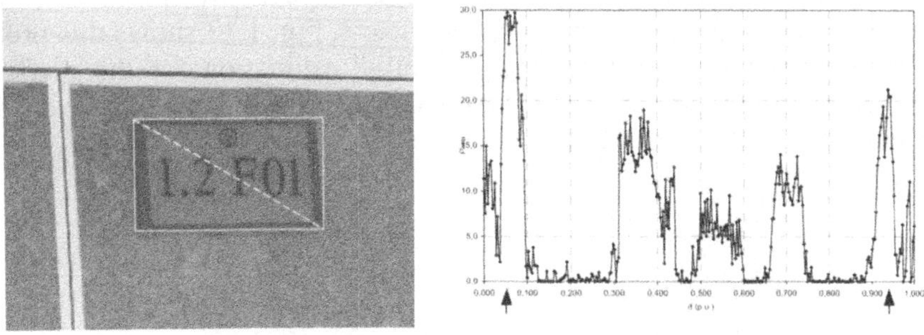

(a) (b)

Fig. 1.13. Pattern-window's position evaluation function

The correlation-windows should be chosen so that each one has a high correlation value in one and only one location inside the target box (for providing good alignment), and low correlation values outside it (to avoid false detections). With this in mind, for each possible value of d_i, the corresponding pattern-window located here is extracted for one of the target boxes. The performance of this pattern-window is evaluated by defining a function with several terms:

1. A positive term with the window's correlation in a very small neighborhood (3-5 pixels) of the theoretical position of the window's center (given by the selected di value over the diagonals of the target boxes).

2. A negative term counting the maximum correlation of the pattern-window inside the target box, but outside the previous theoretical zone.

3. A negative term with the maximum correlation in random zones outside target boxes.

Again, a coarse GA initialization can be easily done in order to decrease training time. Intuitively, the relevant positions where the correlation-windows should be placed are those having strong local variations in the image components (H, L and/or S). A simple method is used to find locations like these. The diagonal lines of the diagonal box of a training image (which will match a theoretical individual's ones) are scanned to H, L and S vectors. Inside these vectors, a local estimate of the derivative is calculated. Then pixels having a high local derivative value are chosen to compute possible initial values for the d_i parameters. Fig. 1.13 shows this process, where the plot represents the derivative estimation for the marked diagonal, starting from the top left corner, while the vertical bars over the plot indicate the selected initial d_i values.

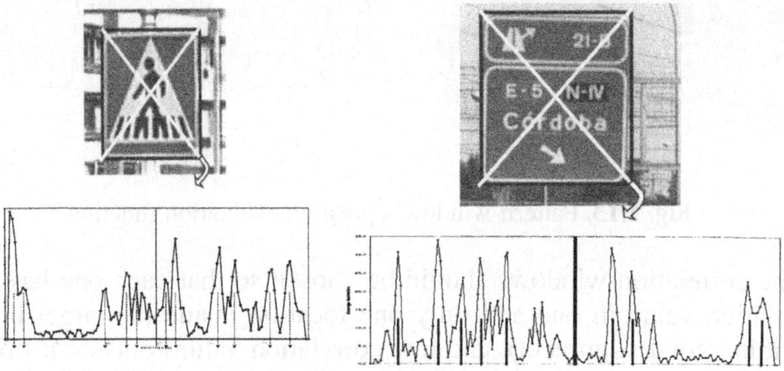

Fig. 1.14. Examples of target box

This function provides a measure for each d_i value; it is evaluated along the diagonals for each target box, and averaged through all target boxes and training images provided, leading to a "goodness" array for each di value. Fig. 1.14 shows this array for one diagonal of two examples of target box. The resulting data is one array for each diagonal. The two pattern-windows over the diagonal are taken in the best peaks from the array. Example pattern-windows selected for some objects are shown (zoomed) in Fig. 1. 15; its real size in pixels can be easily appreciated.

(a) (b)

(c)

Fig. 1.15. Learned pattern-windows for some objects: (**a**) green circle, (**b**) room informative panel, c) pedestrian crossing traffic sign

1.5 System Structure

Pattern search is done using the 2D Pattern Search Engine designed for general application. Once a landmark is found, the related information extraction stage depends on each mark, since they contain different types and amounts of information. However, the topological event (which is generated with the successful recognition of a landmark) is independent from the selected landmark, except for the opportunity of "high level" localization which implies the interpretation of the contents of an office's nameplate. That is, once a landmark is found, symbolic information it could contain, like text or icons, is extracted and interpreted with a neural network. This action gives the opportunity of a "high level" topological localization and control strategies. The complete process is made up by three sequential stages: initialization of the genetic algorithm around regions of interest (ROI), search for the object, and information retrieval if the object is found. This section presents the practical application of the described system. In order to comply with time restrictions common to most real-world applications, some particularizations have been made.

1.5.1 Algorithm Initialization

Letting the GA to explore the whole model's parameters space will make the system unusable in practice, with the available computation capacity at the present. The best way to reduce convergence time is to initialize the algorithm, so that a part of the initial population starts over certain zones of the image that are somehow more interesting than others. These zones are frequently called regions of interest (ROI). If no ROI are used, then the complete population is randomly initialized. This is not a good situation, because algorithm convergence, if the object is in the image, is slow, time varying and so unpractical. Furthermore, if the object is not present in the image, the only way to be sure of that is letting the algorithm run for too long.

The first thing one can do is to use general ROI. There are image zones with presence of borders, lines, etc, that are plausible to match with an object's specific detail. Initializing individuals to these zones increases the probability of setting some individuals near the desired object. Of course, there can be too much zones in the image that can be considered of interest, and it does not solve the problem of deciding that the desired object is not present in the image. Finally, one can use some characteristics of the desired object to select the ROI in the image: color, texture, corners, movement, etc. This will result in few ROI, but with a great probability of belonging to the object searched for. This will speed up the search in two ways: reducing the number of generations until convergence, and reducing the number of individuals needed in the population. If a part of the population is initialized around these ROI, individuals near a correct ROI will have high fitness score and quickly converge to match the object (if the fitness function makes its role); on the other hand, individuals initialized near a wrong ROI will have low fitness score and will be driven away from it by the evolutive process, exploring new image areas. From a statistical point of view, ROI selected using object specific knowledge can be interpreted as object presence hypotheses. The GA search must then validate or reject these hypotheses, by refining the adjustment to a correct ROI until a valid match is generated, or fading away from an incorrect ROI. It has been shown with practical results that, if ROI are properly selected, the GA can converge in a few generations. Also, if this does not happen, it will mean that the desired object was not present in the image. This speeds up the system so it can be used in practical applications.

A simple and quick segmentation is done on the target image, in order to establish Regions of Interest (ROI). A thresholding is performed in the color image following equation (3) and the threshold learned in the training step.These arezones where the selected model has a relevant probability of being found. Then, some morphological operations are carried out in the binary image for connecting interrupted contours. After that, connected regions with appropriate geometry are selected as ROI or object presence hypotheses, these ROIs may be considered as model location hypotheses.

Fig. 1.16 shows several examples of the resulting binary images for indoor and outdoor landmarks. It's important to note that ROI segmentation does not need to be exact, and that there is no inconvenient in generating incorrect ROI. The search stage will verify or reject them.

1.5.2 Object Search

Object search is an evolutionary search in deformable model's parameters space. A Genetic Algorithm (GA) is used to confirm or reject the ROI hypotheses. Each individual's genome is made of five genes (or variables): the individual's Cartesian coordinates (x,y) in the image, its horizontal and vertical size in pixels (ΔX, ΔY) and a measure of its vertical perspective distortion (SkewY).

(a)

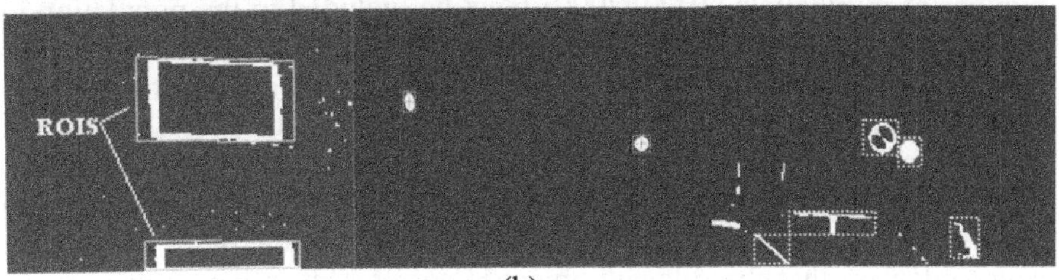

(b)

Fig. 1.16. Example of ROI generation (**a**) original image, (**b**) ROIs

In a general sense, the fitness function can use global and/or local object specific detail. Global details do not have a precise geometric location inside the object, such as statistics of gray levels or colors, textures, etc. Local details are located in certain points inside the object, for example corners, color or texture patches, etc. The use of global details does not need of a perfect alignment between deformable model and object to obtain a high score, while the use of local detail does. Global details allow quickest convergence, but local details allow a more precise one. A trade-off between both kinds of details will achieve the best results.

The individual's health is estimated by the fitness function showed in equation 10b, using the normalized correlation results (on the luminance component of the target image). The correlation for each window ρ_i is calculated only in a very small (about 7 pixels) neighborhood of the pixel in the target image which matches the pattern-window's center position, for real-time computation purpose. The use of four small pattern-windows has enormous advantages over the classical use of one big pattern image for correlation. The relative position of the pattern-windows inside the individual can be modified during the search process. This idea is the basis of the proposed algorithm, as it makes it possible to find landmarks with very different apparent sizes and perspective deformations in the image. Furthermore, the pattern-windows for one landmark does not need to be rotated or scaled before correlation (assuming that only perspective transformation are present), due to their small size. Finally, computation time for one search is much lower for the correlation of the four pattern-windows than for the correlation of one big pattern.

The described implementation of the object detection system will always find the object if it present in the image under the limitations described before. The critical question to be of practical use is the time it takes on it. If the system is used with only random initialization, a great number of individuals (1000~2000) must be included in the population to ensure the exploration of the whole image in a finite time. The selected fitness function evaluation and the individual biasing accelerate convergence once an individual gets close enough to the object, but several tenths and perhaps some hundreds of generations can be necessary for this to happen. Of course there is always a possibility for a lucky mutation to make the job quickly, but this should not be taken into account. Furthermore, there is no way to declare that the selected object is not present in the image, except letting the algorithm run for a long time without any result. This methodology should only be used if it is sure that the object is present in the image, and there are no time restrictions to the search.

When general ROI are used, more individuals are concentrated in interesting areas, so the population can be lowered to 500 ~ 1000 individuals and convergence should take only a few tenths of generations, because the probability of having some deformed models near the object is high. At least, this working way should be used, instead the previous one. However, there are a lot of individuals and generations to run, and search times in a 500 MHz Pentium III PC is still in the order of a few minutes, in 640x480 pixel images. This heavily restricts the applications of the algorithm. And there is also the problem of ensuring the absence of the object in the image.

Finally, if the system with object specific ROI, for example with the representative color segmentation strategy described, things change drastically. In a general real case, there should be only a few ROI; excessively small ones are rejected as they will be noise or objects located too far away for having enough resolution for its identification. From these ROI, some could belong to the object looked for (there can be various instances of the object in the image), and the rest will not. Several objects, about one or two tenth, are initialized scattered around the selected ROI, up to they reach 2/3 of the total population. The rest of the population is randomly initialized to ensure sufficient genetic diversity for crossover operations. If a ROI really is part of the desired object, the individuals close to it will quickly refine the matching, with the help of the slight biasing during fitness evaluation. Here quickly means in very few generations, usually two or three. If the ROI is not part of the object, the fitness score for the individuals around it will be low and genetic drift will move their descendents out. The strategy here is to use only the individuals required to confirm or reject the ROI present in the image (plus some random more); with the habitual number of ROI, about one hundred individuals is enough. Then the GA runs for at most 5 generations. If the object was present in the image, in two or three generations it will be fitted by some deformed models. If after the five generations no ROI has been confirmed, it is considered that the object is not present in the image. Furthermore, if no ROI have been found for the initialization stage, the probabilities of an object to be in the image are very low (if the segmentation was properly learned), and the search process stops here. Typical processing times are 0.2 seconds if no ROI are found, and 0.15 seconds per generation if there are ROI in the image. So, total time for a match is around 0.65 seconds, and less than one second to declare that there is no match (0.2 seconds if no ROI were present). Note that all processing is made by software means, C programmed, and no optimizations have been done in the GA programming –only the biasing technique is non-standard –. In these conditions, mutation has very low probability of making a relevant role, so its computation could be avoided. Mutation is essential only if the search is extended to more generations when the object is not found, if time restrictions allow this.

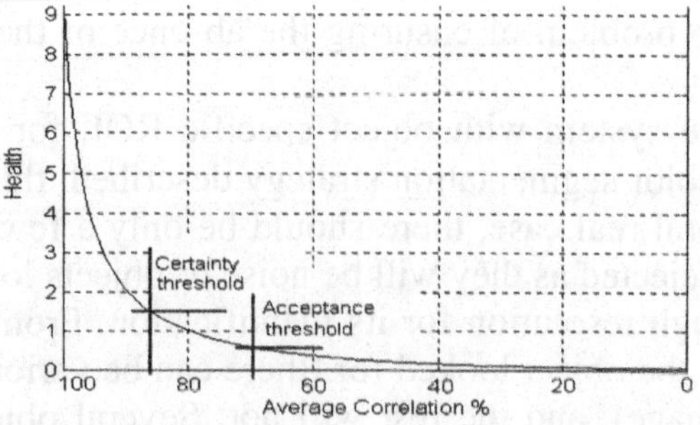

Fig. 1.17. Health vs. average correlation

Fig. 1.17 represents the health of an individual versus the average correlation of its four pattern-windows. Two thresholds have been empirically selected. When a match reaches the certainty threshold, the search ends with a very good result; on the other hand, any match must have an average correlation over the acceptance threshold to be considered as a valid one. The threshold fitness score for accepting a match as valid has been empirically selected. At least 70% correlation in each pattern-window is needed to accept the match as valid (comparatively, average correlation of the pattern-windows over random zones of an image is 25%).

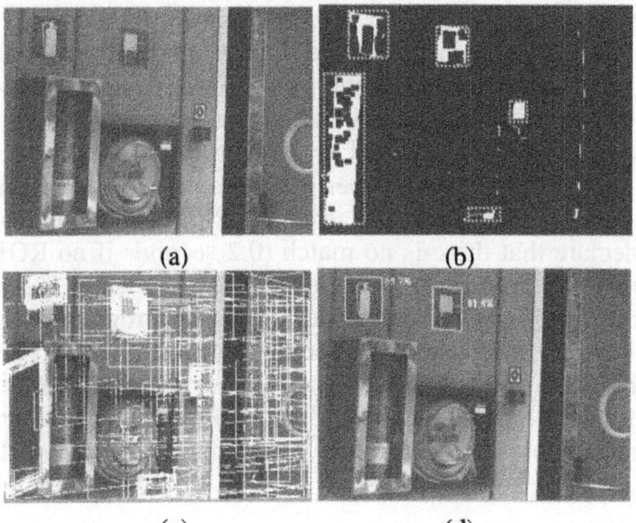

(a) (b)

(c) (d)

Fig. 1.18. (**a**) original images, (**b**) ROIs, (**c**) model search (**d**) Landmarks found

Fig. 1.18 illustrates the full search process one example. Once the search algorithm is stopped, detected objects (if present) are handled by the information extraction stage. Finally, although four pattern-windows is the minimum number which ensures that the individual covers the full extent of the object in the image, a higher number of pattern-windows can be used if needed for more complex landmarks without increasing significantly computation time.

1.5.3 Information Extraction

If the desired object has been found in the image, some information about it shall be required. For topological navigation, often the only information needed from a landmark is its presence or absence in the robot's immediate environment. However, more information may be needed for other navigation strategies, regardless of their topologic or geometric nature. For general application, object location, object pose, distance, size and perspective distortion of each landmark are extracted. Some objects are frequently used for containing symbolic information that is used by humans. This is the case of traffic signs, informative panels in roads and streets, indoor building signs, labels and barcodes, etc. Fig. 1.19 shows some of these objects. All of them have been learned and can be detected by the system, among others. Furthermore, if the landmark found is an office's nameplate, the next step is reading its contents. This ability is widely used by humans, and other research approaches have been done recently in this sense [48]. In our work, a simple Optical Character Recognition (OCR) algorithm has been designed for the reading task, briefly discussed below.

The presented system includes a symbol extraction routine for segmenting characters and icons present into the detected objects. This routine is based in the detection of the background for the symbols through histogram analysis. Symbols are extracted by first segmenting the background region for them (selecting as background the greatest region in the object Luminance histogram), then taking connected regions inside background as symbols, as shown in Fig. 1. 20.

Fig. 1.19. Different objects containing symbolic information

Once the background is extracted and segmented, the holes inside it are considered as candidate symbols. Each of these blobs is analyzed in order to ensure it has the right size: relatively big blobs (usually means some characters merged in the segmentation process) are split recursively in two new characters, and relatively small blobs (fragments of characters broken in the segmentation process, or punctuation marks) are merged to one of their neighbors. Then these blob-characters are grouped in text lines, and each text line is split in words (each word is then a group of one or more blob-characters). Segmented symbols are normalized to 24x24 pixels binary images and feed to a backpropagation neural network input layer. Small deformations of the symbols are handled by the classifier; bigger deformations are corrected using the deformation parameters of the matched model. A single hidden layer is used, and one output for each learned symbol, so good symbol recognition should have one and only one high output. In order to avoid an enormous network size, separated sets of network weights have been trained for three different groups of symbols: capital letters, small letters, and numbers and icons like emergency exits, stairs, elevators, fire extinguishing materials, etc. The weight sets are tried sequentially until a good classification is found, or it is rejected. The final output is a string of characters identifying each classified symbol; the character '?' is reserved for placing in the string an unrecognized symbol. Average symbol extraction and reading process takes around 0.1 seconds per symbol, again by full software processing. This backpropagation network has proved to have a very good ratio between recognition ability and speed compared to more complex neural networks. It has also proved to be more robust than conventional classifiers (only size normalization of the character patterns is done, the neural network handles the possible rotation and skew). This network is trained offline using the quickpropagation algorithm, described in [18]. Fig. 1.21.a shows the inner region of an office's nameplate found in a real image; in b) blobs considered as possible characters are shown, and in c) binary size-normalized images, that the neural network has to recognize, are included. In this example, recognition confidence is over 85% for every character.

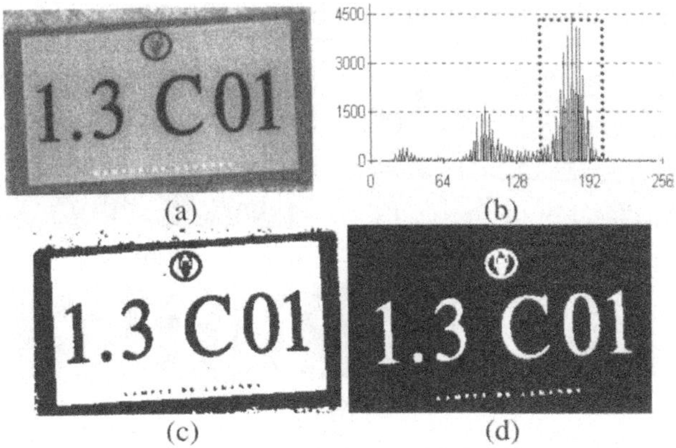

Fig. 1.20. Symbol extraction. **(a)** detected object, **(b)** luminance histogram, **(c)** background segmentation, **(d)** extracted symbols

1.5.4 Learning New Objects

The learning ability makes any system flexible, as it is easy to adapt to new situations, and robust (if the training is made up carefully), because training needs to evaluate and check its progress. In the presented work, new objects can be autonomously learned by the system, as described before. Learning a new object consists in extracting all the needed object-dependent information used by the system. The core of the system, the deformable model-based search algorithm with a GA, is independent of the object. All object-dependent knowledge is localized at three points:

1. Object characteristics used for extraction of ROI (hypotheses generation).

2. Object specific detail to add to the basic deformable model.

3. Object specific symbolic information (if present).

Although on-line training is desirable for its integration ability and continuous update, often an off-line, supervised and controlled training leads to the best results; furthermore, on-line training can make the system too slow to be practical. In the proposed system, off-line training has been used for avoiding extra computing time during detection runs. Learning of symbolic information is done by backpropagation in the neural classifier; this is a classical subject, so it will not be described here.

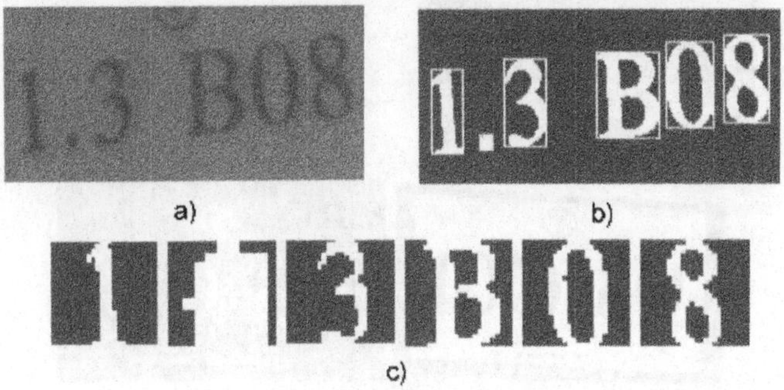

Fig. 1.21. Symbol recognition

1.6 Experimental Results

Experiments have been conducted on a B21-RWI mobile vehicle, in the facilities of the System Engineering and Automation Dept. at the Carlos III University [3] (Fig. 1.22). This implementation uses a JAI CV-M70 progressive scan color camera and a Matrox Meteor II frame grabber plugged in a standard Pentium III personal computer mounted onboard the robot. An Ernitec M2, 8-48 mm. motorized optic is mounted on a Zebra pan-tilt platform. The image processing algorithms for the landmark detection system runs in a standard 500MHz ADM K6II PC. This PC is located inside the robot, and is linked with the movement control PC (also onboard) using a Fast Ethernet based LAN.

Fig. 1.22. RWI B-21 test robot and laboratories and computer vision system

Within the Systems Engineering and Automation department in Carlos III University, an advanced topological navigation system is been developing for indoor mobile robots. It uses a laser telemeter for collision avoidance and door-crossing tasks, and a color vision system for high level localization tasks [34]. The robot uses the Automatic-Deliberative architecture described in [5]. In this architecture, our landmark detection system is an automatic sensorial skill, implemented as a distributed server with a CORBA interface. This way, the server can be accessed from any PC in the robot's LAN. A sequencer is the program that coordinates the robot's skills that should be launched each time, like following a corridor until a door is detected, then crossing the door, and so on.

Experiments have been carried out in the Department installations. It is a typical office environment, with corridors, halls, offices and some large rooms. Each of the floors of the buildings in the campus is organized in zones named with letters. Within each zone, rooms and offices are designated with a number. There is office nameplates (Fig. 1. 23) located at the entrance of room's doors. These landmarks are especially useful for topological navigation for two reasons:

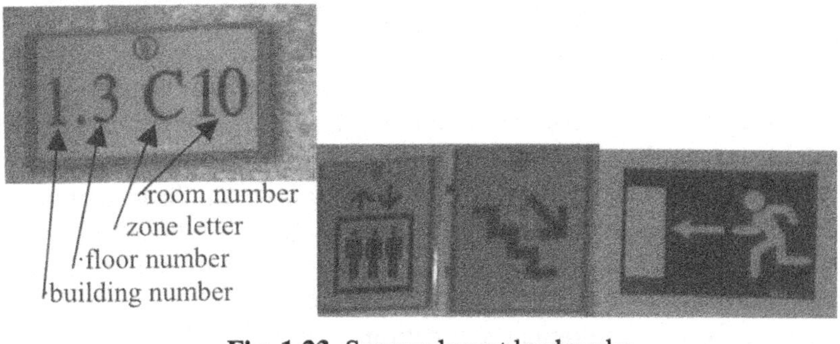

Fig. 1.23. Some relevant landmarks

1. They indicate the presence of a door. If the door is opened, it is easily detected with the laser telemeter, but it can not be detected with this sensor when it is closed. So the detection of the nameplate handles this limitation.

2. The system is able to read and understand the symbolic content of the landmarks. This allows an exact "topological localization", and also confirms the detection of the right landmark.

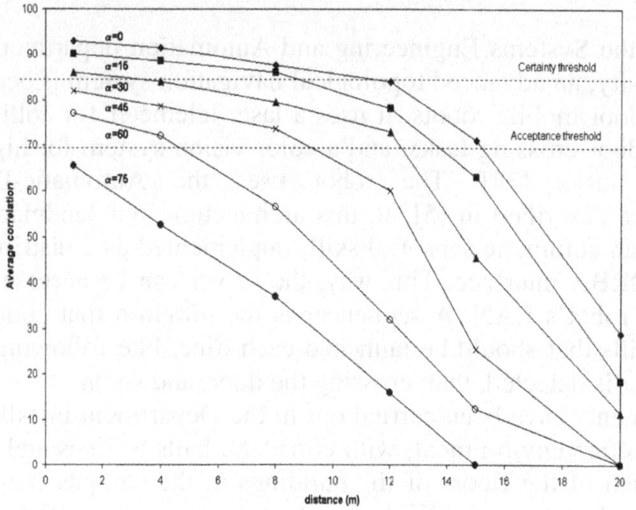

Fig. 1.24. Recognition results

When the office nameplates are available, they offer all the information needed for topological navigation. When they are not, the rest of the landmarks are used. Also, there are other "especially relevant" landmarks: those alerting of the presence of stairs or lifts, since they indicate the ways for moving to another floor of the building. Finally, emergency exit signs indicate ways for exiting the building. Thinking on these examples, it should be noted that some landmarks can be used in two ways. First, its presence or absence is used for robot localization in a classic manner. Second, the contents of the landmark give high level information which is naturally useful for topological navigation, as mentioned before. This is allowed by the symbol reading ability included in our landmark detection system. The experimental results will show its usefulness.

Table 1.1. Recognition results

angle (°) \ distance (m)	1	4	8	12	15	20
0	*93*	*91,5*	*87,5*	*84*	*71*	29
15	*90*	*88,5*	*86*	*78*	63	18,5
30	*86*	*84*	*79,5*	*73*	45,5	11,5
45	*82*	*78,5*	*73,5*	60	25,5	0
60	*77*	*72*	56,5	32	12,5	0
75	65,5	52,5	37	16	0	0

1.6.1 Robot Localization Inside a Room

The pattern recognition stage has shown good robustness with the two landmarks tested in a real application. Table 1.I and Fig. 1.24 summarizes some of the test results. The curves show the average correlation obtained with tested landmarks situated at different distances and angles of view from the robot, under uncontrolled illumination conditions. A "possible recognition" zone in the vicinity of any landmark can be extracted from the data on this plot. This means that there is a very good chance of finding a landmark if the robot enters inside the oval defined by angles and distances over the acceptance threshold line in the graph. Matches above the certainty thresholds are good ones with a probability over 95% (85% for the acceptance threshold). These results were obtained using a 25 mm fixed optic. When a motorized zoom is used with the camera, it is possible to modify the recognition zone at will. The robot is able to localize itself successfully using the standard University's nameplates, and using the artificial landmarks placed in large rooms (Fig. 1.25). The ability of reading nameplates means that there is no need for the robot initial positioning. The robot can move around searching for a nameplate and then use the text inside to realize its whereabouts in the building ("absolute" position). The system can actually process up to 4 frames per second when searching for a landmark, while the text reading process requires about half a second to be completed (once the plate is within range). Since the nameplates can be detected at larger distances and angles of view than those minimum needed for successfully reading their contents, a simple approach trajectory is launched when the robot detects a plate. This approach trajectory does not need to be accurate since, in practice, the text inside plates can be read with angles of view up to 45 degrees. Once this approach movement is completed, the robot tries to read the nameplate's content. If the reading is not good enough, or the interpreted text is not any of the expected, a closer approach is launched before discarding the landmark and starting a new search. In Fig. 1.25.a a real situation is presented. Nine artificial landmarks are placed inside room 1 and four natural landmarks are situated along the hall. The frame captured by the camera (25 mm focal distance and 14.5° horizontal angle of view) is shown in Fig. 1.26, where two artificial landmarks are successfully detected after only one iteration of the genetic search. Fig. 1.25.b illustrates the case where both kinds of landmarks were present in the captured image; in this case two runs of the algorithm were needed to identify both landmarks.

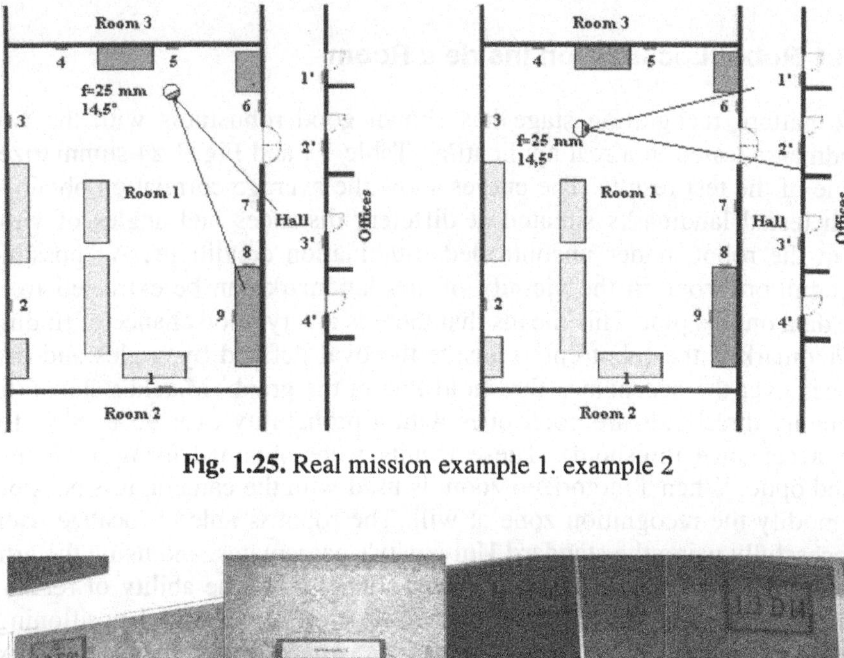

Fig. 1.25. Real mission example 1. example 2

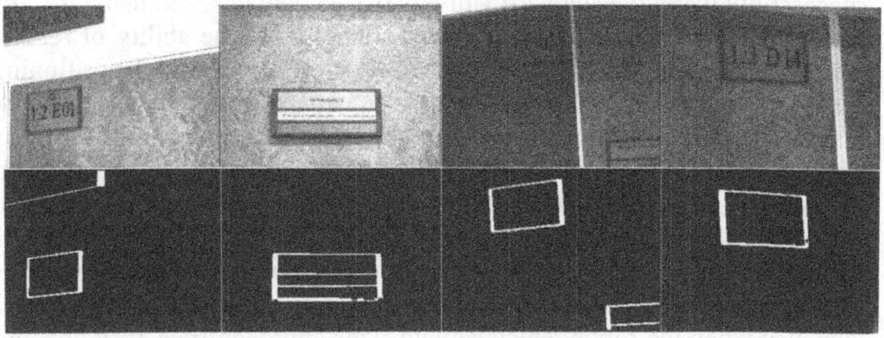

Fig. 1.26. Learnt segmentation results 1

1.6.2 Room Identification

The first high-level skill developed for the robot is the topological identification of a room, using the landmarks detected inside it. This skill is really useful when the robot does not know its initial position during the starting of a mission. Other applications are making topological landmark maps of rooms, and confirming that when the robot enters a room this is truly the expected one. This makes topological navigation more robust, since it helps avoiding the robot to get lost. The philosophy used for this skill is as follows. When the robot is in a room, it uses a basic skill for going towards the room center (coarsely), using a laser telemeter (this only pretends put the robot away from walls and give it a wide field of view). Then, the robot alternates the "rotate left" and the developed "landmark detection" skills to accomplish a full rotation over itself while trying to detect the landmarks present, as follows. The robot stops, search for all the possibly present landmarks (in our case, green circles, fire system signs and emergency exit signs) and stores the detected ones. Then rotates a certain angle (calculated with the focal distance to cover the full scene), stops, stores detected landmarks, makes a new search, and so on. Symbolic content of landmarks having it is extracted and also stored. The result is a detected landmark sequence, with relative rotation angles between them, which is the "landmark signature" of the room. This signature can be compared with the stored ones to identify the room, or to establish that it is a unknown one and it can be added to the navigation chart.

As an example, let us consider the room 1.3C13, shown in Fig. 1.27. There is only one natural landmark that has been learned by the system, a fire extinguisher sign (indicated by a black square), so artificial landmarks (green circles, indicated as black ovals) were added to the room to increase the length of the landmark signature. Images captured by the robot during a typical sweep are presented in Fig. 1.27, where the image sequence is from right to left and top to bottom. All landmarks have been detected, marked in the figure with dotted black rhombus. Note that the fire extinguisher sign is identified first as a generic fire system one, and then confirmed as a fire extinguisher one by interpreting its symbolic content (the fire extinguisher icon).

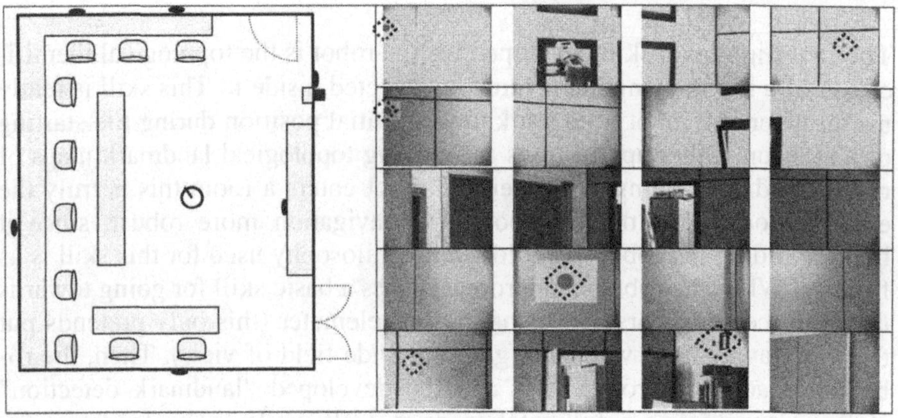

Fig. 1.27. Landmark map of room 1.3C13 and room sweep with detected landmarks

GC	*50*	GC	*85*	GC	*120*	GC	*50*	FE GC	*35*	GC	*20*

Fig. 1.28. Landmark signature for room 1.3C13

The obtained landmark sequence (room signature) is presented in Fig. 1.28, where GC stands for "green circle" and FE for "fire extinguisher" signs. Rotated relative angles (in degrees) between detections are included, since there is no absolute angle reference.

The detection system is designed in such a way so it has a very low probability of false positives, but false negatives can be caused by occlusion by moving obstacles or a robot position very different to the one from where the landmark signature was stored. So the signatures matching algorithm for room identification must manage both relative angle variations and possible lack of some landmarks. A custom algorithm is used for that.

1.6.3 Searching for a Room

The second high-level skill developed is room searching. Here, the robot has to move through the corridors looking for a specific room, indicated by a room nameplate. As an example, the robot must search for the room named 1.3C08. To accomplish that, the robot has to detect a room name plate, and read its content. This is not a new idea (see for example [48]), but it has been applied in practice in very few cases, and only in geometrical navigation approaches. The name of the rooms contains a lot of implicit information. The first number identifies the building, the second number (after the dot) identifies the floor of the building, the letter is the zone of that floor, and the last two digits are the room number. So the reading of a room nameplate allows several critical high-level decisions in topological navigation:

1. If the building number does not match the required one, the robot has to exit the building and enter another one.
2. If the floor number does not match, the robot must search for an elevator to change floor.
3. If the zone letter is wrong, the robot has to follow the corridors, searching for the right letter (see Fig. 1.25).

Once the desired zone is reached, the robot must follow several nameplates until the right number is found. These are correlatives, so it is easy to detect if the desired nameplate is lost, and allows the robot to know the right moving direction along a corridor. Furthermore, the reading of a nameplate at any time implies an absolute topological localization of the robot, since it knows where it is. This avoids the robot to get lost.

Actually, our robot can not use elevators, so experiments are limited to floor 3. Fig. 1.29 shows a zone map of this floor. The robot uses a topological navigation chart resembling the relations between zones [16], so it can know how to go from one to another zone.

The room search skill follows these steps:

1. Follow any corridor (using a laser telemeter-based "corridor following" skill) searching for a room nameplate.
2. Once detected and read, move again and search for a second one. With these two readings, the robot knows the zone where it is and the moving direction along the corridor (room numbers are correlative).
3. Follow the corridors until the right zone is reached (uses navigation chart of Fig. 1.29), checking room nameplates along the way to avoid getting lost.

4. Once in the right zone, follow the corridor until the desired room is reached. Check room numbers for missing ones.

The image sequence "seen" by the robot once the right zone is reached is shown in Fig. 1.29. It exemplifies the standard navigation along a corridor. The robot ends its mission once 1.3C08 nameplate is read. Note that only nameplates containing the room number are read; nameplates with the name of the people who occupies the room are not (the characters are too small). Of course, they can be read if needed for any task.

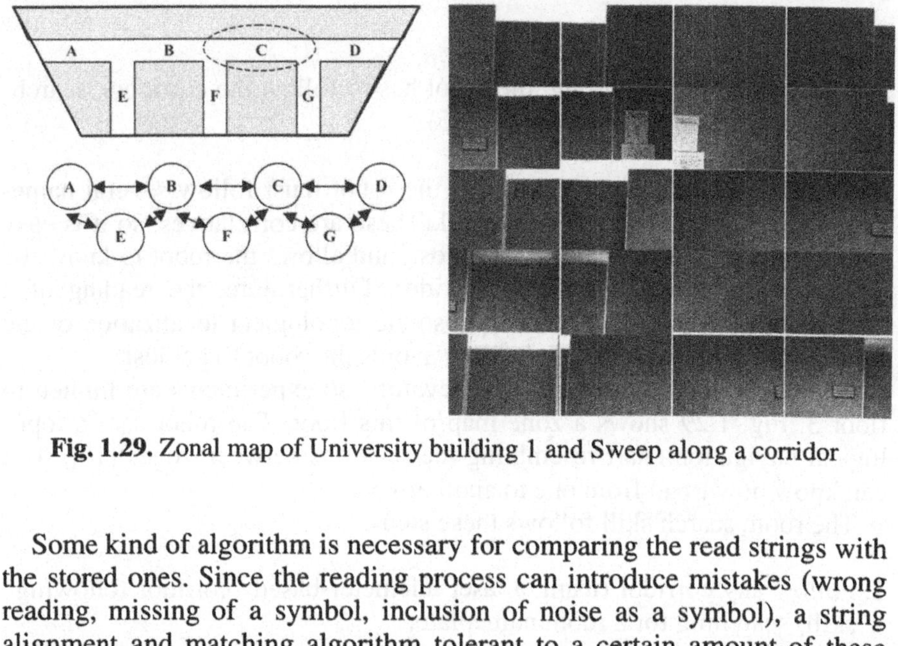

Fig. 1.29. Zonal map of University building 1 and Sweep along a corridor

Some kind of algorithm is necessary for comparing the read strings with the stored ones. Since the reading process can introduce mistakes (wrong reading, missing of a symbol, inclusion of noise as a symbol), a string alignment and matching algorithm tolerant to a certain amount of these mistakes should be used. There is dedicated literature on this topic ([36] among others); however, our database is relatively small, so we use a home-made comparative suboptimal algorithm.

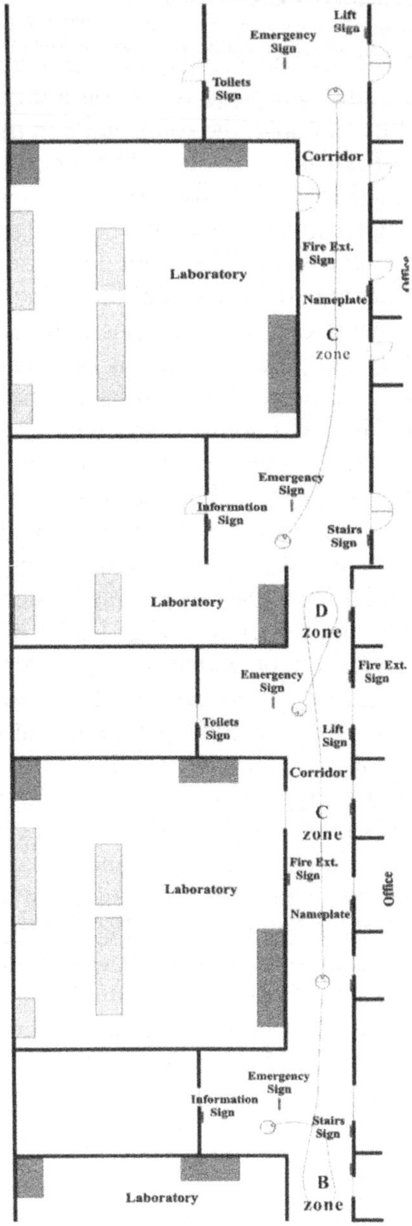

(a) (b)

Fig. 1.30. navigation examples (**a**) test 1 (**b**) test 2

1.6.4 Corridor Navigation Example

A more complex unconstrained navigation problem is presented now. The robot starts in a unknown point of the building, and it must reach a specific location. In this example, the robot starts in the hall between zones B and C, on the third floor of building 1. The robot does not know any of this, and is told to reach room 1.2D01. Fig. 1.30.a presents the landmark distribution and the approximate trajectory (there is no need for odometric measures) described by the robot. The robot does not know its initial position, so it tries to find and read a room nameplate landmark. If it can achieve this, then immediately knows its position (building, zone and office it stands at). In this case, it can't find any one. Then, the "room identification from landmark signature" ability is used. The robot tries to find all the landmarks around it, and compares the obtained landmark sequence with stored ones. Fig. 1.31.a shows an image of this location, taken with the robot's camera. In this example, again this is not enough, because there are several halls with a very similar landmark signature. The last strategy considered by the robot is entering a corridor (using the laser telemeter) and trying again to read a nameplate. Now this is successful, and the robot reads "1.3C01" in the image shown in Fig. 1.31.b. Once located, the desired action sequence until the objective room is reached is generated. The robot is in the right building, but in the third floor, so it must search for a lift to go down one floor. The topological map indicates it has to follow the C zone corridor, then enter a hall, and search here for a "lift" sign. It follows the corridor, and tries to read the nameplates for avoiding getting lost. If some are missed, it is not a problem, since reading any of the following ones relocates the robot. If desired, other landmarks present in the corridors (like fire extinguisher ones) can be used as an additional navigation aid. When the corridor ends in a new hall (Fig. 1.31.c), the robot launches the room identification ability to confirm that. The hall's landmark signature includes the lift sign. When this landmark is found and read (Fig. 1.31.d), the robot finishes its path in this floor, and knows that entering the lift lobby is the way to second floor. Our robot is not able to use the lifts, so the experiment ends here.

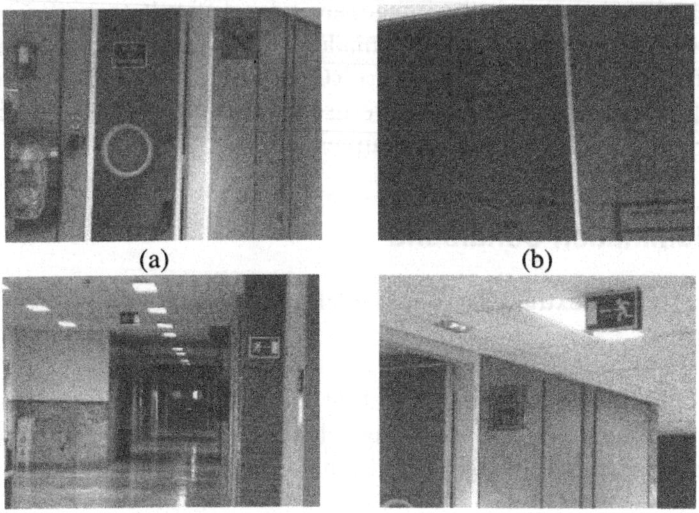

(a) (b) (c) (d)

Fig. 1.31. Some frames in the robot's path

A more complex situation is tested in a second part of the experiment. The robot is initially headed so it will start moving in the wrong direction (entering zone B instead C, see Fig. 1.30.b). When the robot reads the first nameplate in B zone ("1.3B12") realizes the wrong direction and heads back to C zone corridor, and then follows it like before. Furthermore, this time several landmarks (including the lift one) have been occluded for test purposes. The robot can not recognize the hall, so it heads for the new corridor, corresponding to D zone. When a nameplate is read, the robot knows it has just passed the desired hall and heads back for it. The experiment ends when the robot assures it is in the right hall, but unable to find the occluded lift sign.

1.7 Practical Limitations through Experiments

Exhaustive tests have been done to the system to evaluate its performances and limitations. All tests have been carried out with real 640x480 color images, without illumination control. The following points present some limitations to the object detection. If the object in the image complies with these limitations, it will surely be detected. The detection will fail if the limitations are exceeded. On the other hand, false positives (detecting an object that is not present in the image) are very difficult to occur, as a consequence of the particularizations made and the autonomous training with real images. No search is tried if no ROI are detected, and restrictive conditions for accepting the results are used. Unless otherwise specified, the failure conditions are for false negatives.

1.7.1 Illumination Conditions

The system is extremely robust to illumination conditions, as a consequence of:

1. HSL color space is used, separating luminance component from color. Color segmentation is done using relaxed intervals learned from illumination-affected real images. Furthermore, it does not need to be perfect.

2. Normalized correlation minimizes lightning effect in search stage.

3. All related processing thresholds are dynamically selected or have been learned.

Illumination is the main cause of failure only in extreme situations, like strongly saturated images or very dark ones (saturation goes to zero in both cases, and all color information is lost), because no specific ROI are segmented and the search is not launched. This can be handled, if needed, by running the search with general ROI detection, although computation time is severely increased, as established. Strong backlighting can cause failure for the same reason, and so metallic brightness. Fig. 1.32 shows several cases where the object is found in spite of difficult lightning conditions, and Fig. 1.33 shows failures. A white circle indicates the presence of the object when not clearly visible.

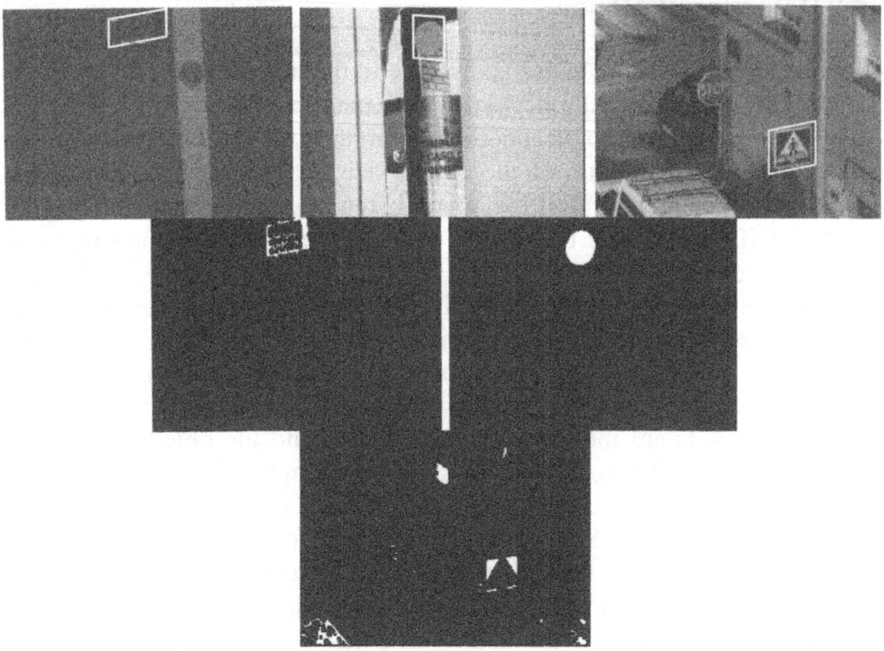

(a) (b) (c)

Fig. 1.32. Object found in difficult illumination conditions: (**a**) poor, (**b**) excessive, (**c**) night

(a) (b) (c)

Fig. 1.33. Failures due to extreme illumination conditions: (**a**) darkness, (**b**) dense mist, (**c**) backlight

1.7.2 Detection Distance

The most frequent failure cause is distance to the object. If the object is too far from the camera, it will occupy too few pixels in the image. A minimal object size in the image is needed for distinguishing it. The maximum detection distance is function of the object size and the camera optic focal distance. On the other hand, if the object is too close to the camera, usually part of it will fall outside the image. The consequences are the same that for partial occlusion (section 1.7.3). There is another source for failure. The correlation between the details included in the pattern-windows and the object decreases slowly as the object details became bigger or smaller that the pattern-window captured details. This decrease will make the correlation values fall under the security acceptance thresholds for the detection. Some details are more robust than others, and the object can be detected over a wider range of distances. Relative angle of view between the object and the optical axis translates into perspective deformation (vertical skew), handled with the SkY parameter of the deformable model. This deformation also affects to the object details, so the correlation will decrease as the vertical deformation increases, too. The pattern-windows are taken on a frontal-view image of the object, so detection distance will be maximal in frontal views, and will decrease as angle of view increases. Fig. 1.34 illustrates this: the average correlation of the four patter-windows for the green circle is painted against the camera position respect to the object in the horizontal plane (the green circle is attached to the wall). The circle is 8 cm diameter, and a 8-48 mm motorized zoom has been used. The effect of visual angle can be reduced if various sets of pattern-windows are used, and switched accordingly to model deformation.

1.7.3 Partial Occlusion

ROI segmentations is barely affected by partial occlusion, it will only change its size. The subsequent search will adjust the deformed model parameter later. The search stage can or can not be affected, depending on the type of occlusion. If the object details used for the matching are not occluded, it will have no effect (Fig. 1.35.b). If one of the four detail zones is occluded, global correlation will descend; depending on the correlation of the other three pattern-windows, the match will be over the acceptance thresholds (Fig. 1.35.a), or will not. Finally, if at least two detail zones are occluded, the search will fail (Fig. 1.35.c), street naming panel).

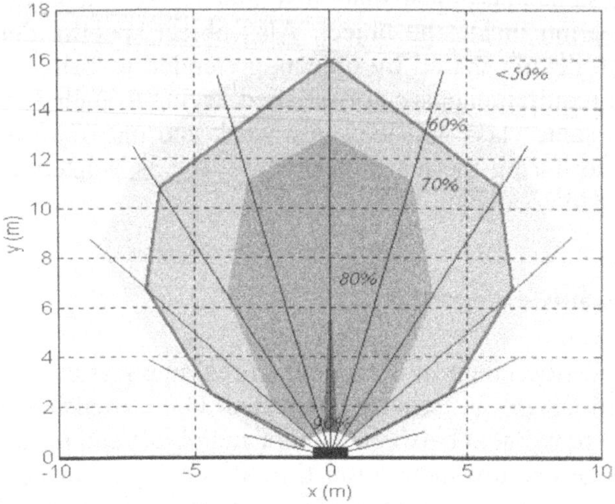

Fig. 1.34. average pattern-window correlation with distance and angle of view for the green circle. Values under 70% are not sufficient for accepting the detection

(a) (b) (c)

Fig. 1.35. Different situations under partial occlusion

1.7.4 Object Morphology

The morphology of the objects to detect is limited by the particularizations made to achieve practical time requirements for the system. The object must be planar (or at least with a relatively small third dimension), or a face of a 3D object. The suppression of the rotation degree of freedom causes that only objects appearing always with the same orientation are detected (although some rotation can be handled by the vertical deformation d.o.f.). Object shape has no restrictions, since the base deformable model only encloses the object; particular shape features will be used for the object search process. Color segmentation requires that objects one wants to include in the same class must share a similar color, independent of its extension or location inside the object. Also object specific detail requires some common details shared by objects pretended to belong to the same class. If these requirements are not satisfied, trying to include too different objects in the same class will lead to a weak and uncertain learning; this can be detected during the learning process (the associated scoring functions will have low values).

1.7.5 Defocusing

Defocusing must be taken into account in real applications, where image capture conditions are not strictly controlled. Optic focusing can be inexact, or relative movement between camera and object can make it to appear blurred if image capture integration time is too high; furthermore, interlaced CCD video cameras capture odd and even fields in different time instants, so they also are affected by movement. A high gain, progressive scan CCD color camera, model CV-M70 from JAI, has been used for the system evaluation to minimize movement effects, for example if the camera is mounted onboard a vehicle (one of the potential application fields). Defocusing only affects color segmentation by changing segmented contours, but this is corrected by the genetic object search. The correlation used for the searching process can be affected under severe defocusing, especially if the learned pattern-windows contain very thin and precise details, which can be destroyed by blur. However, the learning process along a wide set of real examples of the objects tries to minimize this effect (excessive thin details are not always present in the images).

1.8 Conclusions and Future Works

A practical oriented, general purpose deformable model-based object detection system is proposed. Evolutionary algorithms are used for both object search and new object learning. Although the proposed system can handle 3D objects, some particularizations have been done to ensure computation times low enough for real applications. 3D extension is discussed. The system includes a symbolic information reading stage, useful for a wide set of informative panels, traffic signs and so on. The system has been developed and tested using real indoor and outdoor images, and several example objects have been learned and detected. Field experiments have proven the robustness of the system for illumination conditions and perspective deformation of objects, and applicability limits have been explored. Potential application fields are industrial and mobile robotics, driving aids and industrial tasks. Actually it is being used for topological navigation of an indoor mobile robot and for a driver assistance system [17]. There are several related works in the literature in the line exploited in the present article, showing this is an active and interesting one. Aoyagi and Asakura [1] developed a traffic sign recognition system; circular signs are detected with a GA and a NN classifies it as speed sign or other; a 3 d.o.f. circle is matched over a luminance-binarized image for the sign detection. Although seriously limited, includes several interesting concepts. GA initialization or time considerations are not covered. Minami, Agbanhan and Asakura [32] also uses a GA to optimize a cost function evaluating the match between a 2D rigid model of an object's surface and the image, considering only translation and rotation. Cost function is evaluated over a 128x120 pixel grayscale image. It is a very simple model, but the problem of where to select the object specific detail over the model is addressed, concluding that inner zones of the model are more robust to noise and occlusion. In our approach, detail location inside the basic model is autonomously learned over real images. Mignotte et al. [35] uses a deformable model, similar to our 2D presented one, to classify between natural or man-made objects in high-resolution sonar images. The model is a cubic B-spline over control points selected by hand, that is tried to adjust precisely over sonar cast-shadows of the objects. This is focused as the maximization of a PDF relating the model and the binarized (shadow or reverberation) image by edges and region homogeneity. Various techniques are compared to do this: a gradient-based algorithm, simulated annealing (SA), and an hybrid GA; the GA wins the contest. Unfortunately, the application is limited to parallelepipedal or elliptical cast shadows, are multiple object presence is handled by launching a new search. Furthermore, using a binary image for cost function evaluation is always segmentation-dependant; in our approach, correlation in grayscale image is used instead. This chapter shows the usefulness of this new landmark detection and reading systems in topological navigation tasks. The ability of using a wide spread of natural landmarks gives great flexibility and robustness. Furthermore, the landmark reading ability allows high level behaviors for topological navigation, resembling those used by humans. As the examples have shown the robot need not to know its initial position in the environment, it can recover of initial wrong direction and landmark occlusion to reach the desired destination. A new color vision-based landmark learning and recognition system is presented in this chapter. The experiments carried out have shown its utility for both artificial and natural landmarks; furthermore, they can contain written text. This text can be extracted, read and used later for any task, such as high level localization by relating written names to places. The system can be adapted easily to handle new landmarks by learning them, with very little human intervention (only providing a training image set). Different text styles can be read using different sets of neural classifier weights; these sets can be loaded from disk when needed. This generalization ability is the relevant advantage from classical rigid methods. The system has been tested in an indoor mobile robot navigation application, and proved useful. The types of landmark to use are not limited a-priori, so the system can be applied to indoor and outdoor navigation tasks. The natural application environments of the system are big public and industrial buildings (factories, stores, etc.) where the preexistent wall signals may be used, and outside environments with well-defined landmarks such as streets and roads. This chapter presents some high-level topological navigation applications of our previously presented visual landmark recognition system. Its relevant characteristics (learning capacity, generality and text/icons reading ability) are exploited for two different tasks. First, room identification from inside is achieved through the landmark signature of the room. This can be used for locating the robot without any initialization, and for distinguishing known or new rooms during map generation tasks. The second example task is searching for a specific room when following a corridor, using the room nameplates placed there for human use, without any information about distance or location of the room. The textual content of the nameplates is read and used to take high-level control decisions. The ability of using preexistent, human-use designed landmarks, results in a higher degree of integration of mobile robotics in everyday life.

References

1 Aoyagi Y., Asakura, T., (1996) “A study on traffic sign recognition in scene image using genetic algorithms and neural networks”. International Conference on Industrial Electronics, Control and Instrumentation, pp.1838-1843.

2 Argamon-Engelson, S. (1998) “Using image signatures for place recognition”. Patter Recognition Letters 19, pp. 941-951.

3 Armingol J.M., de la Escalera, A., Salichs, M.A., (1998) "Landmark perception planning for mobile robot localization". IEEE International Conference on Robotics and Automation, vol. 3, pp. 3425-30.
4 Balkenius, C. (1998) "Spatial learning with perceptually grounded representations". Robotics and Autonomous Systems, vol. 25, pp. 165-175.
5 Barber R., Salichs, M.A. (2001) "Mobile robot navigation based on events maps". 3rd International Conference on Field and Service Robots, pp. 61-66.
6 Beccari, G.; Caselli, S.; Zanichelli, F. (1998) "Qualitative spatial representations from task-oriented perception and exploratory behaviors". Robotics and Autonomous Systems, vol. 25, pp. 165-175.
7 Betke, M., Makris, N., (2001) "Recognition, resolution, and complexity of objects subject to affine transformations", International Journal of Computer Vision, vol.44, n° 1, pp. 5-40.
8 Bhandarkar, S. M.; Koh, J.; Suk, M., (1997) "Multiscale image segmentation using a hierarchical self-organizing map". Neurocomputing, vol. 14, pp. 241-272.
9 Bin-Ran; Liu, H. X.; Martonov, W., (1998) "A vision-based object detection system for intelligent vehicles". Proceedings of the SPIE- the International Society for Optical Engineering, vol. 3525, pp. 326-337.
10 Blaer, P., Allen, P. (2002) "Topological mobile robot localization using fast vision techniques". IEEE International Conference on Robotics and Automation, pp. 1031-1036.
11 Borenstein, J. and Feng. L., (1996) "Measurement and correction of systematic odometry errors in mobile robots". IEEE Journal of Robotics and Automation, vol. 12, n° 6, pp. 869-880.
12 Colin, V. and Crowley, J., (2000) "Local appearance space for recognition of navigation landmarks". Robotics and Autonomous Systems, vol. 31, pp. 61-69.
13 Cootes, T.F., Taylor, C.J., Lanitis, A., Cooper, D.H., Graham, J. (1993) "Building and using flexible models incorporating gray level information". International Conference on Computer Vision, pp.242-246.
14 Dubuisson M.P., Lakshmanan S., and Jain A.K. (1996) "Vehicle segmentation and classification using deformable templates", IEEE Transaction on Pattern Analysis and Machine Intelligence, vol.18, n° 3, pp.293-308.

15 Edelman S., Bulthoff H. and Weinshall D. (1989) "Stimulus familiarity determines recognition strategy for novel 3D objects", technical report 1138, Massachussets Institute of Technology, Artificial Intelligence Laboratory.
16 Egido, V., Barber, R., Salichs, M.A., (2002) "Self-generation by a mobile robot of topological maps of corridors". IEEE International Conference on Robotics and Automation, pp. 2662-2667.
17 Escalera A. de la, Armingol J. M. and Mata M. (2003) "Traffic sign recognition and analysis for intelligent vehicles", Image and Vision Computing, vol. 21, pp. 247-258.
18 Fahlman, S. E. (1998) "An empirical study of learning speed in back-propagation networks". CMU-CS-88-162.
19 Franz, Matthias O. (1998) "Learning view graphs for robot navigation". Autonomous robots, vol. 5, pp. 111-125.
20 Fukuda, T., Nakashima, M., Arai, F., Hasegawa, Y. (2002) "Generalized facial expression of character face based on deformation model for human-robot communication". International Workshop on Robot and Human Interactive Communication, pp. 331-336.
21 Gaskett, C., Fletcher, L., Zelinsky, A., (2000) "Reinforcement learning for vision based mobile robot". International Conference on Intelligent Robots and Systems, vol. 2 pp. 403-409.
22 Ghita, O., Whelan, P. (1998) "Eigenimage analysis for object recognition", technical report, Vision Systems Laboratory, School of Electronic Engineering, Dublin City University.
23 Iida, M., Sugisaka, M., Shibata, K., (2002) "Application of direct-vision based reinforcement learning to a real mobile robot". International Conference on Neural Information Processing, vol. 5 pp. 2556-2560.
24 Kervrann, C., Heitz, F., (1999) "Statistical deformable model-based segmentation of image motion", IEEE Transactions on Image Processing, vol.8, nº 4, pp.583-8.
25 Kreucher C., Lakshmanan S. (1999) "LANA: a lane extraction algorithm that uses frequency domain features", IEEE Transactions on Robotics and Automation, vol.15, nº 2, pp.343-50.
26 Kubota, N., Hashimoto, S., Kojima, F. (2001) "Genetic programming for life-time learning of a mobile robot". IFSA World Congress and 20th NAFIPS International Conference, vol. 4, pp. 2422-2427.
27 Launay, F., Ohya, A., Yuta, S. (2002) "A corridors lights based navigation system including path definition using topologically corrected map for indoor mobile robots". IEEE International Conference on Robotics and Automation, pp. 3918-3923.

28 Lijun Y., Basu A. (1999) "Integrating active face tracking with model based coding", Pattern Recognition Letters, vol.20, nº 6, pp.651-7.
29 Liu, L., Sclaroff, S., (2001) "Medical image segmentation and retrieval via deformable models". International Conference on Image Processing, vol. 3, pp. 1071-1074.
30 Liu, Y.; Yamamura, T.; Ohnishi, N.; Surgie, N., (1998) "Character-based mobile robot navigation". 1998 IEEE International Conference on Intelligent Vehicles, pp. 563-568.
31 Luo, R. C.; Potlapalli, H., (1994) "Landmark recognition using projection learning for mobile robot navigation". IEEE International Conference on Neural Networks, vol. 4, pp. 2703-2708.
32 Minami, M., Agbanhan, J., Asakura, T. (2001) "Robust scene recognition using a GA and real-world raw-image", Measurement, vol. 29, pp.249-267.
33 Mahadevan, S.; Theocharous, G., (1998) "Rapid concept learning for mobile robots". Machine learning, vol. 31, pp. 7-27.
34 Mata, M.; Armingol, J.M.; Escalera, A.; Salichs, M.A. (2001) "Mobile robot navigation based on visual landmark recognition" International Conference on Intelligent Autonomous Vehicles, pp. 197-192.
35 Mignotte, M., Collet, C., Perez P., Bouthemy, P. (2000) "Hybrid genetic optimization and statistical model-based approach for the classification of shadow shapes in sonar imaginery", IEEE Transactions on Pattern Analysis and Machine Intelligence, vol.22, nº 2, pp.129-141.
36 Myers, E. W., Oliva, P., Guimarães, K.S.(1998) "Reporting Exact and Approximate Regular Expression Matches". Combinatorial Pattern Matching, 9th Annual Symposium CPM'98. pp. 91-103
37 Ohyama, T; (1995) "Neural network-based regions detection". IEEE International Conference on Neural Networks. Proceedings, vol.3, nº 2; pp. 222-302.
38 Perez, F.; Koch, C. (1994) "Toward color image segmentation in analog VLSI: algorithm and hardware". International Journal of Computer Vision, vol. 12, nº 1 pp. 17-42.
39 Poupon F., Mangin J. F., Hasboun D., Poupon C., Magnin I., Frouin V. (1998)"Multi-object deformable templates dedicated to the segmentation of brain deep structures", Medical Image Computing and Computer Assisted Intervention, First International Conference, pp.1134-43.
40 Rosenfeld A., (2000) "Image analysis and computer vision 1999 [survey]". Computer Vision and Image Understanding, vol. 78 nº 2, pp 222-302.

41 Rue H. and Husby O.K. (1998) "Identification of partly destroyed objects using deformable templates". Statistics and Computing, vol.8, n° 3, pp.221-228.
42 Salichs, M.A., Moreno, L. (2000) "Navigation of mobile robots: open questions". Robotica, vol.18, pp. 227-234.
43 Selinger A., Nelson R. C. (1999) "A Perceptual grouping hierarchy for appearance-based 3D object recognition", technical report, Department of Computer Science, University of Rochester.
44 Sim, R., Dudek, G., "Mobile robot localization from learned landmarks". International Conference on Intelligent Robots and Systems, vol. 2. pp. 1060-1065.
45 Takahashi, Y., Asada, M., (2000) "Vision-guided behavior acquisition of a mobile robot by multi-layered reinforcement learning". Conference on Intelligent Robots and Systems, vol. 1, pp. 395-402.
46 Tarr M.J., Bülthoff H.H. (1998) "Image-based object recognition in man, monkey and machine", Cognition, vol.67, pp. 1-20.
47 Thompson, S., Zelinsky, A., (2002) "Accurate local positioning using visual landmarks from a panoramic sensor". IEEE International Conference on Robotics and Automation, pp. 2656-2661.
48 Tomono M., Yuta, S. (2000) "Mobile robot navigation in indoor environments using object and character recognition". IEEE International Conference on Robotics and Automation, pp. 313-320.
49 Tsang, C.K., Fu-Lai Chung., (1998) "Development of a structural deformable model for handwriting recognition ". 14th Conference on Pattern Recognition, vol. 2 pp. 1130-1133.
50 Uchida, S., Sakoe, H., (2003) "Handwritten character recognition using elastic matching based on a class-dependent deformation model". 7th International Conference on Document analysis and Recognition, pp. 163-167.
51 Ullman, S. (1998) "Three-dimensional object recognition based on the combination of views", Cognition, vol.67, pp.21-44.
52 Valveny E., Marti E. (1999) "Application of deformable template matching to symbol recognition in handwritten architectural drawings". 5th International Conference on Document Analysis and Recognition, pp. 483-486.
53 Walker, M., Messom, C.H., (2002) "A comparison of genetic programming and genetic algorithms for auto-tuning mobile robot motion control". IEEE International Workshop on Electronic Design, pp. 507-509.
54 Yu, Z., Jain, A.K., (2000) "Object localization using color, texture and shape", Pattern Recognition, vol.33, n° 4, pp. 671-84.

55 Yuille, A., Halliman, P., Cohen, D., (1992) "Feature extraction from faces using deformable models", International Journal of Computer Vision, vol.8, nº 2, pp.99-111.
56 Yung, N., Lai, A., (1998) "Detection of vehicle occlusion using a generalized deformable model". IEEE International Symposium on Circuits and Systems, vol. 4, pp. 154-157.

2 Foveated Vision Sensor and Image Processing – A Review

Mohammed Yeasin[1], Rajeev Sharma[2]

1. Department of Electrical and Computer Engineering, University of Memphis, TN 38152-3180
   Email: myeasin@memphis.edu

2. Department of Computer Science and Engineering The Pennsylvania State University, University Park, PA-16802

Abstract. The term foveated vision refers to sensor architectures based on smooth variation of resolution across the visual field, like that of the human visual system. The foveated vision, however, is usually treated concurrently with the eye motor system, where fovea focuses on regions of interest (ROI). Such visual sensors expected to have wide range of machine vision applications in situations where the constraint of performance, size, weight, data reduction and cost must be jointly optimized. Arguably, foveated sensors along with a purposefully planned acquisition strategy can considerably reduce the complexity of processing and help in designing superior vision algorithms to extract meaningful information from visual data. Hence, understanding foveated vision sensors is critical for designing a better machine vision algorithm and understanding biological vision system.

This chapter will review the state-of-the-art of the retino-cortical (foveated) mapping models and sensor implementations based on these models. Despite some notable advantages foveated sensors have not been widely used due to the lack of elegant image processing tools. Traditional image processing algorithms are inadequate when applied directly to a space-variant image representation. A careful design of low level image processing operators (both the spatial and frequency domain) can offer a meaningful solution to the above mentioned problems. The utility of such approach was explefied through the computation of optical flow on log-mapped images.

Key words Foveated vision, Retino-cortical mapping, Optical flow, Stereo disparity, Conformal mapping, and Chirp transform.

2.1 Introduction

The amount of data that needs to be processed to extract meaningful in-formation using uniform sampling cameras is often enormous and also

M. Yeasin and R. Sharma: *Foveated Vision Sensor and Image Processing A Review*, Studies in Computational Intelligence (SCI) **7**, 57–98 (2005)
www.springerlink.com

redundant in many machine vision applications. For example, in case of autonomous navigation [1, 2], vergence control [3, 4, 5], estimation of time-to-impact [6, 7, 8], object recognition [9, 10, 11] and object tracking [12, 13, 14], one usually needs a real-time coordination between sensory perception and motor control [15]. A biologically motivated sensor along with purposefully planned acquisition strategy can considerably reduce the complexity of processing. Hence, the main theme behind developing a space-variant sensor is to establish an artificial vision and sensory-motor coordination. The aim could also be to understand how the brain of living systems sense the environment also transform sensory input into motor and cognitive functions by implementing physical models of sensory-motor behaviors.

Studies on primate visual system reveal that there is a compromise which simultaneously provides a wide field of view and a high spatial resolution in the fovea. The basis of this compromise is the use of variable resolution or Foveated vision system [16]. The term Foveated vision refers to sensor architectures based on smooth variation of resolution across the visual field, like that of the human visual system. Like the biological retina, sensor with a high resolution fovea and a periphery whose resolution decreases as a function of eccentricity can sample, integrate, and map the receptor input to a new image plane. This architecture is an efficient means of data compression and has other advantages as well. The larger receptive fields in the periphery integrate contrast changes, and provide a larger separation for sampling the higher velocities. Their elegant mathematical properties for certain visual tasks also motivated the development of Foveated sensors. Foveated architectures also have multi-resolution property but is different from the *pyramid architecture* [17]. Despite some notable advantages space-variant sensors have not been widely used due to the lack of elegant image processing tools.

Nevertheless, the use of space-variant visual sensor is an important factor when the constraint of performance, size, weight, data reduction and cost must be jointly optimized while preserving both high resolution and a wide field of view. Applications scenarios of such sensors include:

- Image communication over limited bandwidth channels such as voice-band telephony [18] and telepresence [19, 20].
- Surveillance applications for public spaces (e.g., intelligent highway applications, factories, etc.) [21] and private spaces (e.g., monitoring vehicles, homes and etc.)[22].
- Applications in which visible or infra-red camera system is used to analyze a large work area [23], and communicate the scene interpretation to a human observer via non-visual cues.

- Field applications (for example, agriculture, forestry, and etc.) in which identification and classification from a wide field of view must be performed by a small, low power portable system and communicated to the human user.
- Autonomous and tele-operated vehicle control.

The broad range of applications mentioned above is by no means exhaustive, rather, an indication of the potential advantages that a biologically motivated design can offer to the large segment of machine vision and image communication. Although many difficult problems are confronted in the application of a space-variant sensor, one is motivated by the example of biological vision, the useful geometric characteristics and elegant mathematical properties of the sensor mapping, favorable space-complexity and synergistic benefits which follows from the geometry as well as from the mapping.

The physiological and perceptual evidence indicates that the log-map image representations approximates the higher vertebrate visual system quite well and have been investigated by several researchers during the past several decades (for example, [24, 25, 26, 27, 28]). Apart from these, a variety of other space-variant sensors has been successfully developed (for example, ESCHeR [29, 30]), and has been used for many machine vision tasks with a proven record of success and acceptable robustness [31, 32]. The problem of image understanding takes a new form with foveated sensor as the translation symmetry and the neighborhood structure in the spatial domain is broken by the non-linear logarithmic mapping. A careful design of low level image processing operator (both the spatial and frequency domain) can offer a meaningful solution to the above problems. Unfortunately, there has been little systematic development of image understanding tools designed for analyzing space-variant sensor images.

A major objective of this chapter is (i) to review the state-of-the-art of foveated sensor models and their practical realizations, and (ii) to review image processing techniques to re-define image understanding tools to process space-variant images. A review catadioptric sensor [33, 34, 35, 36, 37] and panoramic camera [38, 39] which also share similar characteristics i.e., variable resolution and wide field of view were not included. The rest of the chapter is organized as follows. Section 2 review the retino-cortical mapping models reported in the literature. The synergistic benefits of log-polar mapping were presented in Section 3.Following this; Section 4 presents the sensor implementations to date to provide a picture of the present state-of-the-art of the technology. Subsequently, discussions on the space-variant form of the spatial and frequency-domain image processing operators to process space-variant images were presented in Section 5 Section 6 presents the space-variant form of classic vision algorithms (for example, optical flow on log-mapped image plane). The utility of the biologically motivated sensors were discussed in Section 7 and finally, Section 8 concludes the chapter with few concluding remarks.

2.2 A Review of Retino-cortical Mapping Models

The visual system has the most complex neural circuitry of all sensory systems. The flow of visual information occurs in two stages [40]: first from the retina to the mid-brain and thalamus, then from thalamus to the primary visual cortex. Although, the primate eye has components serving functions similar to those of standard video cameras – the eye's light transduction component, the retina, differs greatly from its electronic counterpart. Primate visual field has both binocular and monocular zones. Light from the binocular zone strikes the retina in both eyes, whereas light from the monocular zone strikes the retina only in the eye on the same side. The retina responds to the light intensities over a range of at least 5 orders of magnitude which is much more then standard cameras. Structurally, the retina is a three layer membrane constructed from six types of cells (for details please see [41]). The light transduction is performed at the photoreceptors level, and the retinal output signals are carried by the optic nerve which consists of the ganglion cell axons. The ganglion cell signals are connected to the first visual area of the cortex (V1) via an intermediary body.

The investigation of the space-variant properties of the mammalian retino-cortical mapping dates back to the early 1940s. In 1960s Daniel et. al. [42] introduced the concept of cortical magnification factor μ_c, measured in millimeters of cortex per degree of visual angle, in order to characterize the transformation of visual data for retinal coordinates to primary visual cortex. The magnification factor is not constant across the retina, but rather varies as a function of eccentricity. Empirically, the cortical magnification factor has been found to be approximated by [43]

$$\mu_c(\rho) = \frac{C_1}{1 + C_2 \rho}, \tag{1}$$

where ρ is the retinal eccentricity measured in degrees, and C_1 and C_2 are experimentally determined constants related to the foveal magnification and the rate at which magnification falls off with the eccentricity, respectively. Integrating Equation (1) yields a relationship between the retinal eccentricity and cortical distance r

$$r(\rho) = \int_0^{\rho} \frac{C_1}{1 + C_2\rho} d\rho = \frac{C_1}{C_2} \log(1 + C_2\rho). \tag{2}$$

To obtain an understanding of the variable resolution mechanism involved in the retina-to-cortex data reduction, one needs to understand the different aspects of the primate visual path ways (see [40] for details). Researchers from inter-disciplinary fields have been investigating this issue for quite some time and Schwartz [43] has pointed out that the retino-cortical mapping can be conveniently and concisely expressed as a conformal transformation[1], i.e., the log(z) mapping. This evidence does not by any means paint a complete picture about the processing and extent of data reduction performed by the retina. Nevertheless, it lays the foundation for the retino-cortical mapping models reviewed in this paper. Conceptually, the log(z) retino-cortical model consists of considering the retina as a complex plane with the center of fovea corresponding to the origin and the visual cortex as another complex plane. Retinal positions are represented by a complex variable z, and the cortical position, by a complex variable Ω. The correspondence between these two planes is dictated by the function $\Omega = \log(z)$. The mapping model $\Omega = \log(z)$, has a singularity at the origin i.e. at $z = 0$, which complicates the sensor fabrication.

To avoid the singularity at origin and to fabricate a physical sensor, Sandini et. al. [27, 44, 45] have proposed a separate mapping models for the fovea and the periphery. These mappings are given by equations (3) and (4) for continuous and discrete case, respectively:

$$\begin{cases} \eta = q\theta, \\ \xi = \log_a \dfrac{\rho}{\rho_0} - \rho, \end{cases} \tag{3}$$

$$\begin{cases} \eta = q\theta_j & j \in [1,....,N_{ang}], \\ \xi = \log_a \dfrac{\rho_1}{\rho_2} & i \in [1,.....,N_{circ}] \end{cases} \tag{4}$$

where (ρ, θ) are the polar coordinates and (ξ, η) are the log-polar coordinates. In the above expressions ρ_0 is the radius of the innermost circle, $1/q$ corresponds to the minimum angular resolution of the log-polar layout,

[1] A conformal mapping is a function of complex variable which has the property of preserving relative angles. Mathematically, a function $\Omega = f(z)$, where Ω and Z are complex variables, is conformal at the point Z if it is analytic at point z and its derivative at z is non-zero.

and p, q and a are constants determined by the physical layout of the CCD sensor that are related to the conventional Cartesian reference system by: $x = \rho\cos\theta$ and $y = \rho\sin\theta$. Though this method provides an easy way to construct a physical sensor, the fovea-periphery discontinuity is a serious drawback. In addition, the mapping is not conformal over the range of the sensor, which is an important factor in developing tools to process space-variant images.

Alternatively, Schwartz [46] proposes a modified mapping, $\Omega = \log(z+a)$ and shows that by selecting an appropriate value for a (a is real number in the range of 0.3 − 0.7 [47]), a better fit to retino topic mapping data of monkeys and cats can be obtained [48]. As opposed to the $\log(z)$ model $\log(z+1)$ provides a single output image. With modified mapping, the singularity problem, the need for uniform resolution patch in the fovea and the fovea-periphery boundary problems, is eliminated. To perform the mapping, the input image is divided into two half-planes along the vertical mid-line. The mapping for the two hemi-fields can be concisely given by the equation

$$\Omega = \log(z+ka) - \log(a)\,, \tag{5}$$

where $z = x+iy$ is the retinal position and $\Omega = \xi + i\eta$ is the corresponding cortical point, while $k = \mathrm{sgn}\quad x = \pm 1$ indicates left or right hemisphere. The combined mapping is conformal within each half plane[2]. In a strict mathematical sense, the properties of scale and rotation invariance are not present in the mapping. However, if $|z| >> a$, then $\log(z+a) \cong \log(z)$, and therefore, these properties hold. Also, since the $\log(z+a)$ template has a slice missing in the middle, circles concentric with and rays through the foveal center do not map to straight lines. To the best of our knowledge, no physical sensor exists which exactly mimics this model, but there are emulated sensor that approximates this model [24].

Another attempt to combine peripheral and foveal vision has been reported in [49] using specially designed lens. The lens characteristics are principally represented by the *projection curve* expressed in Equation (6), which maps the incident angle Θ of a sight ray entering the camera to r(Θ), the distance of the projected point on the image plane from the image center. This curve has been modeled in three distinct parts to provide wide and

[2] Note that this is similar to the anatomy of the brain: The two sides of this mapping are in direct correspondence with the two hemispheres of the brain.

high resolution images: a standard projection in the fovea, a spherical one in the periphery and a logarithmic one to do a smooth transition between the two:

$$r(\Theta) = \begin{cases} f_1 \tan\Theta, & 0 \le \Theta \le \Theta_1 \\ \log_a(f_2\Theta) - p, & \Theta_1 \le \Theta \le \Theta_2 \\ f_3\Theta + q, & \Theta_2 \le \Theta \le \Theta_{max} \end{cases}, \qquad (6)$$

where *q, p* and *a* are constants computed by solving continuity conditions on zeroth and first order derivatives, *f1, f2* and *f3* are the respective focal length (in pixels) of the three projections and Θ_1, Θ_2 and Θ_{max} are angular bounds.

It combines a wide field of view of 120 degree with a very high angular resolution of 20 pixels per degree in the fovea. Those properties were achieved by carefully assembling concave and convex optics sharing the same axis. Despite the complexity of its optical design, the physical implementation of the lens is very light and compact, and therefore suitable for active camera movement such as saccade and pursuit.

2.3 Synergistic Benefits

There are a number of synergistic benefits which follows from a biologically motivated (i.e., complex log-mapping, log-polar, etc.) sensor. Like the human eye, a foveated sensor does not require a high quality optics off-axis, as conventional cameras do, since peripheral pixels are in effect low-pass filters. The complex log-mapping also provides a smooth multi-resolution architecture [47] which is in contrast with the truncated pyramid architecture[3] that is common in machine vision [17]. The scale and rotation invariant properties of the mapping simplifies the calculation of radial optical flow of approaching objects, allowing the system to quickly calculate the time to impact. The selective data reduction is helpful in reducing the computation time and is useful in many image analysis and computer vision application.

A mentioned earlier, retino-cortical mapping model provides a scale and rotation invariant representation of an object. The scale and rotation invariance of the transformation is illustrated (see Fig 2.1) by mapping bars of various size and orientation from standard Cartesian representation to a

[3] The truncated pyramid architecture provides a data structure which is coarsely sampled version of the image data.

cortical representation. Figure 2.1 shows the results of the on center bars and the off-center bars. Clearly, the mapping (cortical representation) produce results which is independent of the size and orientation of the bar. It is important to note that the above properties hold if the rotation and scaling are centered about the origin of the complex-plane. This is due to the fact that the inertial axis is not unique, and can be ambiguous. The scale and rotation invariance property is of paramount importance and can be used to improve form invariant shape/object recognition. Traditional shape/object recognition schemes (i.e., template-matching and etc.) suffer from the variance of the size and the orientation of an object. The retinotopic mapping model of the form-invariant shape recognition approach may help in recognition of two-dimensional shapes independently from their position on the visual field, spatial orientation, and distance from the sensing device. The complex log-mapping has some favorable computational properties. It embodies a useful isomorphism between multiplication in its domain and addition in its range. It has line-circle duality[4], which may be an interesting property for finding invariant features in the processing of space-variant images. For an image sensor having a pixel geometry given by $\Omega = \log(z)$, image scaling is equivalent to radial shifting and image rotation is equivalent to annular shifting. Let us assume that the image is scaled by some real amount *S*, which can be written as $\rho e^{j\theta} \rightarrow S.\rho e^{j\theta}$ Applying the log-mapping one would obtain,

$$\log(S.\rho e^{j\theta}) = \log S + \log \rho + j\theta . \tag{7}$$

Similarly, rotating the image by an angle α can be written as $re^{j\theta} \rightarrow re^{j(\theta+\alpha)}$. A log-mapping leads to a relation,

$$\log(re^{j(\theta+\alpha)}) = \log r + j(\theta + \alpha). \tag{8}$$

From equations (7) and (8) it is clear that scaling and rotation produces a shift along the radial and the annular directions, respectively.

[4] The log-mapping transforms lines and circles onto each other.

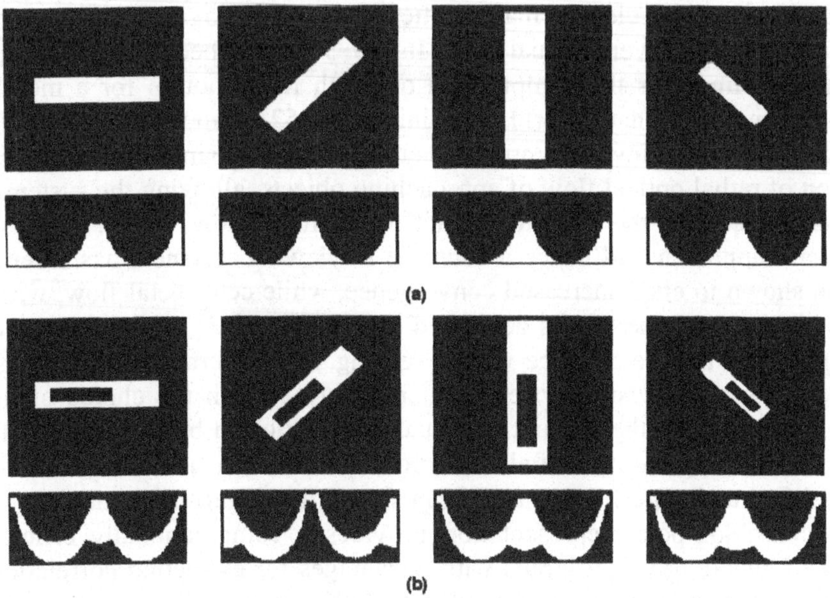

Fig. 2.1: Scale and rotation invariance properties of log-mapping: Log-polar mapping of (**a**) on-center bars and (**b**) off-center bars with various size and orientation. Upper row corresponds to the Cartesian representation and the bottom row is corresponding cortical representation

To illustrate further, a geometrical interpretation of the above concepts is shown in Fig.2.2. Consider a circle that is originating at the center of the fovea (see Fig. 2.2(a)), maps on to a straight vertical line in the peripheral grid (see Fig. 2.2(b)). An increase of the radius of the circle in the Figure 2.2(a) resulted in a shift in the Figure 2.2(b). Rotating a ray about the origin (see Fig. 2.2(c)) produces a shift as shown in Fig. 2.2(d). These properties of the log-mapping have been successfully utilized with regard to the computations in a moving visual field [7]. These properties can also be exploited for the detection of general straight lines, line segments, and circles through the foveation point.

While the use of the log-mapping greatly simplifies the rotation and scale invariant image processing, it significantly complicates the image translation (see Fig. 2.3). The vertical contours representing horizontal translation in the input image Fig. 2.3(a) result curved contours in the log-polar image shown in Fig. 2.3(b). Similarly, Figs. 2.3(c) and 2.3(d) exemplify the effect of vertical translation. It is evident that spatial neighborhood structure in the spatial domain is broken by the space-variant properties of the sensor. Traditional image processing techniques do not hold when applied directly to a space-variant image representation.

Apart from the elegant mathematical properties, logarithmic mapping greatly simplify several visual tasks. In [50, 51] it has been shown how the mapping simplifies the computation of depth from motion for a moving camera in a stationary world. Sandini et. al [52] demonstrated how the scale and rotation invariant properties of the mapping simplifies the calculation of radial optical flow of approaching objects, allowing the system to quickly calculate the time to impact. Centrifugal flow, which signals a forward approach and hence a decrease in viewing distance, has recently been shown to elicit increased convergence, while centripetal flow, which signals the converse, elicits decreased convergence [53]. In [3] Capuro et. al. proposed the use of space-variant sensing, as an alternative imaging geometry for robot vision systems. Interestingly enough the choice of this geometry reduces the amount of visual information to be processed without constraining the visual field size, nor the resolution, and allow for more simplified techniques. It has also been shown that logarithmic mapping, in particular, log-polar representation provides a computationally efficient way of encoding visual inputs with advantages for extracting correlations between binocular images without the need to derive disparity explicitly [3, 54]. In [5], it has been shown that applying correlation techniques on log-polar images produce much better results than standard Cartesian images. It has been argued that the correlation between two log-polar images corresponds to the correlation of Cartesian images weighted by the inverse distance to the image center. Hence, the characteristic of the implicit weighting function (dominance of the areas close to the image center) provides a measure of focus of attention. Space-variant sensors implicitly enhances objects that happen to lie close to the fixation point and through this provides a pre-categorical, fast selection mechanism which requires no additional computation [53].

In a recent study [54] by Sandini et. al. it has been suggested that a reciprocal interaction between biologists and computer vision scientists on a common ground may highlight more synergies. For an example, in a recent study on gaze stabilization mechanisms in primates that deal with the problems created by translational disturbances of the observer were introduced in the context of robotic control. It was found that robots have benefited from inertial sensors that encode the linear as well as angular accelerations of the head just as the human occulomotor does.

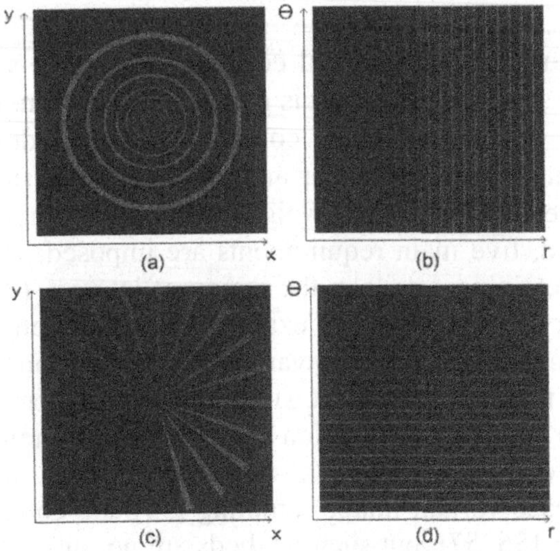

Fig. 2.2: Duality of log-mapping: (**a**) and (**c**) shows retinal image (complex image representation, i.e., $z = x + jy = re^{j\theta}$) while (**b**) and (**d**) shows cortical images (i.e., log-mapped images).Circles centered at the origin as shown in (**a**) maps onto lines in (**b**). Rotating a ray about the origin of (**c**) results in a shift in (**d**)

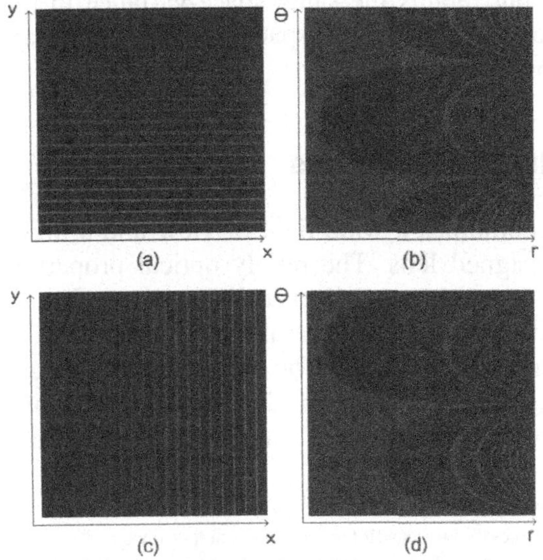

Fig. 2.3: Translation properties of log-mapping: of similar image representation as shown in Fig. 2.2. (**a**) horizontal translation, (**b**) the corresponding log-polar image, (**c**) and (**d**) shows similar images for vertical translation

2.4 Space-variant Vision Sensor

With the introduction of biological concepts, the space-variant vision architecture and issues related to this is gaining momentum, especially in the field of robotic vision and image communication. In designing a visual sensor for an autonomous robot or any other visual system in which the constraints of performance, size, weight, data reduction and cost must be jointly optimized, five main requirements are imposed: (1) a high resolution fovea for obtaining details in the region of interest, (2) a wide field of view useful for many tasks, for example, interest point determination, (3) fast response time[5], (4) smooth variation of resolution across the visual work space and finally, (5) the cost, size and performance[6] of the sensor.

To develop a space-variant vision sensor, several attempts to combine peripheral and foveal vision have already been made in the past decades. For example, space-variant image sampling [55] and combination of wide and tele cameras [56, 57], but such methods are not discussed in this chapter as they do not fit in to the context of our discussion. Studies reveal that there are mainly two types of space-variant sensors available and the clarification of this issue will go a long way towards clarifying several basic issues. First, one could work in 'cortical' plane, which has fundamentally different geometry than the 'retina', but in which the size of the pixels increases towards the periphery. Second, one in which the image geometry is still Cartesian, but retains the same space-variance in the pixel structure. Successful efforts in developing space-variant sensors are summarized in subsequent subsections.

2.4.1 Specially Designed Lens

This approach combines a wide field of view and a high resolution fovea by specially designed lens. The purely optical properties of this method avoid most of the problems involved in space-varying sensor design and implementation, e.g., co-axial parallelism, continuity, hardware redundancy and computational cost. The foveated wide-angle lenses to build space-varying sensors reported in [29] follow the design principles proposed in [58], while improving visual acuity in the fovea, and providing a

[5] Low space-complexity i.e. a small fast to process output image. The space complexity of a vision system is a good measure of the computational complexity, since the number of pixel which must be processed is the space-complexity. Thus, even though the space-complexity does not entirely determine the computational complexity (which depends on many factors and specification of the algorithm), it is believed that the computational complexity is likely to be proportional to space-complexity.

[6] The sensor must preserve the translational and rotational invariance property.

constant, low image compression rate in the periphery which can be useful for periphery-based motion detection. Drawbacks associated with this kind of approach include low photo-sensitivity in the periphery and strong optical deformations in the images that can be challenging for object recognition algorithms. We describe an instance of a complete system in the next subsection to introduce the reader to the related development.

It is important to note that getting an space-varying sensor itself does not solve the problem, the very spirit it has been chosen. It is important to place them strategically (like human visual system), i.e., we need proper platform to make the information extraction process easier. One such architecture is **ESCHeR**, an acronym for **E**tl **S**tereo **C**ompact **H**ead for **R**obot vision, is a custom designed high performance binocular head [59]. Its functionalities are very much inspired by the physiology of biological visual systems. In particular, it exhibits many characteristics similar to human vision: Certainly, the most distinguishing and unique feature of ESCHeR lies in its lenses. Although rarely found in robotic systems, foveal vision is a common characteristic of higher vertebrates. It is an essential tool that permits both a global awareness of the environment and a precise observation of fine details in the scene. In fact, it is also responsible to a great extent for the simplicity and robustness of the target tracking.

ESCHeR was one of the first binocular heads that combines high dynamic performance, in a very compact and light design, with foveal and peripheral vision. The lenses provide the ability to globally observe the environment and precisely analyze details in the scene, while the mechanical setup is capable of quickly redirecting the gaze and smoothly pursue moving targets.

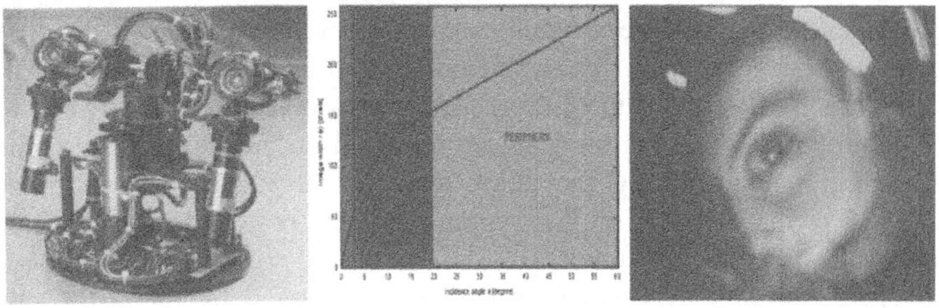

Fig. 2. 4: **ESCHeR, a high performance binocular head**. A Picture of ESCHeR (*left*), the lens projection curve (*middle*), and an image of a face (*right*) taken with its foveated wide-angle lenses (adopted from [60])

2.4.2 Specially Designed Chips

This foveated sensor has been designed by several groups from the University of Genoa, Italy, University of Pennsylvania, USA, Scoula Superore S Anna of Pisa, and has been fabricated by IMEC in Leuven, Belgium [61, 62, 63]. It features a unique concept in the VLSI implementation of a vision chip. The foveated chip, which uses a CCD process, mimics the physically foveated retina of human. This approach in designing a foveated sensor adopts a distribution of receptors of size gradually increasing from the center to the periphery. The chip has a foveated rectangular region in the middle with high resolution and a circular outer layer with decreasing resolution. This mapping provides a scale and rotation invariant transformation. The chip has been fabricated using a triple-poly buried channel CCD process provided by IMEC. The rectangular inner region has 102 photo-detectors. The prototype has the following structure: the pixels are arranged on 30 concentric circles each with 64 photosensitive sites. The pixel size increases from 30 micron × 30 micron to 412 micron × 412 micron from the innermost circle to the outermost circle. The total chip area is 11 mm × 11 mm. The video acquisition rate is 50 frames per second. The total amount of information stored is less than 2 Kbytes per frame. Thus, the chip realizes a good trade-off between image resolution, amplitude of the visual field and size of the stored data. In references [61, 62, 63] other aspects of the design, such as read-out structures, clock generation, simple theories about the fovea, and hardware interface to the chip are described.

The foveated CMOS chip designed by by the IMEC and IBIDEM consortium [Ferrari et al. 95b, Ferrari et al. 95a, Pardo 94], and dubbed "FUGA", is similar to the CCD fovea described in Section 2.11 [van der Spiegel et al. 89]. The rectangularly spaced foveated region in the CCD retina has been replaced by reconfiguring the spatial placement of the photo-detectors. As a result of this redesign, the discontinuity between fovea and the peripheral region has been removed. In the CCD retina a blind sliced region (for routing the clock and control signals) exists. In the FUGA18 retina the need to this region has been removed by routing the signals through radial channels.

A latest version of the sensor has been designed by the IMEC and IBIDEM consortium using [64, 62] the CMOS technology, without compromising the main feature of the retina-like arrangement. In the CCD retina a blind sliced region (for routing the clock and control signals) exists but in CMOS version this has been removed by routing the signals through radial channels. Several versions of the FUGA chip with different sizes have been designed and manufactured by IMEC. Most recent version of this sensor 30, 000 pixels, a figure allowing a 3 to 4 times increase with respect to the old CMOS chip which has 8, 013 pixels. The color version of the chip was obtained by micro-deposition of filters over the monochromatic layout. The pixel's layout is the same as the IBIDEM retina and is composed of 8, 013 pixels.

Wodnicki et. al. [65, 66] have also designed and fabricated a foveated CMOS sensor. The fovea photo-detectors are uniformly spaced in a rectangle and the periphery photo-detectors are placed in a circular array. The pixel pitch in the fovea is 9.6μm in a 1.2μm process. This degree of resolution requires substrate biasing connection to be located outside of the sensor matrix. Photo-detectors have been realized using circular parasitic well diodes operating in integrating mode. Biasing is accomplished with a ring of p4 diffusion encircling the sensor matrix. The area of the photo detector in the circular outer region increases exponentially, resulting in the log-polar mapping. The chip has been fabricated in a 1.2μm CMOS process. It has 16 circular layers in the periphery. The chip size is 4.8 mm × 4.8 mm.

2.4.3 Emulated Chips

Apart from the above, few other emulated sensor implementation has been reported in the literature. For example, the AD2101 and TI320C40 are DSP's used in cortex I and cortex II [67], respectively, with conventional CCD (e.g., Texas-Instruments TC211 CCD used in Cortex-I) to emulate log-map sensor. The $log(z + a)$ mapping model has been used instead of mapping the foveal part with polar mapping and periphery with logarithmic mapping. This ensures the conformality of the mapping at the cost of managing a discontinuity along the vertical-midline. In a similar manner, another log-map sensor using an overlapping data reduction model has been reported in [41]. Next section focuses on image understanding tools to analyze space-variant images; in particular, the log-mapped images.

2.5 Space-variant Image Processing

This chapter discusses the space-variant image processing in a deterministic framework. Humans are accustomed in thinking of an image as a rectangular grid of rectangular pixel where the connectivity and adjacency are well defined. The scenario is completely different for a space-variant image representation. Consequently, image processing and pattern recognition algorithms become much more complex in space-variant systems than in standard imaging systems. There are several reasons for this, namely, the complex neighborhood connectivity and the lack of shift invariant processing. It is important to keep in mind that there are two types of space variance and the clarification of this issue will go a long way towards clarifying several basic issues. First, one could work in a 'retinal plane' in which the image geometry is still Cartesian, but the size of the pixels increases towards the periphery. Second, one could work in a 'cortical' plane, which has a fundamentally different geometry than the 'retina', but retains the same space-variance in the pixel structure. Fig. 2.5 shows an example of a log-polar mapped image. From Fig. 2.5 it is readily seen that image feature changes size and shape as it shifts across the field of a space-variant sensor. The frequency-domain and the spatial domain image processing techniques to process such a complicated image are reviewed in subsequent subsections.

2.5.1 Space-variant Fourier Analysis

As mentioned earlier, the shift-invariant property of the Fourier transform does not hold since translation symmetry in the spatial domain is broken by the space-variant properties of the map. It has been shown in [68, 69] that it is indeed possible to solve the seemingly paradoxical problem of shift invariance on a strongly space variant architecture. The following subsections will systematically discuss the related developments.

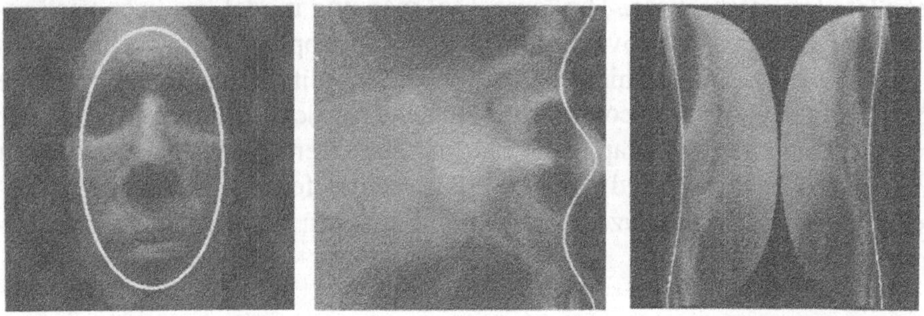

(a) (b) (c)

Fig. 2.5: Illustration of complex neighborhood: (**a**) standard camera image, (**b**) retinal plane representation of log-mapped image, (**c**) cortical plane representation of the log-mapped image. The white line shows how the oval shape maps in log-mapped plane

2.5.1.1 The Generalized Chirp Transform

Given a one dimensional signal *f(x)* and an invertible mapping or transformation $\Omega : x \rightarrow \xi, \Omega \in C^1$, the Fourier transform of *f(x)*

$$F(f) = \int_{-\infty}^{\infty} f(x) e^{-j2\pi fx} dx . \tag{9}$$

By using the Jacobian in the ξ space and by changing the notation one can obtain,

$$\mathrm{E}(f) = \int_{-\infty}^{\infty} f(x(\xi)) \frac{\partial x(\xi)}{\partial \xi} e^{-j2\pi fx(\xi)} d\xi . \tag{10}$$

Defining a kernel as $\kappa(f,\xi) = \frac{\partial x(\xi)}{\partial \xi} e^{-j2\pi fx(\xi)}$, and rewriting equation (10) one can get,

$$\mathrm{E}(f) = \int_{-\infty}^{\infty} f(x(\xi)) \kappa(f,\xi) d\xi . \tag{11}$$

The integral equation (11) is called the exponential chirp transform. A close look at this equation reveals that the transform is invariant up to a phase under translation in the x domain. This follows from the Fourier shift theory which is simply transformed through the map function.

2.5.1.2 1-D Continuous Exponential Chirp Transform (ECT)

Let us consider the 1-D transformation[7] [7]of the following form:

$$\xi(x) = \begin{cases} \log(x+a) & x \geq 0, \\ 2\log(a) - \log(-x+a), & x < 0. \end{cases}$$

For which the kernel as in equation (11) is

$$\begin{matrix} e^{\xi - j2\pi f(e^{\xi} - a)} & \xi \geq \log(a) \\ a^2 e^{-\xi - j2\pi f(a - a^2 e^{-\xi})} & \xi < \log(a) \end{matrix} . \tag{12}$$

[7] This represents a logarithmic mapping in which the singularity at the origin is removed by defining two separate branches, using some finite positive ‘a’ to provide a linear map for ||*x*|| >> *a*.

This kernel is reminiscent of a chirp with the exponentially growing frequency and magnitude. Hence, aliasing must be carefully handled, due to the rapidly growing frequency of the kernel.

2.5.1.3 2-D Exponential Chirp Transform

Given a 2-D function f(x, y) and an invertible and differentiable transform $\Omega:(x,y)\rightarrow(\xi,\eta)$, the 2-D ECT is defined by the following integral transform:

$$\mathrm{E}(k,h)=\iint_{-\infty}^{+\infty}(x(\xi,\eta),y(\xi,\eta))\kappa(\xi,\eta,k,h)d\xi d\eta\,, \tag{13}$$

where k and h are the respective Fourier variables. The ECT in equation (13) can be written as

$$\mathrm{E}(k,h)=\iint_{D}f(\xi,\eta)e^{2\xi}e^{-2\pi j(k(e^{\xi}\cos(\eta)-a)+he^{\xi}\sin(\eta))}d\xi d\eta\,, \tag{14}$$

where D is over the range $-\infty\leq\xi<\infty$ and $\frac{3\pi}{2}\leq\eta<\frac{\pi}{2}$. From equation (14) it is readily seen that the integral transform can be evaluated directly with a complexity of $O(M^2N^2)$, where M and N are the dimension of the log-mapped image.

2.5.1.4 Fast Exponential Chirp Transform

The ECT in equation (14) can be written as

$$\mathrm{E}(k,h)=e^{j2\pi ak}\iint_{D}f(\xi,\eta)e^{2\xi}e^{-j2\pi(k(e^{\xi}\cos(\eta)))+(he^{\xi}\sin(\eta))}d\xi d\eta\,. \tag{15}$$

By introducing the log-mapping in frequency, centered on the frequency origin, it has been shown in [68] that the above equation can be written as

$$\mathrm{E}(r,\theta)=e^{j2\pi ak(r,\theta)}\iint_{D}(f^*(\xi,\eta)e^{2\xi-j2\pi be^{\xi}\cos\eta})*e-j^{2\pi e^{(r+\xi)}\cos(\theta+\eta)}d\xi d\eta\,, \tag{16}$$

where b is a real number and the superscript * stands for a complex conjugate of the function. From equation (16) it is simple to see that the ECT can be computed as a complex correlation. The numerical implementation of equation (16) is referred as the FECT[8]. The inverse FECT (IFECT), 2D discrete ECT and their implementation details can be found in [68].

2.5.1.5 Antialiasing Filtering

When a signal is not sampled at a sufficiently high rate, aliasing error occurs in the reconstructed signal. In order to anti-alias, one must filter out the set of samples from the exponential chirp kernel that do not satisfy the following inequality:

$$\frac{\log(R+1)}{N} \geq N_\xi v_\xi$$
$$\frac{\log(2\pi)}{M} \geq N_\eta v_\eta$$

Where v_ξ, v_η are the 2-D instantaneous frequencies of the complex kernel, N_ξ and N_η are the Nyquist factors, and N and M are the length of the vectors ξ_n and η_m, respectively. Antialiasing filtering can be achieved by multiplying the kernel by the 2-D Fermi function

$$\Phi\left[\frac{\log(R+1)}{N} - N_\xi v_\xi, \frac{2\pi}{M} - N_\eta v_\eta\right].$$

This function can be incorporated in the chirp transform and in equation (16), giving the following cross-correlation *(b = 0)*:

$$e^{j2\pi ah(\gamma,\theta)} \iint_D (f(\xi,\eta)e^{2\xi})e^{j2\pi e^{(\gamma+\xi)}\cos(\theta+\eta)} . \Phi\left[\frac{\log(R+1)}{N} - N_\xi v_\xi, \frac{2\pi}{M} - N_\eta v_\eta\right] d\xi d\eta \cdot (17)$$

The ECT described in this section has been used in [68] for image filtering, cross-correlation. It is simple to see that the ECT discussed in this section can used for the frequency domain analysis of space-variant images. As the ECT guarantees the shift invariance hence it is straightforward to adopt the ECT for phase-based vision algorithms (for example, phase-based disparity and phase-based optical flow and etc.).

[8] This is a slightly different usage than, for example, the FFT where the fast version of the DFT produces result identical to the DFT. The FECT produces results which are re-sampled versions of the DECT due to the log-map sampling in the frequency. Although the FECT is a homeomorphism of the log-mapped image (i.e. invertible and one to one), the DECT and FECT are not numerically identical.

2.5.2 Space-variant Metric Tensor and Differential Operators

This section discusses the space-variant form of the Δ operator, which yields the space-variant form of the gradient, divergence, curl and Laplacian operator.

2.5.2.1 Metric Tensor of the Log-mapping

The metric tensor is a multi-linear map which describes what happens to an infinitesimal length element under the transformation. A useful way to understand the effects of the log-mapping on the standard Cartesian operators is in terms of the metric tensor of the complex log domain. As the coordinate transform is space-variant, so does the metric tensor as a function of the log coordinate. Formally, the metric tensor T of a transformation z from a coordinate system *(ξ, η)* in to another coordinate system *(x, y)* is given by

$$T = \begin{bmatrix} \langle z_\xi, z_\xi \rangle & \langle z_\xi, z_\eta \rangle \\ \langle z_\xi, z_\eta \rangle & \langle z_\eta, z_\eta \rangle \end{bmatrix} = \begin{bmatrix} x_\xi x_\xi + y_\xi y_\xi & x_\xi x_\eta + y_\eta y_\xi \\ x_\xi x_\eta + y_\eta y_\xi & x_\eta x_\eta + y_\eta y_\eta \end{bmatrix} = \begin{bmatrix} e^{2\xi} & 0 \\ 0 & e^{2\xi} \end{bmatrix}, \qquad (18)$$

where $\langle z_i, z_j \rangle$ stands for the inner product of the vectors. The diagonal form of T is a direct consequence of conformal mapping. That is, the metric tensor of any conformal mapping has the form $T = A\delta_{ij}$ (with equal elements on the diagonal). From equation (18) it is apparent that as distance from the fovea increases, the Cartesian length of the log-domain vector is scaled by e^ξ. Conversely, the length of a Cartesian vector mapped into the log-plane is shrinked by a factor of $e^{-\xi}$ due to the compressive logarithmic non-linearity.

2.5.2.2 Space-variant form of ∇ Operator

A conformal mapping insures that the basis vector which are orthogonal in the (ξ, η) space remains orthogonal when projected back to the Cartesian space. Since the gradient is the combination of directional derivatives, one is assured that the gradient in the log-space is of the form

$$\nabla f = A(\xi,\eta)\left(\frac{\partial f}{\partial \xi} e_\xi + \frac{\partial f}{\partial \eta} e_\eta \right), \qquad (19)$$

where e_ξ and $e\eta$ define the orthonormal basis, and $A(\xi,\eta)$ is the term that accounts for the length scaling of a vector under the log mapping. It may be noted that equation (20) holds for any conformal mapping with the specifics of the transformation expressed in the co-efficient function A. By using the invariance of the magnitude of the gradient under a change of coordinates it has been shown that the the space-variant form of ∇ is given by [47]:

$$\nabla = e^{-\xi}\left(\frac{\partial}{\partial \xi}e_{\xi} + \frac{\partial}{\partial \eta}e_{\eta}\right), \tag{20}$$

which allows the direct computation of quantities such as derivative, divergence, curl and Laplacian operator in a log-mapped plane. It may be noted here that this derivation does not account for the varying support of each log-pixel. As one moves towards the periphery of the log-mapped plane, each log-pixel is typically generated by averaging a larger region of the Cartesian space, both in the mammalian retina and in machine vision systems. The averaging is done to avoid aliasing in the periphery, and to attenuate high frequency information, partially offsetting the need for a negative exponential weighting to account for varying pixel separation. It is simple to see that the space-variant gradient operator defined in this section will prove useful for performing low level spatial domain vision operations. Next section presents classic vision algorithms (space-variant optical flow, stereo disparity, anisotropic diffusion, corner detection and etc.) on space-variant images.

2.6 Space-variant Vision Algorithms

As discussed in the previous sections, the elegant mathematical properties and the synergistic benefits of the mapping allows us to perform many visual tasks with ease. While the implementation of vision algorithms on space-variant images remains a challenging issue due to complex neighborhood connectivity and also the lack of shape invariance under translation. Given the lack of general image understanding tools, this section will discuss the computational issues of representative vision algorithms (stereo-disparity and optical flow), specifically designed for space-variant vision system. In principle, one can use the spatial and the frequency domain operators discussed in previous sections to account for the adjustment one needs to make to process space-variant images.

2.6.1 Space-variant Optical Flow

From a biologist's point of view, optical flow refers to the perceived motion of the visual field results from an individual's own movement through the environment. With optical flow the entire visual field moves, in contrast to the local motion of the objects. Optical flow provides two types of cues: information about the organization of the environment and information about the control of the posture. In computer vision, optical flow has commonly been defined as the apparent motion of image brightness patterns in an image sequence. But the common definition of optical flow as an image displacement field does not provide a correct interpretation[9] when dealing with light source motion or generally dominant shading effects. In a most recent effort to avoid this problem a revised definition of optical flow has been given in [70]. It is argued that the new representation, describing both the radiometric and the geometric variations in an image sequence, is more consistent with the common interpretation of optical flow. The optical flow has been defined as a three-dimensional transformation field, $v=[\delta x,\delta y,\delta I]^T$, where $[\delta x,\delta y]$ are the geometric component and δI is the radiometric component of the flow field. In this representation, optical flow describes the perceived transformation, instead of perceived motion, of brightness patterns in an image sequence.

The revised definition of optical flow permits the relaxation of the brightness constancy model (BCM) where the radiometric component δI is explicitly used to be zero. To compute the optical flow, the so-called generalized dynamic image model (GDIM) has been proposed; which allows the intensity to vary in the successive frames. In [70] the GDIM was defined as follows:

$$I_2(x+\delta x)=M(x)I_1(x)+C(x). \tag{21}$$

The radiometric transformation from $I_1(x)$ to $I_2(x+\delta x)$ is explicitly defined in terms of the multiplier and the offset fields *M(x)* and *C(x)*, respectively. The geometric transformation is implicit in terms of the correspondence between points x and $x+\delta x$. If one writes M and C in terms of variations from one and zero, respectively, $M(x)=1+\delta m(x)$ and $C(x)=0+\delta c(x)$ one can express GDIM explicitly in terms of the scene brightness variation field

---

[9] For example, a stationary viewer perceives an optical flow when observing a stationary scene that is illuminated by a moving light source. Though there is no relative motion between the camera and the scene, there is a nonzero optical flow because of the apparent motion of the image pattern

$$I_2(x+\delta x)-I_1(x)=\delta I_1(x)=\delta m(x)I_1(x)+\delta c(x). \tag{22}$$

Where $\delta m=\delta c=0$, the above model simplifies to the BCM. Despite a wide variety of approaches to compute optical flow, the algorithms can be classified into three main categories: gradient-based methods [71], matching techniques [72], and frequency-based approaches [73]. But a recent review [74] on the performance analysis of different kinds of algorithm suggests that the overall performances of the gradient-based techniques are superior. Hence, in this chapter will discuss the gradient-based method to compute the optical flow.

Though there are several implementations to compute the optical flow in the log-polar images (i.e., [14, 7]), but most of the algorithm fails to take into account some very crucial issues related log-polar mapping. Traditionally, the optical flow on space-variant images has been computed based on the BCM using the Cartesian domain gradient operator. On the contrary, the use of GDIM and employ the space-variant form of gradient operator (see the previous section) to compute optical flow on log-mapped image plane [75].

Using the revised definition of the optical flow and by requiring the flow field to be constant within a small region around each point, in [76, 77], it was shown that the optical flow on a log-mapped image plane can be computed by solving system of equations

$$\sum_W \begin{bmatrix} I_\xi^2 & I_\xi I_\eta & -I_\xi I & -I_\xi \\ I_\xi I_\eta & I_\eta^2 & -I_\xi I & -I_\eta \\ -I_\xi I & I I\eta & I^2 & I \\ -I_\xi & I_\eta & I & I \end{bmatrix} \begin{bmatrix} v_\xi \\ v_\eta \\ \delta m \\ \delta c \end{bmatrix} = \sum \begin{bmatrix} -I_\xi I_t \\ -I_\eta I_t \\ I I_t \\ I_t \end{bmatrix}, \tag{23}$$

where W is a neighborhood region. Please note that in a log-mapped image this neighborhood region is complicated and variable due to the nonlinear properties of the logarithmic mapping. A notion called a variable window (see Fig. 2.6) i.e., a log-mapped version of the standard Cartesian window, to preserve the local neighborhood on a log-mapped image is used to address the above problem. From Fig. 2.6(c) it is very easy to see that the size and shape of the window varies across the image plane according to the logarithmic mapping. Also the use of space-variant derivative operator was used to compute the derivative on log-mapped plane. The use of space variant form of the derivative operator is important for a better numerical accuracy as the mapping preserves the angles between the vectors, but not the magnitude.

By solving equations (23) one can compute the optical flow directly on log-mapped images. The GDIM-based model permits us to relax the brightness constancy model (BCM) by allowing the intensity to vary in the successive frames. If one explicitly set the radiometric component I to zero the GDIM models boils down to the BCM. In other words, the BCM assumption holds where the multiplier field $m = 0$ and the offset field $c = 0$. The multiplier and the offset field can become discontinuous at iso lated boundaries, just as image motion is discontinuous at occluding or motion boundaries. As a result, the estimated radiometric and geometric components of optical flow may be inaccurate in these regions. Erroneous

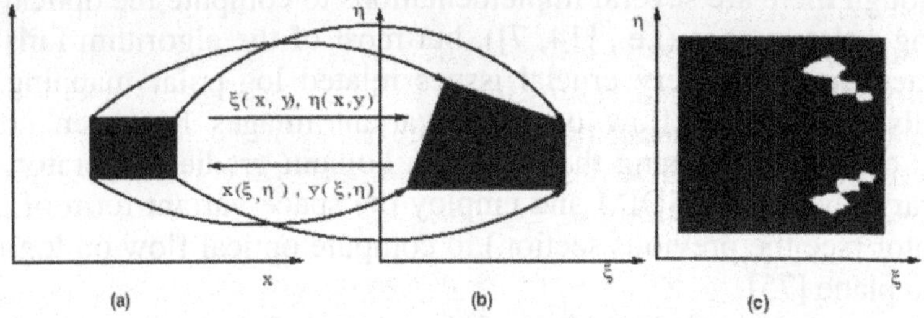

Fig. 2.6: An illustration of variable window: **(a)** A Cartesian window, **(b)** log-mapped window and **(c)** computed shape of windows across the image plane result may be detected by evaluating the residual squared-error. It has been shown that the inclusion of the above features significantly enhances the accuracy of optical flow computation directly for the log-mapped image plane (please see [75, 77])

2.6.2 Results of Optical Flow on Log-mapped Images

As mentioned earlier, the log-mapping is conformal, i.e., it preserves local angles. Empirical study were conducted with both the synthetic and the real image sequences. For real image sequences, indoor laboratory, an outdoor and an underwater scene were considered to show the utility of the proposed algorithm. Synthetically generated examples include the computed image motion using both the BCM and GDIM-based method to demonstrate the effect of neglecting the radiometric variations in an image sequence. In order to retain this property after discretization, it is wise to keep identical discretization steps in radial and angular directions.

2.6.2.1 Synthetic Image Sequences

The first image is that of a textured 256 × 256 face image (see Fig. 2.7(a)). Using a known motion (0.4 pixel horizontal motion in the Cartesian space which corresponds to 0–30 pixel image motion in the log-mapped image) and a radiometric transformation field (a Gaussian distribution of radiometric transformation field (δm) in the range between 0.8 − 1.0 and $\delta c = 0$), were used to compute the second image. The third image was derived from the first image using the above radiometric transformation only. Two sequences using frame 1–2 and 1 − 3 are considered. Fig. 2.7(b) shows a sample transformed log-mapped image of (derived from Fig. 2.7(a)).

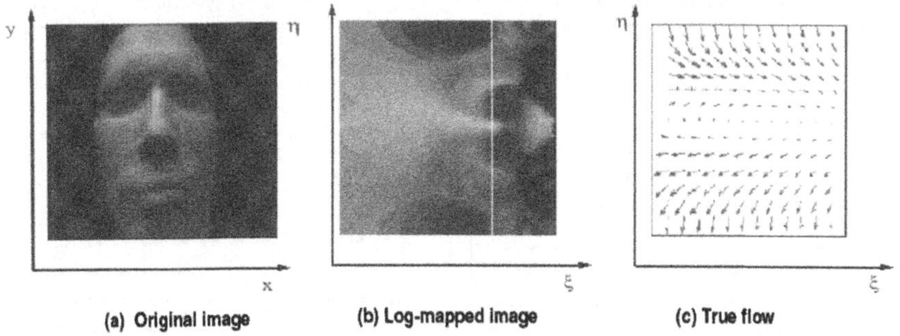

Fig. 2.7: Simulated optical flow: (**a**) a traditional uniformly sampled image, (**b**) log-map representation of the uniformly sampled image, and (**c**) true image motion used to generate synthetic image sequences

The peripheral part of the image i.e., the portion of the log-mapped image right to the white vertical line for the computation of optical flow (see Fig. 2.7(b)). The idea of using the periphery stems from biological motivation and also to increase the computational efficiency. It may be noted that the same algorithm will hold incase of computation of the optical flow for the full frame. It is also important to recognize that the computation of optical flow on the peripheral part is hard as the resolution decreases towards the periphery.

To analyze the quantitative performance the error statistics for both the BCM and GDIM methods are compared. The error measurements used here are the root mean square (RMS) error, the average relative error (given in percentage), and the angular error (given in degrees). The average relative error in some sense gives the accuracy of the magnitude part while the angular error provides information related to phase of the flow field. Compared are, at a time, the two vectors (u, v, 1) and (ˆu, ˆv, 1), where (u, v) and (ˆu, ˆv) are the ground truth and estimated image motions, respectively. The length of a flow vector is computed using the Euclidean norm. The relative error between two vectors is defined as the difference of length in percentage between a flow vector in the estimated flow field and the corresponding reference flow field:

$$\frac{\sum \| (\hat{u}-u, \hat{v}-v) \|_2}{\sum \| (u,v) \|_2} .100 . \tag{24}$$

The angular error between two vectors is defined as the difference in degrees between the direction of the estimated flow vector and the direction of the corresponding reference flow vector.

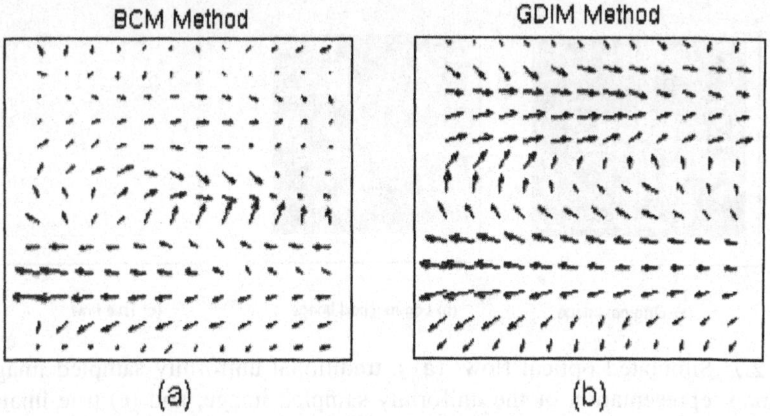

Fig. 2.8: Computed optical flow in case of both geometric and radiometric transformations. Figures 2.8(**a**) and 2.8(**b**) represents the computed flow field using BCM and GDIM methods, respectively.

Synthetically generated images with ground truth were used to show both the qualitative and the quantitative performance of the proposed algorithm. Figure 2.7(c) shows the true log-mapped image motion field which has been used to transform the image sequence 1 − 2. Figures 2.8(a) and 2.8(b) show the computed image motion as Quiver diagram for the sequence 1 − 2 using the BCM and GDIM, respectively. The space-variant form of gra-dient operator and variable window were used to compute the optical flow for both the GDIM-based and BCM-based method. A visual comparison of the Fig. 2.6(c) with Figs. 2.8(a) and 2.8(b) reveals that the image motion field estimated using GDIM method is similar to that of the true image motion field, unlike the BCM method. This result is not surprising as the BCM method ignores the radiometric transformation. To provide a quantitative error measure and to compare the performance of the proposed algorithm with the traditional method; the average relative error, which in some sense reflects the error in estimating the magnitude of the flow field were used. It was found that the average relative error 7.68 and 6.12 percent for the BCM and GDIM, respectively.

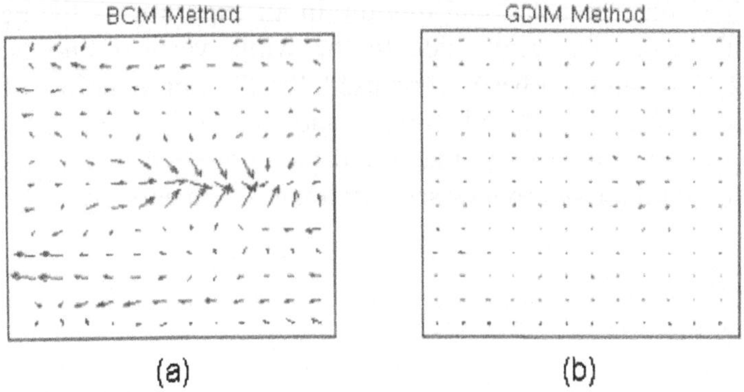

Fig. 2.9: Computed optical flow in case of radiometric transformation. Figures 2.9(**a**) and 2.9(**b**) represents the computed flow using BCM and GDIM, respectively

To provide more meaningful information about the error statistics the average angular error which in some sense reflects the error in estimating the phase of the flow field were also computed. The average angular error was found to be 25.23 and 5.02 degree for the BCM and GDIM, respectively. The RMS error was found to be 0.5346 and 0.1732 for the BCM and the GDIM method, respectively. The above error statistics clearly indicates that the performance of the proposed GDIM-based method is superior to the BCM method. Figs. 9(a) and 9(b) displays the computed optical flow using sequence 1–3, where there is no motion (only the radiometric transformation has been considered to transform the image). It is clear from the Fig. 2.9(a), when employing BCM; one obtains the erroneous interpretation of geometric transformation due to the presence of radiometric variation. On the contrary, the proposed GDIM-based method shows no image motion (see Fig. 2.9(b)), which is consistent with ground truth. Figs. 10(a)- 10(d) shows the mesh plot of the true and computed $\delta\xi$ and $\delta\eta$ components of the image motion, respectively. From Figs. 10(a)-10(d) it is evident that the proposed method estimated the spatial distribution of the image motion quite accurately.

2.6.2.2 Real Image Sequences

To further exemplify the robustness and accuracy of the proposed method, empirical studies were conducted using real sequence of images captured under both the indoor and the outdoor a well as using under water camera by fixing the camera parameters. The motion for the under water and the outdoor sequence of images were dominantly horizontal motion, while the motion for the indoor laboratory was chosen to be the combination of rotation and horizontal translational motion. In all experiments the peripheral portion of images i.e., right side to the white vertical line (see Figs. 2.11(b), 2.12(b) and 2.13(b)) were used for the computation of optical flow. Figures 2.11(a)–(c), 2.12(a)–(c) and 2.13(a)–(c) shows a sample frame, log-polar transformed image and the computed image motion for under water, outdoor and indoor scenery images, respectively.

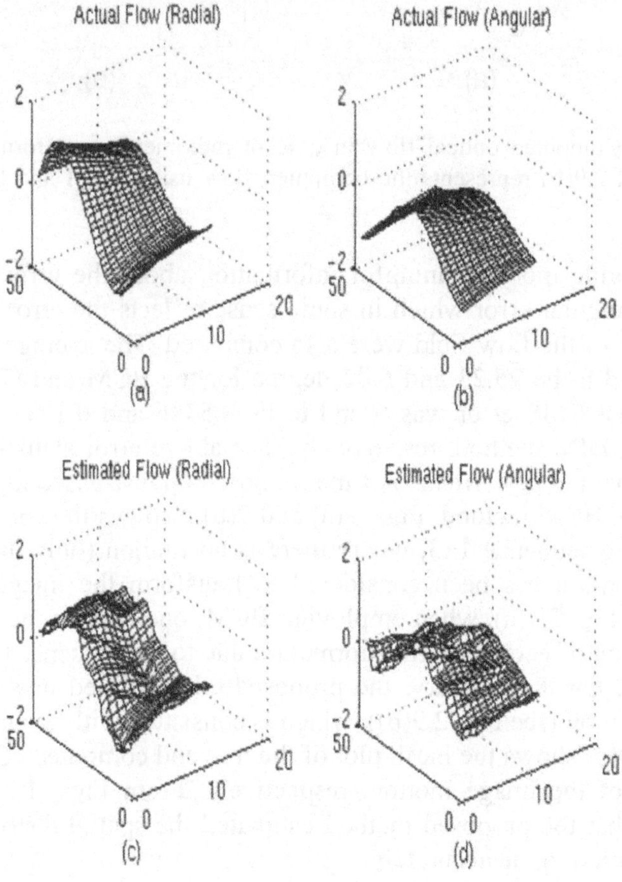

Fig. 2.10: Quantitative comparison of true flow and computed flow using GDIM method. (**a**) and (**b**) shows the true flow and (**c**) and (**d**) shows the computed flow in the radial and angular directions, respectively

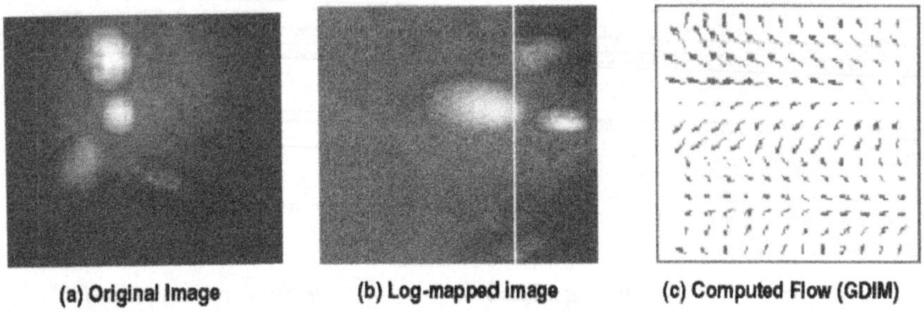

Fig. 2.11: Optical flow computation using an under water scene. (**a**) sample image from the under water scene; (**b**) the log-mapped transformed image and, (**c**) the computed image motion using GDIM-based method

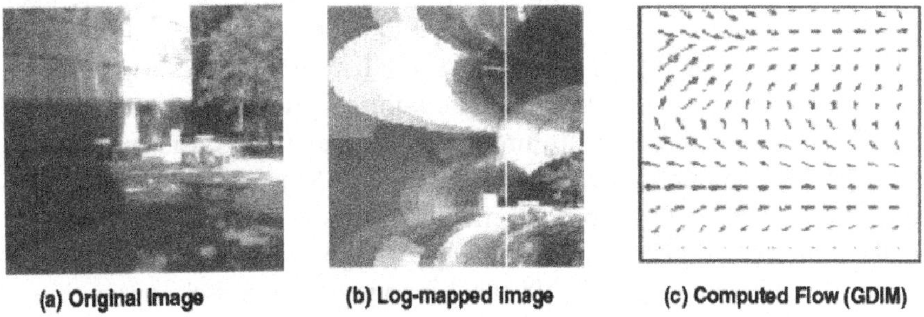

Fig. 2.12: Similar results as shown in Fig. 2.11 using an outdoor scene

From Figs. 2.11(c) and 2.12(c) it is clear that the flow distributions for the underwater and outdoor scenery images are similar to that of the Fig. 2.7(c), as expected. But, the flow distribution of the indoor laboratory sequence (see Fig. 2.13(c)) is different from that of the Fig. 2.7(c), due to the different motion profile. As mentioned earlier, the rotation in the image plane produces a constant flow along the radial direction. Hence, the flow distribution of Fig. 2.13(c) can be seen as the superposition of the flow distribution of the translational flow and that of the constant angular flow.

These results show the importance of taking into account the radiometric variation as well as the space-variant form of the derivative operator for log-mapped images by providing a accurate image motion estimation and unambiguous interpretation of image motion. It is clear from the results that the proposed method is numerically accurate, robust and provides consistent interpretation. It is important to note that the proposed method has error in computing optical flow. The main source of error is due to the non-uniform sampling.

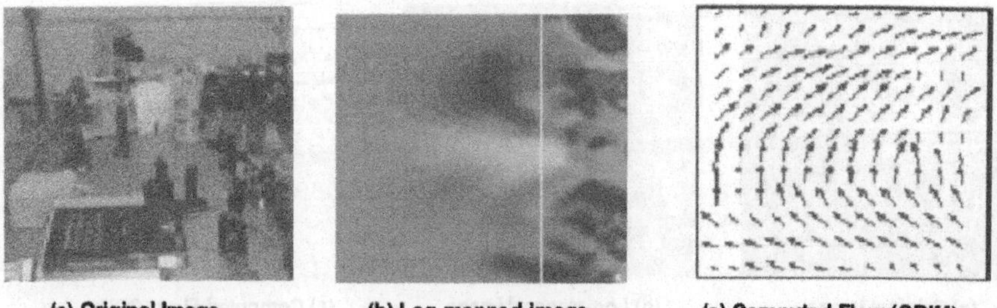

Fig. 2.13: Similar results shown in Fig. 2.11 using an indoor laboratory scene

2.6.2.3 Aperture Problem

P. Stumpf is credited (as translated in [78]) with first describing the aperture problem in motion analysis. The aperture problem arises as a consequence of the ambiguity of one-dimensional motion of a simple striped pattern viewed through an aperture. The failure to detect the true direction of motion is called the aperture problem. In other words, the motion of a homogeneous contour is locally ambiguous [79-81], i.e., within the aperture, different physical motions are indistinguishable.

In the context of primate vision, a two-stage solution to the aperture problem was presented in [82]. In machine vision literature, application of some form of smoothness constraint has been employed to overcome the aperture problem in devising techniques for computing the optical flow (for example, [83-84]). The aperture problem is critical in case of log-polar mapped images. As shown in section 3 straight lines are mapped into curves. Since the aperture problem appears only in the case of straight lines for the Cartesian images, the log-polar mapping seems to eliminate the problem. This of course is not true. It may be noted that a circle in the Cartesian image mapped on to a straight line in log-polar mapped image. This means that the aperture problem appears at points in the log-polar plane where the aperture problem does not occur in the corresponding points in the Cartesian image. Alternatively, it is possible to compute optical flow at points in the log-polar plane where the corresponding Cartesian point does not show curvature. Of course, the superficial elimination of the aperture problem produces optical flow values that show large error regarding the expected motion field. The problem is much more complex with GDIM model. If one assume, $\delta m = \delta c = 0$, the above model simplifies to the BCM. Mathematically, one of the two fields, say M is sufficient to describe the radiometric transformation in an image sequence if this is allowed to vary arbitrarily from point to point and one time instant to the next. In this case the multiplier field is typically a complicated function of several scene events that contribute to the radiometric transformation, each of which may vary sharply in different isolated regions [70]. This is not desirable since it then becomes very difficult to compute optical flow due to the generalized aperture problem (please see [70] for details regarding the generalized aperture problem).

2.6.3 Stereo Disparity

When a moving observer looks in the direction of heading, radial optical flow is only one of several cues which indicate the direction of speed of heading. Another cue, which is very significant at generating vergence at ultra short latencies is binocular disparity [54]. The pattern of retinal binocular disparities acquired by a fixating visual system depends on both the depth structure of the scene and the viewing geometry. In some binocular machine vision systems, the viewing geometry is fixed (e.g., with approximately parallel cameras) and can be determined once and for all by a calibration procedure. However, in human vision or any fixating vision system, the viewing geometry changes continually as the gaze is shifted from point to point in the visual field. In principle, this situation can be approached in two different ways: either a mechanism must be provided which continuously makes the state of the viewing geometry available to the binocular system, or invariant representations that fully or partially side-step the need for calibration of the viewing geometry must be found. For each approach a number of different techniques are possible, and any combination of these may be used as they are not mutually exclusive.

The viewing geometry could in principle be recovered from extra-retinal sources, using either in-flow or out-flow signals from the occulomotor and/or accommodation systems. The viability of this approach has been questioned on the ground that judgments of depth from occulomotor/accommodation information alone are poor [85, 86, 87, 40]. Alternatively, viewing geometry can be recovered from purely visual information, using the mutual image positions of a number of matched image features to solve for the rotation and translation of one eye relative to the other. This is often referred to as the "relative orientation" [88]. For normal binocular vision the relative orientation problem need not be solved in its full generality since the kinematics of fixating eye movements is quite constrained. These constraints lead to a natural decomposition of the disparity field into a horizontal and vertical component, which carries most of the depth information, and a vertical component, which mainly reflect the viewing geometry.

Apart from few exceptions [3, 89], most active vision researchers use Cartesian image representations. For tracking, the main advantage of the log-polar sensor is that objects occupying the central high resolution part of the visual field become dominant over the coarsely sampled background elements in the periphery. This embeds an implicit focus of attention in the center of the visual field where the target is expected to be most of the time. Furthermore, with Cartesian images, if the object of interest is small, the disparity of the background can lead to erroneous estimate. In [54], it has been argued that a biologically inspired index of fusion provides a measure of disparity.

Disparity estimation on space-variant image representation has not been fully explored. A cepstral filtering method is introduced in [90] to calculate stereo disparity on columnar image architecture architecture for cortical image representation [91]). In [92], it has been shown that the performance of cepstral filtering is superior then phase-based method [93]. In [5] correlation of log-polar images has been used to compute the stereo disparity. It has been argued that correlation based method works much better in log-polar images than for Cartesian images. It has been shown that correlation between log-polar images corresponds to the correlation in Cartesian images weighted by the inverse distance to the image center. To account for the translation in Cartesian domain (in the log-polar domain the translation is very complicated) a global search for the horizontal disparity has been proposed which minimizes the SSD.

It is believed that stereo disparity on a space-variant architecture can be conveniently estimated using phase-based technique by computing the local phase difference of the signals using the ECT. As mentioned earlier, ECT preserves the shift invariant property hence standard phase-disparity relation holds. To cope with the local characteristics of disparity in stereo images, it is standard practice to compute local phase using a complex band-pass filters (for example, [94, 95]). It is important to note that one needs to take a proper account of the aliasing and quantization issues to compute the phase of the signals using the ECT discussed in the previous section. A possible computational approach could be,

Step 1: Obtain the phase of left and right camera images using the ECT-based method.

Step 2: Calculate the stereo disparity using the standard phase-disparity relationship.

For most natural head motions the eyes (cameras) are not held on precisely the same lines of sight, but it is still true that the angular component of disparity is approximately independent of gaze. Weinshall [96] treated the problem of computing a qualitative depth map from the disparity field in the absence of camera calibration. Rather than decomposing disparity vectors into horizontal and vertical components, Wienshall used a polar decomposition and showed that two different measures derived from the angular component alone contains enough information to compute an approximate depth ordering. It has also been established that a numerical simulations showing that the pattern of polar angle disparities can be used to estimate the slope of a planar surface up to scaling by fixation distance, and that this pattern is affected by unilateral vertical magnification. In summary, eccentricity- scaled log-polar disparity, which can be computed from a single pair of corresponding points without any knowledge of the viewing geometry, directly indicates relative proximity.

2.7 Discussions

Biological and artificial systems that share the same environment may adopt similar solution to cope with similar problems. Neurobiologists are interested in finding the solutions adopted by the biological vision systems and machine vision scientists are interested in which of technologically feasible solutions that are optimal or suited of building autonomous vision based systems. Hence, a meaningful dialogue and reciprocal interaction between biologists and engineers with a common ground may bring fruitful results. One good example could be finding a better retino-cortical mapping model for sensor fabrication. It is believed that research in this front will help in designing a much more sophisticated sensor which preserves complete scale and rotation invariance at the same time maintains the conformal mapping. Another fundamental problem with space-variant sensor arises from their varying connectivity across the sensor plane. Pixels that are neighbors in the sensor are not necessarily neighbors ones computer reads data into array, making it difficult or impossible to perform image array operations. A novel sensor architecture using a 'connectivity graph' [97] or data abstraction technique may be another avenue which potentially can solve this problem.

Sensor-motor integration, in one form commonly known as eye-hand coordination, is a process that permits the system to make and test hypotheses about objects in the environment. In a sense, nature invented the scientific method for the nervous system to use as a means to predict and prepare for significant events. The motor component of perception compensates for an uncooperative environment. Not only does the use of effectors provide mobility, but it alters the information available, uncovering new opportunities to exploit. The development of purposive movement allows the host to judiciously act in the environment and sample the results. Prediction forms the basis of the judgment to act, and the results are used to formulate new predictions. Hence an action-sensation-prediction-action chain is established through experience and conditioned learning.

One behavioral piece of evidence for the action-sensation-prediction sequence is the scan path. The scan path is a sequence of eye (or camera) saccades that sample a target in a regular way to collect information. The scan path after learning became more regular and the inter-saccade interval get reduced compared to the naive state. It is believed that an invariant recognition can be achieved by transforming an appropriate behavior. For example, a scan path behavior to an image at different sizes, the saccade amplitudes must be modulated. This could be accomplished by the use of the topographical mapping that permits a natural rescaling of saccade amplitude based upon the locus of activity on the output map. To change the locus of activity, it is only necessary to match the expectation from the associative map with the available sensor information.

2.8 Conclusions

Anthropomorphic visual sensor and the implication of logarithmic mapping offer the possibility of superior vision algorithms for dynamic scene analysis and is motivated by the biological studies. But the fabrication of space-variant sensor and implementation of vision algorithms on space-variant images is a challenging issue as the spatial neighborhood connectivity is complex. The lack of shape invariance under translation also complicates image understanding. Hence, the retino-cortical mapping models as well as the state-of-the-art of the space-variant sensors were reviewed to provide a better understanding of foveated vision systems. The key motivation is to discuss techniques for developing image understanding tools designed for space-variant vision systems. Given the lack of general image understanding tools for space-variant sensor images, a set of image processing operators both in the frequency and in the spatial domain were discussed. It is argued that almost all the low level vision problems (i.e., shape from shading, optical flow, stereodisparity, corner detection,surface interpolation, and etc.) in the deterministic framework can be addressed using the techniques discussed in this article. For example, ECT discussed in section 5.1 can be used to solve the outstanding bottleneck of shift invariance while the spatial domain operators discussed in section 5.2 paves the way for easy use of traditional gradient-based image processing tools. In [68], convolution, image enhancement, image filtering, template matching was done by using ECT. The computational steps to compute the pace-variant stereo disparity was outlined in section 6.3 using ECT. Also operations like anisotropic diffusion [47] and corner detection [98], on a space-variant architecture was done using the space-variant form of differential operator and the Hessian of the intensity function $(I_{\xi\xi}I_{\eta\eta} - I_{\xi\eta}^2)$, respectively. A GDIM-based method to compute the optical flow which allows image intensity to vary in the subsequent images and that used the space-variant form of the derivative operator to calculate the image gradients was reported in [77, 75]. It is hypothesized that the outline of classical vision algorithms based on space-variant image processing operators will prove invaluable in the future and will pave the way of developing image understanding tools for space-variant sensor images. Finally, the problem of 'attention' is foremost in the application of a space-variant sensor. The vision system must be able to determine where to point its high-resolution fovea. A proper attentional mechanism is expected to enhance image understanding by strategically directing fovea to points which are most likely to yield important information.

Acknowledgments

This work was partially supported by NSF ITR grant IIS-0081935 and NSF CAREER grant IIS-97-33644. Authors acknowledge various personal communications with Yasuo Kuniyoshi.

References

1. A.C. Bovik W.N. Klarquist, "FOVEA: a foveated vergent active stereo vision system for dynamic three-dimensional scene recovery", IEEE Transactions on Robotics and Automation, vol. 5, pp. 755 –770, 1998.
2. N.C. Griswold and C.F. Weinman, "A modification of the fusion model for log polar coordinates", in SPIE- Intelligent robot and computer vision VIII: Algoritms and techniques,, 1989, pp. vol 938, pp.854–866, Bellingham,WA.
3. C. Capurro, F. Panerai and G. Sandini, "Dynamic vergence using log-polar images", Intl. Journal on Computer Vision, vol. 24, no.1, pp. 79–94, 1997.
4. J. Dias, H. Araujo, C. Paredes and J. Batista, "Optical normal flow estimation on log-polar images: A solution for real-time binocular vision ", RealTime Img., vol. 3, pp. 213–228, 1997.
5. A. Bernardino and Jose Santos-victor, "Binocular tracking: Integrating perception and control", IEEE Tran. on Robotics and Automation, vol. 15, no.6, pp. 1080–1094, 1999.
6. C. Silva and J. Santos-Victor, "Egomotion estimation using log-polar images", in Proc. of Intl. Conf. on Computer Vision, 1998, pp. 967–972.
7. M. Tistarelli and G. Sandini, "On the advantage of log-polar mapping for estimation of time to impact from the optical flow", IEEE trans. on Patt. Analysis and Mach. Intl., vol. 15(4), pp. 401–410, 1993.
8. M. Tistarelli and G. Sandini, "Ddynamic aspects in active vision ", CVGIP:Image understanding, vol. 56(1), pp. 108–129, 1992.
9. S.S Young, P.D. Scott and C. Bandera, "Foveal automatic target recognition using a multiresolution neural network", IEEE Transactions on Image Processing, vol. 7, 1998.
10. J.C. Wilson and R.M. Hodgson, "Log-polar mapping applied to pattern representation and recognition", CVGIP, pp. 245–277, 1992.
11. F.L. Lim, G. West and S. Venkatesh, "Investigation into the use of log polar space for foveation and feature recognition", To appear in IEE Proceedings - Vision, image and Signal Processing, 1997.
12. P. Mueller R. Etienne-Cummings, J.Van der Spiegel and Mao-Zhu Zhang, "A foveated silicon retina for two-dimensional tracking", IEEE Trans. on Circuits and Systems II: Analog and Digital Signal Processing, vol. 47 Issue: 6, pp. 504–517, June 2000.
13. C.F. Weinman and R.D. Juday, "Tracking algorithms for log-polar mapped image coordinates", in the SPIE- Intelligent robot and computer vision VIII: Algoritms and techniques, vol 938, pp.138-145, SPIE, Bellingham,WA 1989, 1998.

14. K. Daniilidis, C. Krauss, M. Hansen and G. Sommer, "Real-time tracking with moving objects with an active camera", Journal of Real-time Imaging, Academic Press, 1997.
15. Luc Berthouze, Paul Bakker, and Yasuo Kuniyoshi, Learning of oculomotor control: a prelude to robotic imitation, in Proc., IEEE International conference Intelligent Robots and Systems, Osaka, Japan, November 1996, vol. 1, pp. 376–381.
16. M.D. Levine, "Vision in man and machine", Addision-Wesley, Reading, MA, 1984.
17. P.J. Burt, "Algoritms and architectures for smart sensing", in the Proc. of DARPA Image understanding workshop, 1988, pp. 139–153.
18. L. Sanghoon, C. Podilchuk and A.C. Bovic, "Foveation-based error resilience for video transmission over mobile networks", in Proc. of Multimedia and Expo, 2000, ICME 2000, 2000, pp. 1451–1454.
19. L.Sanghoon and A.C. Bovic, "Very low bit rate foveated video coding for h.263", in IEEE International Conference on Acoustics, Speech, and Signal Processing, 1999, 1999, pp. 3113 –3116 vol.6.
20. G. Bonmassar and E.L. Schwartz, "Real-time restoration of images degraded by uniform motion blur in foveal active vision systems", IEEE Transactions on Image Processing, vol. 12, pp. 1838 –1842, 1999.
21. H. Qian C. Yuntao, S. Samarasckera and M.Greiffenhagen, "Indoor monitoring via the collaboration between a peripheral sensor and a foveal sensor", in IEEE Workshop on Visual Surveillance, 1998, 1998, pp. 2–9.
22. G.A. Baricco, A.M. Olivero, E.J. Rodriguez, F.G. Safar and J.L. CSanz, "Conformal mapping-based image processing: Theory and applications", Journal Vis. Com. And Image. Rend., vol. 6, pp. 35–51, 1995.
23. J.M. Kinser, "Foveation from pulse images", in Proc. of Information Intelligence and Systems, 1999, 1999, pp. 86 –89.
24. A.S. Rojer and E. L. Schwartz, "Design considerations for a space-varying sensors with complex logarithmic geometry", in the Proc. Intl. Conf. on Patt. Rec.,, 1990, pp.278–285.
25. B.R. Friden and C. Oh, "Integral logarithmic transform : Theory and applications", Applied Optics, vol. 15, pp. 1138–1145, March, 1992.
26. J.J. Clark, M.R. Palmer and P.D. Lawrence, "A transformation method for the reconstruction of functions from non-uniformly spaced sensors", IEEE trans. Accoustic, speech and signal processing, vol. 33(4), pp. 1151, 1985.
27. G. Sandini and V. Tagliasco, An anthropomorphic retina-like structure for scene analysis, Computer Graphics and Image Processing, vol. 14, no. 3, pp. 365–372, 1980.
28. B. Dierickx F. Pardo and D. Scheffer, "Space-variant nonorthogonal structure cmos image sensor design", IEEE Trans. on Solid-State Circuits, vol. 6, pp. 842 –849, 1998.
29. Y. Kuniyoshi, N. Kita and K. Sugimoto, "A foveated wide angle lense for active vision", in the Proc. of IEEE intl. Conf. Robotics and Automation, 1995.

30. L. Berthouze, S. Rougeaux, and Y. Kuniyoshi, Calibration of a foveated wide angle lens on an active vision head, in in Proc., IMACS/IEEE-SMC Computational Engineering Systems Applications, Lille, France, 1996.
31. S. Rougeaux and Y. Kuniyoshi, Velocity and disparity cues for robust real-time binocular tracking, in in Proc., IEEE International Conference on Computer Vision and Pattern Recognition , Puerto Rico, 1997, pp. 1–6.
32. S′ebastien Rougeaux and Yasuo Kuniyoshi, Robust tracking by a humanoid vision system, in in Proc., First Int. Workshop Humanoid and Human Friendly Robotics, Tsukuba, Japan, 1998.
33. S. K. Nayar, "Omnidirectional video camera", in Proc. of DARPA Image Understanding Workshop, New Orleans, May 1997.
34. S. K. Nayar, "Ccatadioptric omnidirectional camera", in Proc. of IEEE Conference on Computer Vision and Pattern Recognition (CVPR), Peurto Rico, June 1997.
35. S. K. Nayar and S. Baker, "Catadioptric image formation", in Proc. of DARPA Image Understanding Workshop, New Orleans, May 1997.
36. S. K. Nayar, "Omnidirectional vision", in Proc. of Eight International Symposium on Robotics Research (ISRR), Shonan, Japan, October 1997.
37. S. K. Nayar and S. Baker, "A theory of catadioptric image formation", in Proceedings of the 6th International Conference on Computer Vision, India, Jan., 1998, pp. 35–42.
38. H.Ishiguro, K.Kato and S.Tsuji, "Multiple vision agents navigating a mobile robot in a real world", in Proceedings IEEE Intl. Conf. on Rob. and Automation, USA, May 1993.
39. M.J.Barth and H.Ishiguro, "Distributed panoramic sensing in multiagent robotics", in Proceedings of 1994 IEEE Intl. Conf. on MFI '94, Las Vegas, USA, Oct. 1994.
40. E.R. Kandel, J.H. Schawartz and T.M. Jessell, "Principles of Neural Science, 4/e", McGraw-Hill, New York, 2000.
41. M. Bouldac and M. D. Levine, "A real-time foveated sensor with overlapping receptive fields", Journal of Real-time Imaging, Academic Press, vol. 3, pp. 195–212, 1997.
42. P.M. Daniel and D. Whitridge, "The representation of the visual field on the cereberal cortex of monkeys", Journal of Physiology, vol. 159, pp. 203–221, 1961.
43. E.L. Schwartz, "Spatial mapping in the primate sensory perception: Analytic structure and relevance to percetion", Biol. Cybern., vol. 25, pp. 181–194, 1977.
44. C. Braccini, G. Gambardella and G. Sandini, "A signal theory approach to the space and frequency variant filtering performed by the human visual system", Signal Processing, vol. 3 (3), 1981.
45. C. Braccini, G. Gambardella, G. Sandini and V. Tagliasco, "A model of the early stages of human visual system : functional and topological transformation performed in the periferal visual field", Biological Cybernetics, vol. 44, 1982.

46. E.L. Schwartz, Computational studies of spatial architecture of primate visual cortex, Vol 10, Chap. 9, pp. 359-411, Plenum, New York, 1994.
47. B. Fischl, A. Cohen and E.L. Schwartz, "Rapid anisotropic diffusion using space-variant vision", Int. Journal of Comp. Vision, vol. 28(3), pp. 199–212, 1998.
48. E.L. Schwartz, "Computational anatomy and functional architecture of the strait cortex", Vision research, vol. 20, pp. 645–669, 1980.
49. Y. Kuniyoshi, N. Kita, K. Sugimoto, S. Nakamura, and T. Suehiro, A foveated wide angle lens for active vision, in Proc., IEEE International Conference on Robotics and Automation, Nagoya, Japan, 1995, vol. 3, pp. 2982–2985.
50. R.C. Jain, S.L. Bartlett and N. O'Brian, "Motion stereo using ego-motion complex logarithmic mapping", Tech. Rep., Technical report, Center for research on integrated manufacturing, University of Michigan, February, 1986.
51. R. Jian, L. Bartlett and N. O'Brien, "Motion stereo using ego-motion complex logarithmic mapping", IEEE Tran. on Pattern Anal. and Mach. Intell., vol. 3, pp. 356–369, 1987.
52. M. Tistarelli and G. Sandini, "Estimation of depth from motion using an antropomorphic visual sensor", Image and Vision Computing, vol. 8(4), 1990.
53. C. Busetti, F.A. Miles and R.J. Krauzlis, "Radial optical flow induces vergence eye movements at ultra-short latencies", nature, vol. 390, pp. 512–515, 1997.
54. G. Sandini, F. Panerai and F.A. Miles, "The role of inertial and visual mechanisms in the stabilization of gaze in natural and artificial systems, From: Motion Vision Computational, Neural, and Echological Constraints, Eds. J. M. Zanker and J. Zeli", Springer-Verlag, New York, 2001.
55. William Klarquist and Alan Bovik, Fovea: a foveated vergent active stereo system for dynamic three-dimensional scene recovery, in Proc., IEEE International Conference on Robotics and Automation, Leuven, Belgium, 1998.
56. Albert J. Wavering, Henry Schneiderman, and John C. Fiala, High-performance tracking with triclops, in Proc., Asian Conf. on Comp. Vision, Singapore, 1995, vol. 1.
57. Brian Scassellati, A binocular, foveated, active vision system, Tech. Rep., Massachussetts Institute of Technology, January 1998.
58. Y. Suematsu and H. Yamada, A wide angle vision sensor with fovea - design of distortion lens and the simulated image, in the Proc., of IECON93, 1993, vol. 1, pp. 1770–1773.
59. Y. Kuniyoshi, N. Kita, S. Rougeaux, and T. Suehiro, Active stereo vision system with foveated wide angle lenses, in Proc., 2nd Asian Conference on Computer Vision, Singapore, 1995, vol. 1, pp. 359–363.
60. M. Yeasin and Y. Kuniyoshi, "Detecting and trackning human face and eye using an space-varying sensor and an active vision head", in the Proc.

of computer vision and pattern recognition, South Carolina, USA, June 2000, pp. 168–173.

61. G. Sandini and P. Dario, "Active vision based on space variant sensing", in the Proc. of Intl. Symp. on Robotic Research, 1989.
62. F. Pardo, "Development of a retinal image sensor based on cmos technology", Tech. Rep., LIRA-TR 6/94, 1994.
63. E. Martinuzzi and F. Pardo, "FG II new version of the ccd retinal sensor frame grabber", Tech. Rep., LIRA-TR 1/94, 1994.
64. F. Ferrari, J. Nielsen, P. Questa, and G. Sandini, Space variant imaging, Sensor Review, vol. 15, no. 2, pp. 17–20, 1995.
65. R. Wodnicki, G. W. Roberts and M.D. Levine, "A foveated image sensor in standard cmos technology", in Proc. of custom integrated circuit conference, 1995, pp. 357–360.
66. R. Woodnicki, G.W. Roberts and M. Levine, "Design and evaluation of log-polar image sensor fabricated using standard 1.2 μ m ASIC and CMOS process", IEEE Trans. On solid state circuits, vol. 32, no. 8, pp. 1274–1277, 1997.
67. E. Schwartz, N. Greve and G. Bonmassar, "Space-variant active vision: Definition, overview and examples", Neural Network, vol. 7/8, pp. 1297–1308, 1995.
68. G. Bonmassar and E.L. Schwartz, "Space-variant Fourier analysis : The exponential chirp transform ", IEEE trans. on Patt. Analysis and Mach. Intl., vol. 19(10), pp. 1080–1089, Oct., 1997.
69. J. Portilla, A. Tabernero and R. Navarro, "Duality of log-polar image representations in the space and spatial-frequency domains", IEEE Transactions on Signal Processing, vol. 9, pp. 2469 –2479, 1999.
70. S. Negadharipour, "Revised definition of optical flow: Integration of radio-metric and geometric cues for dynamic scene analysis", IEEE Trans. on Pat. Analysis and Mach. Intl., vol. 20(9), pp. 961–979, 1998.
71. B. Horn and B. Schunck, Determining optical flow, Artificial Intelligence, vol. 17, pp. 185–204, 1981.
72. P. Anandan, Measuring visual motion from image sequences, PhD thesis, University of Massachussetts, Amherst, MA, 1987.
73. A. B. Watson and A. J. Ahumada, Motion: perception and representation, chapter A look at motion in the frequency domain, pp. 1–10, J. K. Tsotsos, 1983.
74. J. L. Barron, D. J. Fleet, and S. S. Beauchemin, Performance of optical flow techniques, International Journal of Computer Vision, vol. 12, no. 1, pp. 43–77, February 1994.
75. M. Yeasin, "Optical flow on log-mapped image plane: A new approach", in the lecture notes on computer science, Springer-Verlag, NY, USA, Feb 2001, pp. 252–260.
76. M. Yeasin, "Optical flow on log-mapped image plane: A new approach", IEEE Trans. Pattern Analysis and Machine Intelligence, vol. to appear.

77. M. Yeasin, "optical flow in log-mapped image plane - A new approach", IEEE Transactions on Pattern Analysis and Machine Intelligence, vol. 24, pp. 125 –131, 2002.
78. D. Todorovic, "A gem from the past: Pleikart stumpf's anticipation of the aperture problem, Reichardt detectors, and perceived motion loss at equiluminance", Perception, vol. 25, pp. 1235–1242, 1996.
79. J. Marroquin, S. Mitter and T. Poggio, "Probabilistic solutions of ill-posed problems in computational vision", Jour. of Amer. Stat. Assoc., vol. 79(387), pp. 584–589, 1987.
80. E.C. Hildereth, "The computation of velocity field", in Proc. of Royal Stat. Soc., London, B221, 1984, pp. 189–220.
81. J.A. Movshon, E.H. Edelson, M.S. Gizzi and W.T. Newsome, "Pattern recognition mechanisms", In the analysis of moving visual patterns: C. Chagas, R. Gattass and C. Gross eds., Springer-Verlag, pp. 283-287, New York, 1985.
82. J.A. Movshon, "Visual processing of moving image", In the Images and Understanding:/Thoughts about images; Ideas about Understanding, H. Burlow, C. Blackmore and M. Weston-Smith eds., pp. 122-137, Cambridge Univ. Press, New York, 1990.
83. B.K.P. Horn and B.G. Schunk, "Determining optical flow", Artificial Intelligence, vol. 17, pp. 185–203, 1981.
84. J.K. Kearney, W.R. Thompson and D.l. Bolly, "Optical flow estimation, an error analysis of gradient based methods with local optimization", IEEE Tran. on Pattern Anal. and Mach. Intell., vol. 14(9), pp. 229–244, 1987.
85. J.M. Foley, "Binocular distance perception", Psychological Review, vol. 87, pp. 411–435, 1980.
86. J.M. Foley, "Binocular distance perception: Egocentric visual task", Journal of experimental Psychology, vol. 11, pp. 133–149, 1985.
87. E.C. Sobel and T.S. Collett, "Does vertical disparity scales the perception of stereoscopic depth?", In the Proc. of Royal society London B, vol. 244, pp. 87–90, 1991.
88. B.K.P Horn, "Relative orientation", International Journal of Computer Vision, vol. 4, pp. 59–78, 1990.
89. G. Salgian and D.H. Ballard, "Visual routines for vehicle control", In the confluence of visual control, D. Kreigman, G. Haggar, and S. Murase Eds.,, Springer-Verlag, New York, 1998.
90. Y. Yeshurun and E. L. Schwartz, "Cepstral filtering on a columnar image architecture: a fast algorithm for binocular stereo segmentation", IEEE Trans. Pattern Analysis and Machine Intelligence, vol. 11, no. 7, pp. 759–767, 1989.
91. Y. Yeshurun and E. L. Schwartz, "Cortical hypercolumn size determines stereo fusion limits", Bio. Cybernetics, vol. 80(2), pp. 117–131, 1999.
92. T. J. Olson and R. D. Potter, Real-time vergence control, Tech. Rep. TR 264, 1988.

93. David J. Fleet, “Stability of phase information”, Transaction on Pattern Analysis and Machine Intelligence, vol. 15, no. 12, pp. 1253–1268, 1993.
94. David J. Fleet, Disparity from local weighted pase-correlation, IEEE International Conference on Systems, Man and Cybernetics, San Antonio, pp. 48–56, October 1994.
95. D. J. Fleet and K. Langley, “Recursive filters for optical flow”, IEEE Transaction on Pattern Analysis and Machine Intelligence, vol. 17, no. 1, pp. 61–67, January 1995.
96. D. Weinshall, “Qualitative depth from stereo”, Comp. Vis., Graph. and Image Proc.,, vol. 49, pp. 222–241, 1990.
97. R. Wallace, P. Ong and E. Schwartz, “Space-variant image processing”, Int. Journal of Comp. Vision, vol. 13(1), pp. 71–90, 1994.
98. B. Fischcl, M. Cohen and E. Schwartz, “The local structure of space-variant images”, Neural Network, vol. 10(5), pp. 815–831, 1997.

3 On-line Model Learning for Mobile Manipulations

Yu Sun[1], Ning Xi[1], and Jindong Tan[2]

1. Department of Electrical and Computer Engineering, Michigan State University, East Lansing, MI 48824, U.S.A.
{sunyu, xin}@msu.edu
2. Department of Electrical and Computer Engineering, Michigan Technological University, Houghton, MI 49931, U.S.A.
jitan@mtu.edu

8.1 Introduction

In control system design, modeling the control objects is the foundation of high performance controllers. In most cases, the model is unknown before the task is performed. To learn the model, therefore, is essential. Generally, real-time systems use the sensors to measure the environment and the objects. The model learning task is particularly difficult when real-time control systems run in a noisy and changing environment. The measurements from the sensors may be contaminated by the non-stationary noise, i.e. it is changing randomly and depends on the environmental uncertainties. The factors, which cause the environmental uncertainties, may include wind, vibration, friction, and so on.

For robotic automation, a manipulator mounted on a mobile platform can significantly increase the workspace of the manipulation and its application flexibility. The applications of mobile manipulation range from manufacturing automation to search and rescue operations. A task such as a mobile manipulator pushing a passive nonholonomic cart can be commonly seen in manufacturing or other applications, as shown in Fig. 3.1.

The mobile manipulator and nonholonomic cart system, shown in Fig. 3.1, is similar to the tracker-trailer system. Tracker-trailer systems generally consist of a steering mobile robot and one or more passive trailer(s) connected together by rigid joints. The tracking control and open loop motion planning of such a nonholonomic system have been discussed in the literature. The trailer system can be controlled to track certain trajectory using a linear controller based on the linearized model [6]. Instead of pulling the trailer, the tracker pushes the trailer to track certain trajectories in

Y. Sun et al.: *On-line Model Learning for Mobile Manipulations*, Studies in Computational Intelligence (SCI) **7**, 99–135 (2005)
www.springerlink.com

the backward steering problem. Fire truck steering is another example for pushing a nonholonomic system and the chained form has been used in the open loop motion planning [3]. Based on the chained form, motion planning for steering a nonholonomic system has been investigated in [16]. Time varying nonholonomic control strategies for the chained form can stabilize the tracker and trailers system to certain configurations [13].

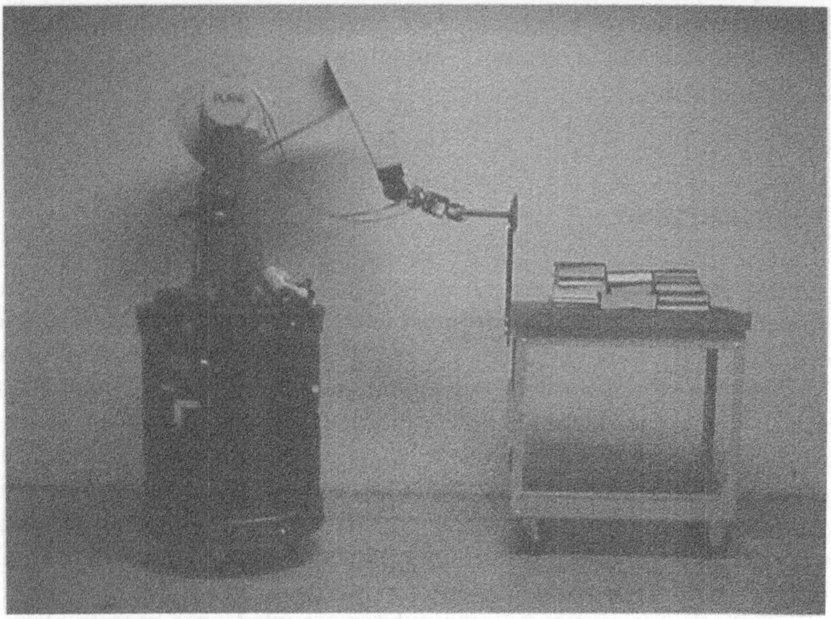

Fig. 3.1. Mobile Manipulation and Noholonomic Cart

In robotic manipulations, manipulator and cart are not linked by an articulated joint, but by the robot manipulator. The mobile manipulator has more flexibility and control while maneuvering the nonholonomic cart. Therefore, the planning and control of the mobile manipulator and nonholonomic cart system is different from a tracker-trailer system. In a tracker-trailer system, control and motion planning are based on the kinematic model, and the trailer is steered by the motion of the tracker. In a mobile manipulator and nonholonomic cart system, the mobile manipulator can manipulate the cart by a dynamically regulated output force.

The planning and control of the mobile manipulator and nonholonomic cart system is based on the mathematic model of the system. Some parameters of the model, including the mass of the cart and its kinematic length, are needed in the controller [15]. The kinematic length of a cart is defined on the horizontal plane as the distance between the axis of the front fixed wheels and the handle.

A nonholonomic cart can travel along its central line and perform turning movement about point C; in this case, the mobile manipulator applies a force and a torque on the handle at point A, as shown in Fig. 3.2. The line between A and C is defined as the kinematic length of the cart, |AC|, while the cart makes an isolated turning movement. As a parameter, the kinematic length |AC| of the cart can be identified by the linear velocity of point A and the angular velocity of line AC. The most frequently used parameter identification methods are the Least Square Method (LSM) and the Kalman filter [8] method. Both have the recursive algorithms for on-line estimation. Generally, if a linear model (linearized model) of a dynamic system can be obtained, the system noise and observation noise are also known. The Kalman filter can estimate the states of the dynamic system through the observations regarding the system outputs. However, the estimation results are significantly affected by the system model and the noise models. Least Square method can be applied to identify the static parameters in absence of accurate linear dynamic models.

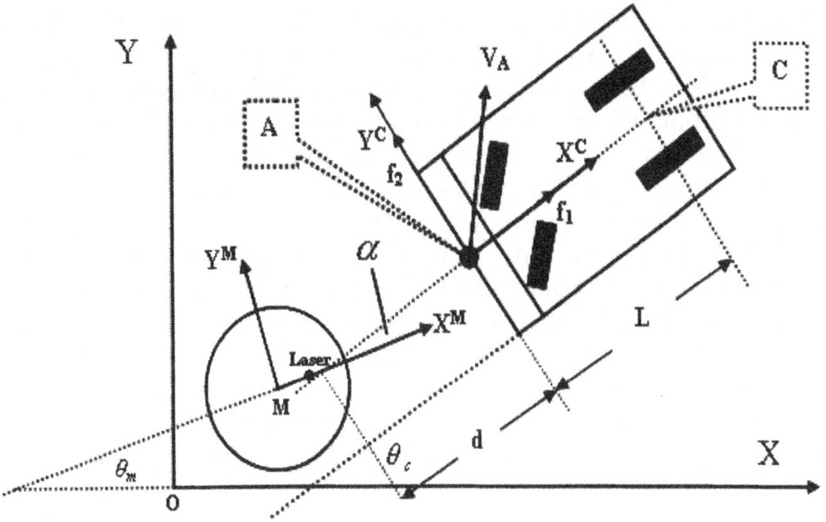

Fig. 3.2. Associated Coordinate Frames of Mobile Manipulation System

Parameter identification has been extensively investigated for robot manipulations. Zhuang and Roth [19] proposed a parameter identification method of robot manipulators. In his work, the Least Square Method is used to estimate the kinematic parameters based on a linear solution for the unknown kinematic parameters. To identify the parameters in the dynamic model of the robot manipulator [12], a least square estimator is applied to identify the parameters of the linearized model. It is easy to see that LSM has been widely applied in model identification as well as in the field of robotics.

With the final goal of real-time online estimation, the Recursive Least Square Method (RLSM) has been developed to save computation resources and increase operation velocities for real time processing [9].For identification, measurement noise is the most important problem. There are two basic approaches to processing a noisy signal. First, the noise can be described by its statistical properties, i.e., in time domain; second, a signal with noise can be analyzed by its frequency-domain properties.

For the first approach, many algorithms of LSM are used to deal with noisy signals to improve estimation accuracy, but they require the knowledge of the additive noise signal. Durbin algorithm and Levinson-Wiener algorithm [2] require the noise to be a stationary signal with known autocorrelation coefficients. Each LSM based identification algorithm corresponds to a specific model of noise [10]. Iserman and Baur developed a Two Step Process Least Square Identification with correlation analysis [14]. But, for on-line estimation, especially in an unstructured environment, relation analysis results and statistical knowledge cannot be obtained. In this case, estimation results obtained by the traditional LSM are very large (Table 2 in section 3.4). The properties of LSM in the frequency domain have also been studied. A spectral analysis algorithm based on least-square fitting was developed for fundamental frequency estimation in [4]. This algorithm operates by minimizing the square error of fitting a normal sinusoid to a harmonic signal segment. The result can be used only for fitting a signal by a mono-frequent sinusoid.

In a cart-pushing system, positions of the cart can be measured directly by multiple sensors. To obtain other kinematic parameters, such as linear and angular velocity of the object, numerical differentiation of the position data is used. This causes high frequency noises which are unknown in different environments. The unknown noise of the signal will cause large estimation errors. Experimental results have shown that the estimation error can be as high as 90%. Therefore, the above least square algorithms can not be used, and eliminating the effect of noise on model identification becomes essential.

This chapter presents a method for solving the problem of parameter identification for a nonholonomic cart modeling where the sensing signals are very noisy and the statistic model of the noise is unknown. When analyzing the properties of the raw signal in frequency domain, the noise signal and the true signal have quite different frequency spectra. In order to reduce the noise, a method to separate the noise signal and the true signal from the raw signal is used to process them in the frequency domain. A raw signal can be decomposed into several new signals with different bandwidths. These new signals are used to estimate the parameters; the best estimate is obtained by minimizing the estimation residual in the least square sense.

Combined with digital subbanding technique [1], a Wavelet based model identification method is proposed. The estimation convergence of the proposed model is proven theoretically and verified by experiments. The experimental results show that the estimation accuracy is greatly improved without prior statistical knowledge of the noise.

8.2 Control and Task Planning in Nonholonomic Cart Pushing

In this section, the dynamic models of a mobile manipulator and a nonholonomic cart are briefly introduced. The integrated task planning for manipulating the nonholonomic cart is then presented.

8.2.1 Problem Formulation

A mobile manipulator consists of a mobile platform and a robot arm. Fig. 3.2 shows the coordinate frames associated with both the mobile platform and the manipulator. They include:

- **World Frame** Σ : XOY frame is Inertial Frame;
- **Mobile Frame** Σ_M : $X^M M Y^M$ frame is a frame attached on the mobile manipulator;
- **Mobile Frame** Σ_A: $X^C A Y^C$ frame is a frame attached on the nonholonomic cart.

The models of the mobile manipulator and the cart are described in a unified frame Σ. The dynamics of a mobile manipulator comprised of a 6 DOF PUMA-like robot arm and a 3 DOF holonomic mobile platform in frame Σ, is described as [15]:

$$M(p)\ddot{x} + c(p,\dot{p}) + g(p) = \tau \tag{3.2.1}$$

where $\tau \in \Re^9$ are the generalized input torques, $M(p)$ is the positive definite mobile manipulator inertia matrix, $c(p,\dot{p})$ are the centripetal and coriolis torques, and $g(p)$ is the vector of gravity term. The vector

$p = \{q_1, q_2, q_3, q_4, q_5, q_6, x_m, y_m, \theta_m\}^T$ is the *joint* variable vector of the mobile manipulator, where $\{q_1, q_2, q_3, q_4, q_5, q_6\}^T$ is the joint angle vector of the robot arm and $\{x_m, y_m, \theta_m\}^T$ is the configuration of the platform in the unified frame Σ. The augmented system output vector x is defined as $x = \{x_1, x_2\}$, where $x_1 = \{p_x, p_y, p_z, O, A, T\}^T$ is the end-effector position and orientation, and $x_2 = \{x_m, y_m, \theta_m\}$ is the configuration of the mobile platform.

Applying the following nonlinear feedback control

$$\tau = M(p)u + c(p, \dot{p}) + g(p) \tag{3.2.2}$$

where $u = \{u_1, u_2, u_3, u_4, u_5, u_6, u_7, u_8, u_9\}^T$ is a linear control vector, the dynamic system can then be linearized and decoupled as

$$\ddot{x} = u \tag{3.2.3}$$

It is seen that the system is decoupled in an augmented task space. The decoupled model is described in a unified frame and the kinematic redundancy could be resolved in the same frame. Given $x^d = \{p_x^d, p_y^d, p_z^d, O^d, A^d, T^d, x_m^d, y_m^d, \theta_m^d\}$ as the desired position and orientation of the mobile manipulator in the world frame Σ, a linear feedback control for model (3.2.3) can be designed.

The nonholonomic cart shown in Fig. 3.2 is a passive object. Assuming that the force applied to the cart can be decomposed into f_1 and f_2, the dynamic model of the nonholonomic cart in frame Σ can be represented by

$$\begin{aligned}
\ddot{x}_c &= \frac{\lambda}{m_c}\sin\theta_c + \frac{f_1}{m_c}\cos\theta_c \\
\ddot{y} &= -\frac{\lambda}{m_c}\cos\theta_c + \frac{f_1}{m_c}\sin\theta_c \\
\ddot{\theta}_c &= \frac{f_2}{I_c}
\end{aligned} \tag{3.2.4}$$

where x_c, y_c and θ_c are the configuration of the cart, m_c and I_c are the mass and inertia of the cart λ is a Lagrange multiplier and θ_c is the cart orientation.

As shown in Fig.3.2, The input forces applied to the cart, f_1 and f_2, correspond to the desired output force of the end-effector, f_x^d, f_y^d. In other words, the mobile manipulator controls the cart to achieve certain tasks by its output force. Therefore, manipulating the cart requires motion planning and control of the mobile manipulator based on the decoupled model (3.2.3), and the motion and force planning of the cart based on its dynamic model (3.2.4). An integrated task planning that combines both is desired.

3.3.2 Force and Motion Planning of the Cart

Due to the large workspace and dexterous manipulation capability, mobile manipulators can be used in a variety of applications. For some complex tasks, such as pushing a nonholonomic cart, the mobile manipulator maneuvers the object by the output forces. Therefore, task planning for manipulating a cart involves three issues: motion and force planning of the cart, trajectory planning and control of the mobile manipulator, and synchronization of the motion planners for the cart and the mobile manipulator.

The cart shown in Fig. 3.2 is a nonholonomic system. Path planning for a nonholonomic system involves finding a path connecting specified configurations that also satisfy nonholonomic constraints. The cart model is similar to a nonholonomic mobile robot except that the cart is a passive object. Therefore, many path planning algorithms for a nonholonomic system such as steering using sinusoids and goursat normal form approach [16] can be used for the motion planning of the cart. In this chapter, the kinematic model is transformed into the chained form and the motion planning is considered as the solution of an optimal control problem. Considering the kinematic model of the cart as

$$\dot{x}_c = v_1 \cos\theta_c \tag{3.2.5}$$

$$\dot{y}_c = v_1 \sin\theta_c \tag{3.2.6}$$

$$\dot{\theta}_c = v_2 \tag{3.2.7}$$

$$\dot{x}_c \sin\theta_c - \dot{y}_c \cos\theta_c = 0 \tag{3.2.8}$$

where v_1 and v_2 are the forward velocity and the angular velocity of the cart respectively, an optimal trajectory connecting an initial configuration and a final configuration can be obtained by minimizing the control input energy $J = \int_{t_0}^{t_f} \frac{1}{2}\|v(t)\|^2 dt$. Given a geometric path, an event-based approach can be used to determine the trajectory [17].

Trajectory, the input forces applied onto the cart, f_1 and f_2, also need to be planned. The input force planning of the cart is equivalent to the output force planning of the mobile manipulator. The control input of the nonholonomic cart is determined by its dynamic model (3.2.4), the nonholonomic constraint (3.2.8) and its desired trajectory $(x_c^d, y_c^d, \theta_c^d)$.

Because a continuous time-invariant stabilizing feedback is lacking, output stabilization is considered in this chapter. Choosing a manifold rather than a particular configuration as the desired system output, the system can be input-output linearized. By choosing x_c, y_c as the system output, the system can be linearized with respect to the control inputs f_1 and v_2. This can be explained by the following derivations. From equation (3.2.5) and (3.2.6), it is easy to see that $v_1 = \dot{x}_c \cos\theta_c + \dot{y}_c \sin\theta_c$, where v_1 is the forward velocity along x' direction. Considering that velocity along y' direction is $\dot{x}_c \sin\theta_c - \dot{y}_c \cos\theta_c = 0$, the following relation can be obtained:

$$\dot{v}_1 = \ddot{x}_c \cos\theta_c + \ddot{y}_c \sin\theta_c$$

$$v_2 = \ddot{\theta}_c$$

Suppose the desired output of interest is $\{x_c, y_c\}$. The following input-output relation can be obtained by the derivative of equations (3.2.5) and (3.2.6):

$$\ddot{x}_c = \frac{1}{m_c}\cos\theta_c f_1 - v_1 \sin\theta_c \cdot v_2$$
$$\ddot{y}_c = \frac{1}{m_c}\sin\theta_c f_1 + v_1 \cos\theta_c \cdot v_2$$

Considering f_1 and v_2 as the control inputs of the system, the input and the output can be formulated in a matrix form:

$$\begin{pmatrix}\ddot{x}_c \\ \ddot{y}_c\end{pmatrix} = G\begin{pmatrix}f_1 \\ v_2\end{pmatrix}$$

where

$$G = \begin{pmatrix}\dfrac{\cos\theta_c}{m_c} & -v_1\sin\theta_c \\ \dfrac{\sin\theta_c}{m_c} & v_1\cos\theta_c\end{pmatrix}$$

The nonholonomic cart can then be linearized and decoupled as

$$\begin{pmatrix}\ddot{x}_c \\ \ddot{y}_c\end{pmatrix} = \begin{pmatrix}w_1 \\ w_2\end{pmatrix}$$

Where $\{w_1, w_2\}^T = G\{f_1, v_2\}^T$. Given a desired path of the cart, x_c^d, y_c^d, which satisfies the nonholonomic constraint, the controller can be designed as:

$$w_1 = \ddot{x}_c^d + k_{px}(x_c^d - x_c) + k_{dx}(\dot{x}_c^d - \dot{x}_c)$$
$$w_2 = \ddot{y}_c^d + k_{py}(y_c^d - y_c) + k_{dy}(\dot{y}_c^d - \dot{y}_c)$$

The angular velocity v_2 can then be obtained by $\{f_1, v_2\}^T = G^{-1}\{w_1, w_2\}^T$. The physical meaning of this approach is that the control input f_1 generates the forward motion and v_2 controls the cart orientation such that x_c^d and y_c^d are tracked. However, control input v_2 is an internal state controlled by f_2, the tangential force applied to the cart. The control input f_1 can then be computed by the backstepping approach based on the design of v_2. Defining $v_2 = \phi(x_c, y_c, \theta_c)$ and $z = \dot{\theta}_c - \phi$, then equation (3.2.7) can be transformed into

$$\dot{z} = -\dot{\phi} + \frac{L}{I_c} f_2 \tag{3.2.9}$$

The control input f_2 can then simply be designed as

$$f_2 = \frac{I_c}{L}(\dot{\phi} - k_\theta(\dot{\theta}_c - \phi)) \tag{3.2.10}$$

It is worth noting that the desired cart configuration $\{x_c^d, y_c^d\}$ is the result of trajectory planning.

8.2 Online Model Learning Method

8.2 Parameter Identification for Task Planning

It is known that L and m_c are two very important parameters when obtaining the control input. While a mobile manipulator starts to manipulate an unknown cart, it is important to identify L and m_c on-line through the initial interaction with the cart.

Figure. 3.2 illustrates the geometric relation between the mobile manipulator and the nonholonomic cart. Point A in Fig. 3.2 on the cart is the point on which the mobile manipulator force and torque are applied, C is the intersection between the rotation axis of the front wheels and the central line of the cart. V_A is the velocity of point A in frame Σ. θ_c, θ_m are the orientations of the cart and the mobile manipulator in frame Σ, respectively. α is the orientation of the cart in frame Σ_A.

For simplification, the cart can be described as a segment of the central line of the cart. The line starts from its intersection with the rotation axis of front wheels (point C) and ends at the point (Point A) handled by the gripper. The length of the line is L. The turning movement of the object can be isolated to study its properties.

The geometric relationship between the angles is:

$$\theta_c = \alpha + \theta_m \tag{3.3.1}$$

The on-board sensors in mobile manipulator include a Laser Range Finder, Encoders, and a Force/Torque Sensor. They can provide real-time sensory measurement of position of the mobile base and the end-effector, orientation of the cart, force and acceleration applied on the end-effector.

Based on the mobile manipulator sensor readings, the position of A can be described as

$$\begin{bmatrix} x_a \\ y_a \end{bmatrix} = \begin{bmatrix} \cos\theta_m & -\sin\theta_m \\ \sin\theta_m & \cos\theta_m \end{bmatrix} \begin{bmatrix} p_x \\ p_y \end{bmatrix} + \begin{bmatrix} x_m \\ y_m \end{bmatrix} \tag{3.3.2}$$

The derivatives of the above equations are:

$$\dot{\theta}_c = \dot{\alpha} + \dot{\theta}_m \tag{3.3.3}$$

$$\begin{bmatrix} \dot{x}_a \\ \dot{y}_a \end{bmatrix} = \begin{bmatrix} \cos\theta_m & -\sin\theta_m \\ \sin\theta_m & \cos\theta_m \end{bmatrix} \begin{bmatrix} \dot{p}_x \\ \dot{p}_y \end{bmatrix} + \begin{bmatrix} \dot{x}_m \\ \dot{y}_m \end{bmatrix}$$

(3.3.4)

$$+\dot{\theta}_m \cdot \begin{bmatrix} -\sin\theta_m & -\cos\theta_m \\ \cos\theta_m & -\sin\theta_m \end{bmatrix} \begin{bmatrix} p_x \\ p_y \end{bmatrix}$$

We define:

$$V_A = \begin{bmatrix} \dot{x}_a \\ \dot{y}_a \end{bmatrix}; \tag{3.3.5}$$

Then $V_{AY}^C = \dot{x}_a \cdot \sin\theta_c - \dot{y}_a \cdot \cos\theta_c$.

According to the characteristics of a nonholonomic cart, the turning movement of the cart can be isolated as an independent motion. The movement can be described as

$$V_{AY}^C = L \cdot \dot{\theta}_c \tag{3.3.6}$$

Equation (3.3.6) is the fundamental equation for identifying the parameter L.

To estimate the mass of the cart, the dynamics of the cart along x_c can be described by the following equation:

$$a_1 = \frac{f_1}{m_c} - \frac{F_f}{m_c}. \tag{3.3.7}$$

Where m_c is the mass of the cart, a_1 is acceleration on X_c axis, f_1 and F_f are the input force and the friction on X_c axis, respectively. a_1 and f_1 can be measured by the $Jr3$ force/torque sensor. Rewrite it in vector form

$$a_1 = \begin{bmatrix} f_1 \\ -1 \end{bmatrix}^T \cdot \begin{bmatrix} \frac{1}{m_c} \\ \frac{F_f}{m_c} \end{bmatrix} \tag{3.3.8}$$

Based on (3.3.8), the mass of the cart can be identified by measurements of a_1 and f_1.

To estimate the kinematics length of a nonholonomic cart, positions, orientations, linear velocities and angular velocities of the cart are needed. The first two can be obtained by mobile manipulator sensor readings and previous work on orientation finding [1]. To find the velocity signals, a numerical differentiation method is used to obtain the derivative of position and orientation information. For a time domain signal, $Z(t)$, this procedure can be expressed by the following equation,

$$\dot{Z}(t) = \frac{Z(t) - Z(t-t')}{t-t'}, t - t' = \Delta t > 0. \tag{3.3.9}$$

Where Δt is the sampling time.

The orientation signal is very noisy due to the sensor measurement error. The noise involved in the orientation signal is significantly amplified after numerical differentiation, and it can not be easily statistically modeled. The unmodeled noise causes a large estimation error in parameter identification.

3.2.2 State Estimation

The state estimation includes the estimation of the cart position and orientation with respect to the mobile platform. The mobile platform is equipped with a laser range finder. Fig. 3.2 shows a configuration of the cart in the moving frame of the mobile platform. The laser sensor provides a polar range map of its environment in the form of (α, r), where α is an angle from $[-\pi/2, \pi/2]$ which is discretized into *360* units. The range sensor provides a ranging resolution of *50mm*. Fig. 3.3 (a) shows the raw data from the laser sensor. Obviously the measurement of the cart is mixed with the environment surrounding it. Only a chunk of the data is useful and the rest should be filtered out. A sliding window, $w = \{\alpha_1, \alpha_2\}$ is defined. The data for $\alpha \notin w$ are discarded. To obtain the size of w, the encoder information of the mobile manipulator should be fused with the laser data. Since the cart is held by the end-effector, the approximate position of x_r, y_r, as shown in Fig. 3.2, can be obtained. Based on x_r, y_r and the covariance of the measurement r, which is known, the sliding window size can be estimated. The filtered data by the sliding window is shown in Fig. 3.3 (b).

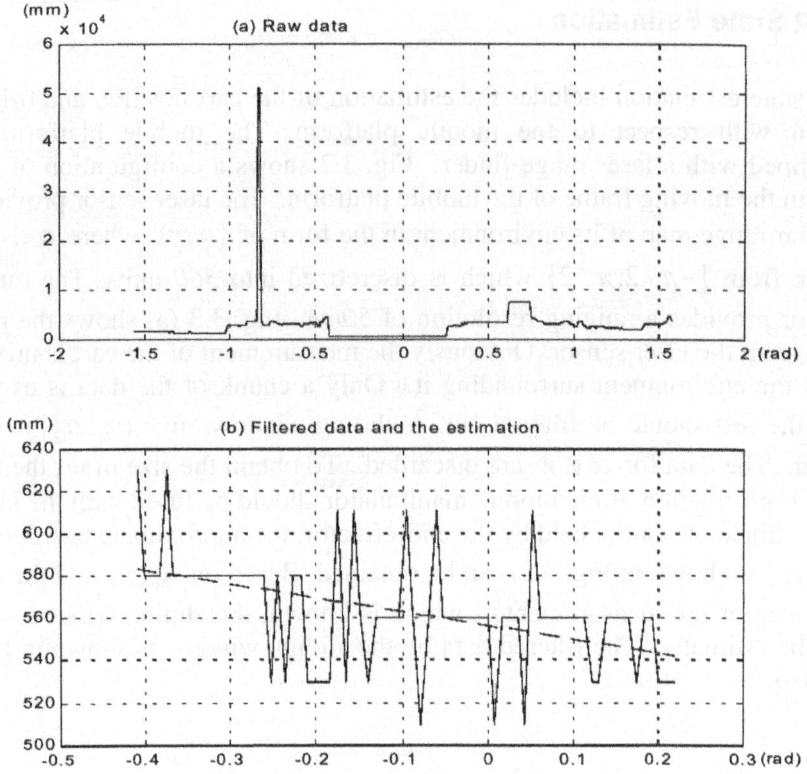

Fig. 3.3. Laser Range Data

In this application, the cart position and orientation (x_c, y_c, θ_c) are the parameters to be estimated. Assuming the relative motion between the cart the robot is slow, the cart position and orientation can be estimated by the standard Extended Kalman Filter (EKF) technique [11]. The cart dynamics are described by the following vector function:

$$\dot{X}_c = F(X_c, f) + \xi \tag{3.3.10}$$

where f is the control input of the cart, which can be measured by the force torque sensor equipped on the end-effector of the mobile manipulator. $X_c = \{x_c, \dot{x}_c, y_c, \dot{y}_c, \theta_c, \dot{\theta}_c\}$. ξ is a random vector with zero mean and covariance matrix σ_x. Considering the output vector as $Z(t) = \{\alpha, r\}^T_i$ as shown in Fig. 3.2, the output function is described as:

$$Z(t) = G(X_c) + \xi \tag{3.3.11}$$

Where ξ is a random vector with zero mean and covariance matrix σ_z. It is worth noting that nonholonomic constraint should be considered in (3.3.10) and the measurement (α, r) should be transformed into the moving frame. By applying the standard ***EKF*** technique to (3.3.10) and (3.3.11), an estimate of the cart state variables $\hat{X}_c$ for X_c can be obtained. The dotted line in Fig 3.3 (b) shows an estimation of the cart configuration in the mobile frame Σ_b. It is worth noting that both the force/torque sensor information and the laser range finder information are used in the estimation of the cart configuration.

The state estimation task can be solved by Least Square Method [9]. The equations of the recursive least square method are:

$$\hat{X}_r = \hat{X}_{r-1} + K_r(Y_r - H_r' X_{r-1}), \tag{3.3.12}$$

$$K_r = P_{r-1} H_r (I + H_r' P_{r-1} H_r)^{-1}, \tag{3.3.13}$$

$$P_r = P_{r-1} - P_{r-1} K_r \tag{3.3.14}$$

3.3.3 Wavelet Transform in Digital Signal Processing

The Fourier Transform is a frequency domain representation of a function. The essence of the Fourier transform of a waveform is to decompose the waveform into a sum of sinusoids of different frequencies. In other words, the Fourier transform analyzes the ``frequency contents" of a signal. The Fourier Transform identifies or distinguishes the different frequency sinusoids, and their respective amplitudes, which combine to form an arbitrary waveform.

The Fourier Transform, with its wide range of applications, has its limitations. For example, this transformation cannot be applied to non-stationary signals that have time varying or spatial varying characteristics. Although the modified version of the Fourier Transform, referred to as Short-Time Fourier Transform (STFT), can resolve some of the problems associated with non-stationary signals, it does not address all the problems.

The wavelet transform is a tool to perform the concept of multiresolution that cuts up data or functions or operators into different frequency components, and then studies each component with a resolution matched to its scale. The wavelet transform has been successfully applied to non-stationary signals for analysis and has provided an alternative to the Windowed Fourier Transform. We will give a brief and concise review on Fourier Transform and Wavelet Transform.

3.3.3.1 *Short-Time Fourier Transform and Wavelet Transform*

The Short-Time Fourier Transform (STFT) replaces the Fourier transform's sinusoidal wave by the product of a sinusoid and a single analysis window which is localized in time. The STFT has a constant time frequency resolution. This resolution can be changed by rescaling the window. It is a complete, stable, redundant representation of the signal.

The STFT takes two arguments: time and frequency. It has a constant time frequency resolution. This resolution can be changed by rescaling the window.

Wavelet Transform is different from STFT. The continuous wavelet transform is defined by

$$STFT_x^{\psi}(\tau,s) = \psi_x^{\psi} = \frac{1}{\sqrt{(|s|)}} \int (x(t)\psi^*(\frac{t-\tau}{s}))dt \tag{3.3.15}$$

As seen in the above equation, the transformed signal is a function of two variables, τ and s, the translation and scale parameters, respectively. $\psi(t)$ is the transforming function, and it is called the mother wavelet. The term mother wavelet gets its name due to two important properties of the wavelet analysis as explained below:

The term wavelet means a small wave. The smallness refers to the condition that this window function is of finite length. The wave refers to the condition that this function is oscillatory. The term "mother"" implies that the functions with different region of support that are used in the transformation process are derived from one main function, or the mother wavelet. In other words, the mother wavelet is a prototype for generating the other window functions.

The term "translation" is used in the same sense as it was used in the STFT; it is related to the location of the window, as the window is shifted through the signal. This term, obviously, corresponds to the time information in the transform domain. However, we do not have a frequency parameter, as we had before for the STFT. Instead, we have scale parameter which is defined as $(frequency)^{-1}$. The term frequency is reserved for the STFT.

3.3.3.2 *Discrete Wavelet Transform*

Wavelet Transform (WT) is a technique for analyzing signals. It was developed as an alternative to STFT to overcome problems related to its frequency and time resolution properties. More specifically, unlike the STFT that provides uniform time resolution for all frequencies. The Discrete Wavelet Transform (DWT) provides high time resolution and low frequency resolution for high frequencies and high frequency resolution and low time resolution for low frequencies.

The DWT is a special case of the WT that provides a compact representation of a signal in time and frequency that can be computed efficiently. It can be explained as a subband coding. The time-domain signal is passed from various highpass and lowpass filters, which filter out either high frequency or low frequency portions of the signal. This procedure is repeated, every time some portion of the signal corresponding to some frequencies is being removed from the signal.

The Continuous Wavelet Transform (CWT) can be thought of as a correlation between a wavelet at different scales and the signal with the scale (or the frequency) being used as a measure of similarity. The continuous wavelet transform is computed by changing the scale of the analysis window, shifting the window in time, multiplying by the signal, and integrating over all times. In the discrete case, filters of different cutoff frequencies are used to analyze the signal at different scales. The signal is passed through a series of high pass filters to analyze the high frequencies, and it is passed through a series of low pass filters to analyze the low frequencies. The subbanding procedure [7, 1] is shown in Fig. 3.4.

In discrete signals, the *Nyquist rate* corresponds to π rad/s in the discrete frequency domain. After passing the signal through a half band lowpass filter, half of the samples can be eliminated according to the Nyquist's rule, since the signal now has a highest frequency of $\pi/2$ radians instead of π radians. Simply discarding every other sample will subsample the signal by two, and the signal will then have half the number of points. The scale of the signal is now doubled. Note that the lowpass filtering removes the high frequency information, but leaves the scale unchanged. Only the subsampling process changes the scale. Resolution, on the other hand, is related to the amount of information in the signal, and therefore, it is affected by the filtering operations. Half band lowpass filtering removes half of the frequencies, which can be interpreted as losing half of the information. Therefore, the resolution is halved after the filtering operation.

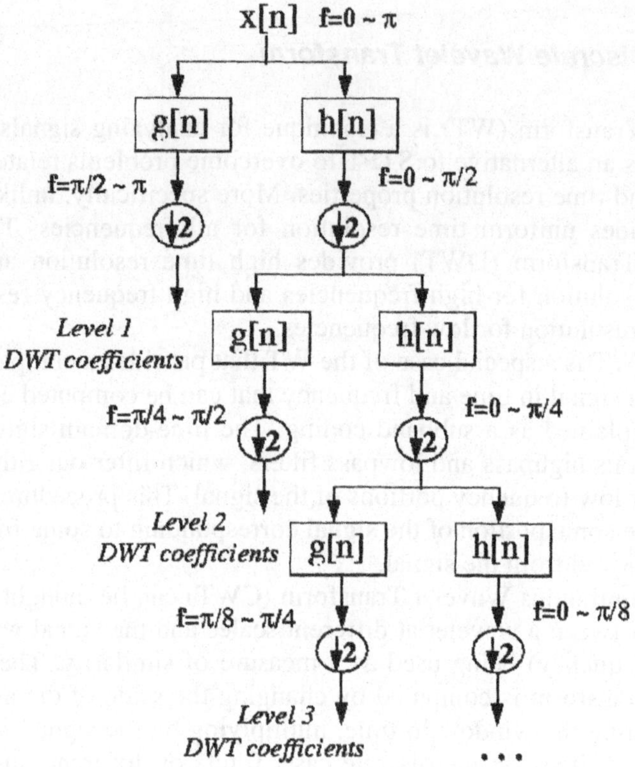

Fig. 3.4. Multilevel Discrete Wavelet Transform

3.3.4 Discrete Wavelet Based Model Identification (DWMI)

The on-line parameter estimation problem can be described by a recursive Least Square Method.

Given a linear observation equation

$$Y_r = H_r X_r + v_r \tag{3.3.16}$$

Y_r is an observation vector at time instance r,

H_r is an observation matrix at time instance *r*.

X_r is an estimated parameter vector.

v_r is the measurement noise.

Corresponding to the estimation equations developed in section 3.2, for estimation of L, Y_r is the measurement sequence of V_{AY}^C, H_r is the measurement sequence of $\dot{\theta}_c$, X is the parameter L; for estimation of the mass, Y_r is the measurement sequence of the acceleration a_1, and H_r is the sequence of the vector of $\begin{bmatrix} f_1 \\ -1 \end{bmatrix}^T$.

Table 3.1. Signal/Noise Distribution

Measurements \ Features	$\dot{\theta}_c$ measurements		V_{AY}^C measurements	
	Signal	Noise	Signal	Noise
Strength	Normal	Strong	Normal	Week
Spectra	Lower frequencies	Higher frequencies	Lower frequencies	

In Table 3.1, the signal analysis of $\dot{\theta}_c$ and V_{AY}^c shows that the former has strong noise in the high frequency range while the signal components for both $\dot{\theta}_c$ and V_{AY}^c reside in the lower frequency range.

To obtain the best estimation of the kinematic length of the cart, we have to find the proper bandwidth, which causes the minimum estimation error. It is known that in the frequency domain the true signal is at low frequencies, the major parts of the noise are at high frequencies. Hence, the measured signal and the true signal have closer spectra at low frequencies than at high frequencies. Furthermore, to extract the proper low frequency part from measured signals will result in accurate estimation.

Daubechies Discrete Wavelet Transform (DWT) [5] is applied to decompose and reconstruct the raw signal in different bandwidth. In this chapter, a raw signal will be sequentially passed through a series of Daubechies filters with imposed response $h_p(n)$ for wavelets with P vanishing moments.

To subband a signal, Discrete Wavelet Transform is used. As shown in Fig. 3.4, $h(n)$ and $g(n)$ are a lowpass filter and a highpass filter, respectively. The two filters can halve the bandwidth of the signal at this level. Fig. 3.4 also shows the DWT coefficients of the higher frequency components at each level.

As a result, the raw signal is preprocessed to have the desired low frequency components. The multiresolution approach from discrete wavelet analysis will be used to decompose the raw signal into several signals with different bandwidths. This algorithm makes the signal, in this case, the raw angular velocity signal passes through several lowpass filters. At each level it passes the filter, and the bandwidth of the signal would be halved. Then the lower frequency component can be obtained level by level.

The algorithm can be described as the following procedures:

(a) Filtering: Passing the signal through a lowpass Daubechies filter with bandwidth which is the lower half bandwidth of the signal at the last level. Subsampling the signal by factor 2, then reconstructing the signal at this level;

(b) Estimating: Using the RLSM to process the linear velocity signal and the angular velocity signal obtained from the step (a) to estimate the kinematic length of the cart.

(c) Calculating: Calculating the expectation of the length estimates and the residual.

(d) Returning: Returning to (a), until it can be ensured that $\widetilde{e}$ is increasing.

(e) Comparing: Comparing the residual in each level, take the estimate of length at a level, which has the minimum residual over all the levels, as the most accurate estimate.

The block diagram of DWMI algorithm is shown in Fig. 3.5.

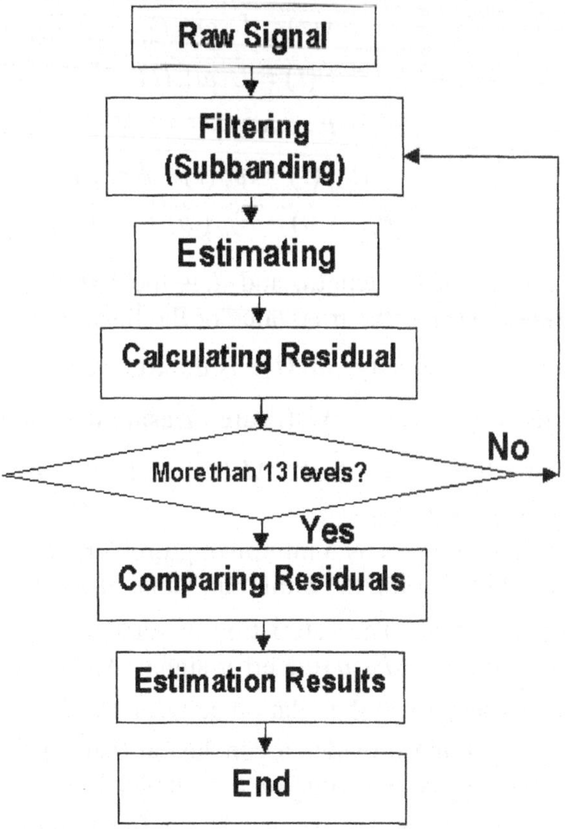

Fig. 3.5. Block Diagram of Model Identification Algorithm

3.4 Convergence of Estimation

In this section, the parameter estimation problem in time domain is analyzed in frequency domain. The estimation convergence means that the estimate of the parameter can approximately approach the real value, if the measurement signal and the real signal have an identical frequency spectrum. First, we convert the time based problem into frequency domain through Fourier Transform.

The least square of estimation residual can be described by

$$\tilde{e} = \int_0^T (\hat{v}(t) - v(t))^2 dt \tag{3.4.1}$$

and the relationships can be designed as follows:

$$v(t) = L \cdot \omega_T(t)\,, \tag{3.4.2}$$

$$\hat{v}(t) = \hat{L} \cdot \omega_M(t)\,, \tag{3.4.3}$$

$$\hat{L} = \Delta L + L\,, \tag{3.4.4}$$

$$\omega_M(t) = \omega_T(t) + \Delta\omega(t), \tag{3.4.5}$$

$$F_{\omega M}(\omega) = F_{\omega T}(\omega) + \Delta F(\omega)\,, \tag{3.4.6}$$

L is the true value of the length, and $\hat{L}$ is the estimate of the length in least square sense. $v(t)$ is the true value of the linear velocity, $\hat{v}(t)$ is the estimate of the linear velocity, $\omega_T(t)$ is the true value of $\dot{\theta}_c$, $\omega_M(t)$ is the measurements of $\dot{\theta}_c$ and $\Delta\omega(t)$ are measurement additive noise signal of $\dot{\theta}_c$, respectively. $F_{\omega_T}(\omega)$, $\Delta F(\omega)$ and $F_{\omega_M}(\omega)$ are their corresponding Fourier Transforms.

Considering the problem as a minimizing problem, the estimation error can be minimized by finding the minimum value of the estimation residual $\tilde{e}$ in least square sense. The estimation residual is in terms of the frequency domain form of $\Delta F(\omega)$ the error signal $\Delta\omega(t)$. Hence, the problem is turned into describing the relation between the $\tilde{e}$ and $\Delta F(\omega)$.

The following lemma provides a conclusion that functions with a certain form are increasing functions of a variable. Based on the lemma, a theorem can be developed to prove that $\tilde{e}$ is a function of $(\frac{\Delta L}{L})^2$ which has the same form as in the lemma. Thus, the estimation error decreases, as the residual is reduced.

Lemma: Let $\Omega : (-\infty, +\infty)$ and

$$X = \left(\frac{\int_\Omega F_{\omega_M}(\omega)\overline{\Delta F(\omega)}d\omega}{\int_\Omega \left|F_{\omega_M}(\omega)\right|^2 d\omega}\right)^2 \tag{3.4.7}$$

$$\tilde{e} = \frac{L^2}{2\pi}\left(\int_\Omega |\Delta F(\omega)|^2 d\omega - \int_\Omega |F_{\omega_M}(\omega)|^2 d\omega \cdot X\right) \tag{3.4.8}$$

If $F_{\omega_M}(\omega)$ is orthogonal to $\Delta F(\omega)$, then $\tilde{e}$ is a strictly increasing function of X.

Proof: First, we try to transfer the problem to real space through simplifying X. Since $\Delta F(\omega)$ is orthogonal to $F_{\omega_M}(\omega)$, i.e.

$$\int_\Omega F_{\omega_M}(\omega)\overline{\Delta F(\omega)} d\omega = 0 \tag{3.4.9}$$

Simplifying the integrals

$$\int_\Omega F_{\omega_M}(\omega)\overline{\Delta F(\omega)} d\omega = \int_\Omega |\Delta F(\omega)|^2 d\omega$$

$$\int_\Omega |F_{\omega_M}(\omega)|^2 d\omega = \int_\Omega |F_{\omega_T}(\omega)|^2 d\omega + \int_\Omega |\Delta F(\omega)|^2 d\omega$$

These two questions can move out some terms in X, it is clear that X is a real function as

$$X = \left(\frac{\int_\Omega |\Delta F(\omega)|^2 d\omega}{\int_\Omega (|F_{\omega_T}(\omega)|^2 d\omega + |\Delta F(\omega)|^2) d\omega}\right)^2$$

It implies

$$\int_\Omega |\Delta F(\omega)|^2 d\omega = \frac{\sqrt{X}}{1-\sqrt{X}} \int_\Omega |F_{\omega_T}(\omega)|^2 d\omega \tag{3.4.10}$$

$\tilde{e}$ can be expressed in teams of X

$$\tilde{e} = \frac{L^2}{2\pi}\left(\int_\Omega |\Delta F(\omega)|^2 d\omega - X \cdot \int_\Omega |F_{\omega_T}(\omega)|^2 d\omega - X \cdot \int_\Omega |\Delta F(\omega)|^2 d\omega\right)$$

$$= \frac{L^2}{2\pi}(1-\sqrt{X})\frac{\sqrt{X}}{1-\sqrt{X}} \int_\Omega |F_{\omega_T}(\omega)|^2 d\omega$$

It can be written as

$$\tilde{e} = \frac{L^2 \int_\Omega |F_{\omega_T}(\omega)|^2 d\omega}{2\pi}\sqrt{X}$$

Let $f(X) = \tilde{e}$, then

$$f'(X) = \frac{L^2 \int_\Omega |F_{\omega_T}(\omega)|^2 d\omega}{4\pi\sqrt{X}} > 0 \qquad (3.4.11)$$

Hence, given $\int_\Omega | F_{\omega_T}(\omega) |^2 d\omega$, $f(X)$ is an increasing function of X.

Finally, $\tilde{e}$ is an increasing function of X. If $\tilde{e} = 0$, $X = 0$.

The lemma provides a foundation to prove $(\frac{\Delta L}{L})^2$ will reach a minimum value when the estimation residual $\tilde{e}$ takes a minimum value.

Theorem: Given $\Delta F : \omega \to C$, C is a complex space, when $\tilde{e}$ takes a minimum value, $(\frac{\Delta L}{L})^2$ also takes a minimum value.

Proof: Consider the continuous case:

$$\tilde{e} = \int_0^T [\hat{v}(t)^2 - \hat{v}(t)v(t) + v(t)^2]dt$$

Given $\Omega : (-\infty, +\infty)$, according to Parseval's Equation,

$$\tilde{e} = \frac{1}{2\pi}\int (|F_{\hat{v}}(\omega)|^2 - 2F_{\hat{v}}(\omega)\overline{F_v(\omega)} + |F_v(\omega)|^2)d\omega$$

From (3.4.3) and Linear properties of Fourier Transform, it can be easily seen that

$$\tilde{e} = \frac{1}{2\pi}\int_{\Omega}(\hat{L}^2|F_{\hat{\omega}}(\omega)|^2 - 2\hat{L}F_{\hat{\omega}}(\omega)\overline{F_{\omega_T}(\omega)})d\omega + \frac{1}{2\pi}\int_{\Omega}|F_v(\omega)|^2 d\omega \tag{3.4.12}$$

$\tilde{e}$ is a function of $\hat{L}$, then based on the least square criterion the following equation in terms of $\hat{L}$ must satisfy

$$\frac{\partial\tilde{e}}{\partial\hat{L}} = \frac{2\hat{L}}{2\pi}\int_{\Omega}(\hat{L}^2|F_{\omega_M}(\omega)|^2 - 2\hat{L}F_{\omega_M}(\omega)\overline{F_v(\omega)})d\omega = 0$$

The above equation implies that the solution of $\hat{L}$ can be expressed as

$$\int_{\Omega}(\hat{L}|F_{\omega_M}(\omega)|^2 - F_{\omega_M}(\omega)\overline{F_v(\omega)})d\omega = 0$$

Using (3.4.2), the above equation implies that the solution of $\hat{L}$ can be expressed as

$$\hat{L} = \frac{\int_{\Omega}F_{\omega_M}(\omega)\overline{F_v(\omega)}d\omega}{\int_{\Omega}|F_{\omega_M}(\omega)|^2 d\omega} = \frac{L\cdot\int_{\Omega}F_{\omega_M}(\omega)\overline{F_{\omega_T}(\omega)}d\omega}{\int_{\Omega}|F_{\omega_M}(\omega)|^2 d\omega} \tag{3.4.13}$$

Let $L^* = \hat{L}$, then substituting (3.4.4),(3.4.6) into (3.4.13) to remove the terms of the linear velocity.

$$\frac{L+\Delta L}{L} = \frac{\int_{\Omega}|F_{\omega_M}(\omega)|^2 - F_{\omega_M}(\omega)\overline{\Delta F(\omega)}d\omega}{\int_{\Omega}|F_{\omega_M}(\omega)|^2 d\omega} \tag{3.4.14}$$

There exists the relation between the estimation error $(\frac{\Delta L}{L})$ in the time domain and the measurement error ($\Delta F(\omega)$) in frequency domain,

$$\frac{\Delta L}{L} = -\frac{\int_{\Omega}F_{\omega_M}(\omega)\overline{\Delta F(\omega)}d\omega}{\int_{\Omega}|F_{\omega_M}(\omega)|^2 d\omega} \tag{3.4.15}$$

Note that if X is defined in the beginning of the section, then $X = (\frac{\Delta L}{L})^2$.

Substituting (3.4.13) into (3.4.12) yields

$$\tilde{e} = \frac{L^2}{2\pi}\left(-\frac{\int_\Omega F_{\omega_m}(\omega)\overline{\Delta F(\omega)}d\omega}{\int_\Omega |F_{\omega_M}(\omega)|^2 d\omega} + \int_\Omega (|F_{\omega T}(\omega)|^2 - |F_{\omega_M}(\omega)|^2 + 2F_{\omega_M}(\omega)\overline{\Delta F(\omega)})d\omega\right) \quad (3.4.16)$$

We define:

$$\Delta e = \int_0 (\Delta\omega(t))^2 dt = \int_0 [\omega_M(t)^2 - 2\omega_M(t)\omega_T(t) + \omega_T(t)^2]dt$$

Applying Parserval's Equation to the error signal $\Delta\omega$ yields

$$\begin{aligned}\int_\Omega |\Delta F(\omega)|^2 d\omega &= \int_\Omega |F_{\omega_T}(\omega)|^2 d\omega + |F_{\omega_M}(\omega)|^2 - 2F_{\omega_M}(\omega)\overline{F_{\omega_T}(\omega)}dw \\ &= \int_\Omega |F_{\omega_T}(\omega)|^2 d\omega + \int_\Omega |F_{\omega_M}(\omega)|^2 d\omega \\ &\quad -2\int_\Omega F_{\omega_M}(\omega)(\overline{F_{\omega_M}(\omega) - \Delta F(\omega)}d\omega\end{aligned}$$

Therefore,

$$\begin{aligned}&\int_\Omega (|F_{\omega T}(\omega)|^2 - |F_{\omega_M}(\omega)|^2)d\omega \\ &= \int_\Omega (|\Delta F(\omega)|^2 - 2F_{\omega_M}(\omega)\overline{\Delta F(\omega)}d\omega\end{aligned} \quad (3.4.17)$$

Substituting (3.4.7), (3.4.8) into (3.4.17) $\tilde{e}$ can be given in terms of X

$$\tilde{e} = \frac{L^2}{2\pi}\left(\int_\Omega |\Delta F(\omega)|^2 d\omega - X \cdot \int_\Omega |F_{\omega_M}(\omega)|^2 d\omega\right) \quad (3.4.18)$$

It can be easily seen that $\tilde{e}$ has the same form as in the lemma, then $\tilde{e}$ is an increasing function of X, for different ΔF, when $\tilde{e}$ takes a minimum value, $(\frac{\Delta L}{L})^2$ also takes a minimum value. Since the minimum value of $\tilde{e}$ is equal to 0, the $(\frac{\Delta L}{L})^2$ will approach 0 as well. The residual of the estimation is convergence and the estimation error goes to 0, as the two frequency spectra are identical.

3.5 Experimental Implementation and Results

The proposed method has been tested using a Mobile Manipulation System consisting of a Nomadic XR4000 mobile robot, and a Puma560 robot arm attached on the mobile robot. A nonholonomic cart is gripped by the end-effector of the robot arm as shown in Fig. 3.1. There are two PCs in the mobile platform, one uses Linux as the operating system for the mobile robot control and the other uses a real time operating system QNX for the control of the Puma560. The end-effector is equipped with a *Jr*3 force/torque sensor.

In order to identify the model of the cart, two types of interaction between mobile manipulator and the cart are planned. First, the robot pushes the cart back and forward without turning the cart. The sensory measurement of the acceleration and the force applied to the cart can be recorded. Second, the cart was turned left and right alternatively to obtain the sensory measurements of the position of the point A and the orientation of the cart. The mass and length estimation are carried out on different carts of varying length and mass.

3.5.1 Mass Estimation

To estimate the mass of the cart, the regular recursive Least Square Method (LSM) is used. The measured acceleration signal and the measured signal of the pushing force contain independent white noise. Hence, the estimation should be unbiased. The estimate of the mass of the cart can be obtained directly by LSM.

Fig. 3.6, 3.7, 3.8 indicate the mass estimation process. At the beginning, the estimation is oscillating, however, a few seconds later, the estimation became stable. The mass estimation results are listed in Table 3.2, which indicates that the mass estimation errors, normally, less than 15%.

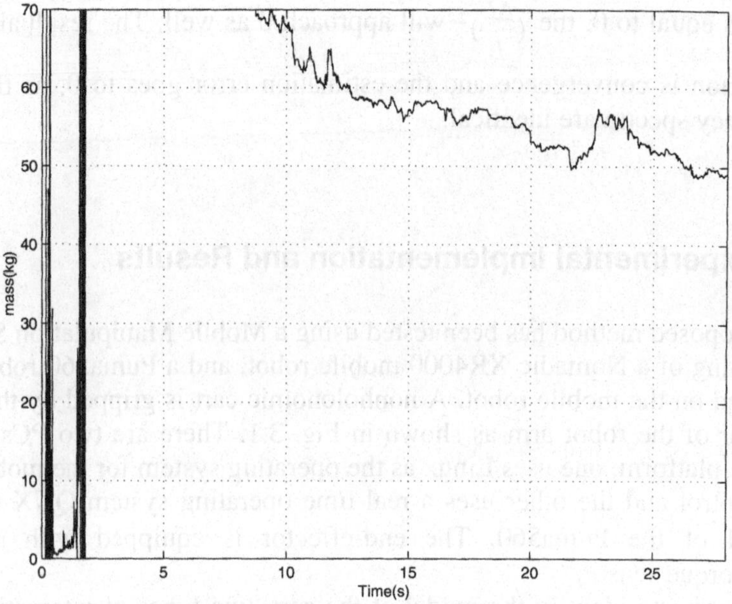

Fig 3.6. Mass Estimation, for M=45kg

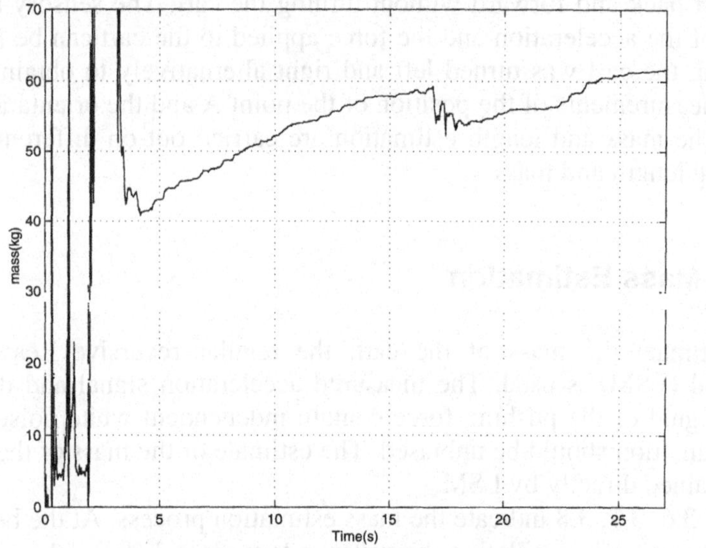

Fig. 3.7. Mass Estimation, for m = 55kg

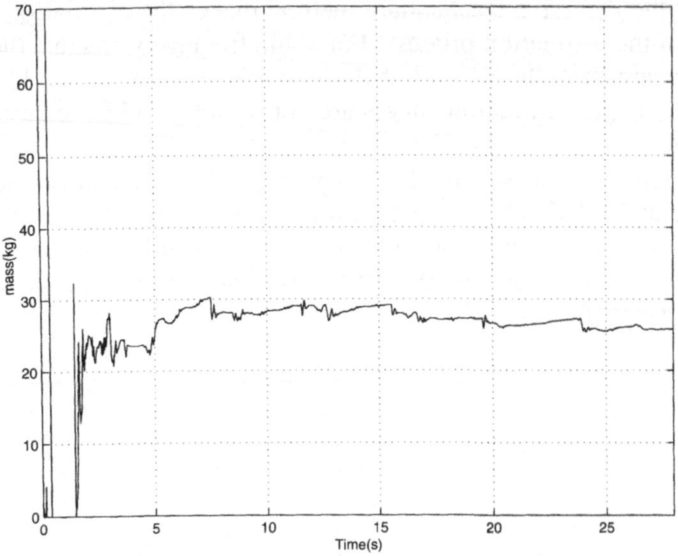

Fig. 3.8. Mass Estimation, for m = 30kg

Table 3.2. Mass Estimation Results

Mass	*Estimate*	*Error(kg)*	*Error(%)*
45.0	49.1	4.1	9.1%
55.0	62.2	7.2	13.1%
30.0	26.8	3.2	10.7%

3.5.1 Length Estimation

According to the proposed method, the algorithm filters the raw signal to have different bandwidths. For different frequency ranges of the signal, recursive Least Square Method is used for parameter identification. The experimental results of length estimation are shown by the graphs below.

Corresponding to the frequency components of the angular velocity signal at different lower ranges, $(0, \pi \cdot (\frac{1}{2})^{level}]$. There are maximally 13 estimation stages in this estimation, therefore the index of the levels ranges from 1 to 13.

Figures 3.9, 3.10, 3.11 and 3.12 show the estimation processes at 9th-12 levels for L=1.31m and L=0.93m. The tends of variance P at all the levels show that the recursive least square method makes the estimation error decreasing in the estimation process. For some frequency ranges, the estimation errors are quite large, and at those levels (For example, 11th and 12th levels), the length estimation curves are not smooth, and have large estimation errors.

For length estimation with L=1.31m, Figs. 3.9, 3.10 show the estimation curve at 9th, 10th, 11th, and 12th level. The estimation result at 10th level provides a smooth estimation, and an accurate result. For L=0.93, Figs. 3.11 and 3.12 indicate a smooth curve of the estimation at 11th level, which results in the best estimate.

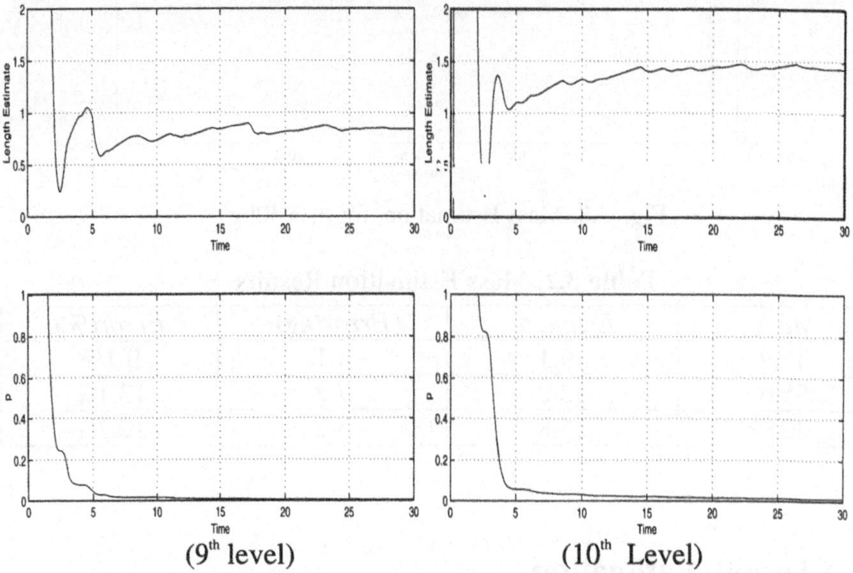

(9th level) (10th Level)

Fig. 3.9. Length Estimate and Variance P at 9th-10th levels for L=1.31m

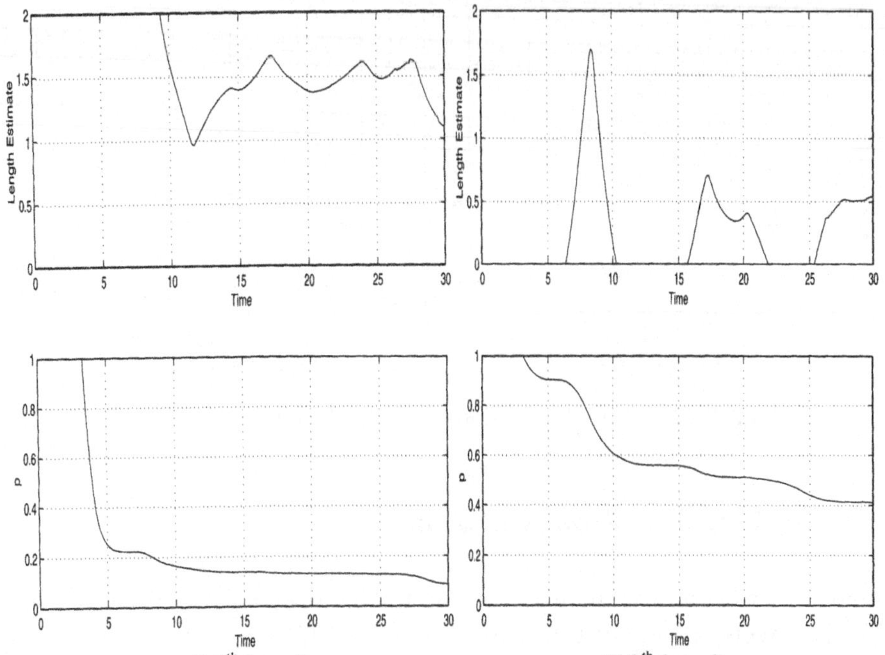

(11th level) (12th level)

Fig. 3.10. Length Estimate and Variance P at 11th-12th levels for L=1.31m

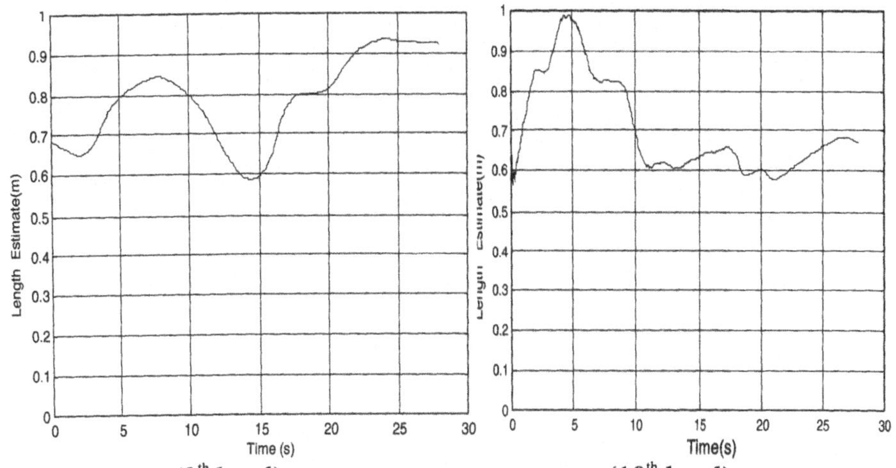

(9th level) (10th level)

Fig. 3. 11. Length Estimation at 9th-10th levels for L=0.93m

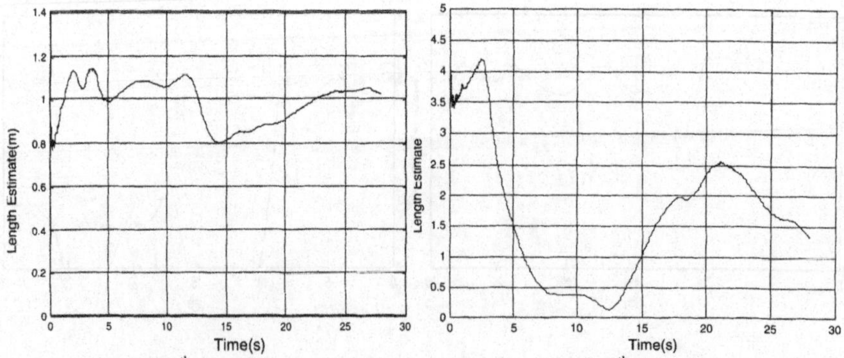

(11th level) (12th level)

Fig. 3. 12. Length Estimation at 11th-12th levels for L=0.93m

3.5.3 Verification of Proposed Method

Figures 3.13, 3.14, 3.15 indicate $\widetilde{e}$ and the parameter estimation errors at different levels, in case of L=0.93m, 1.31m, and 1.46m, respectively.

The horizontal axes represent the index of the estimation level, as shown in Figs. 3.13, 3.14, 3.15. The vertical axes of the figures represent the absolute value of relative estimation error, and the value of $\widetilde{e}$.

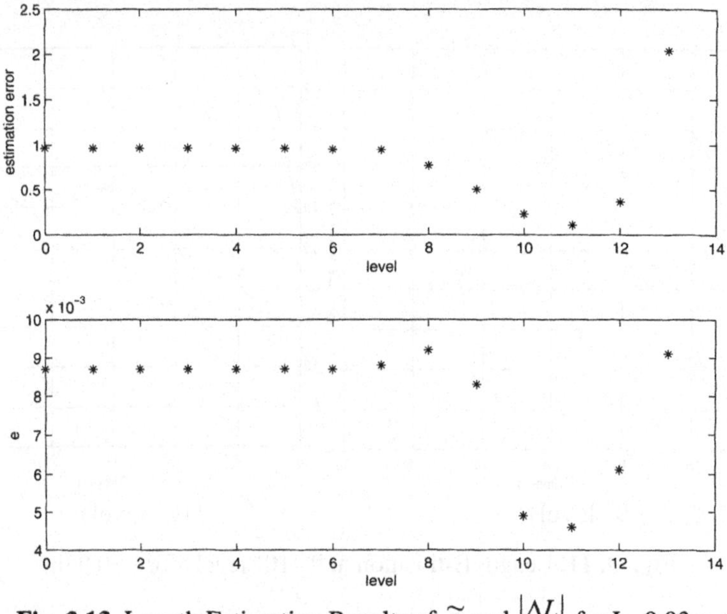

Fig. 3.13. Length Estimation Results of $\widetilde{e}$ and $\left|\frac{\Delta L}{L}\right|$ for L=0.93m

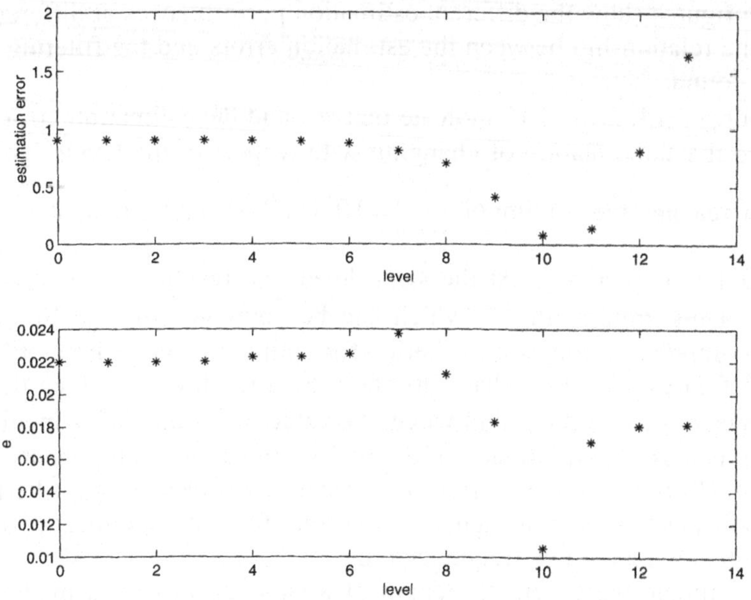

Fig . 3.14. Length Estimation Results of $\tilde{e}$ and $\left|\frac{\Delta L}{L}\right|$ for L=1.31m

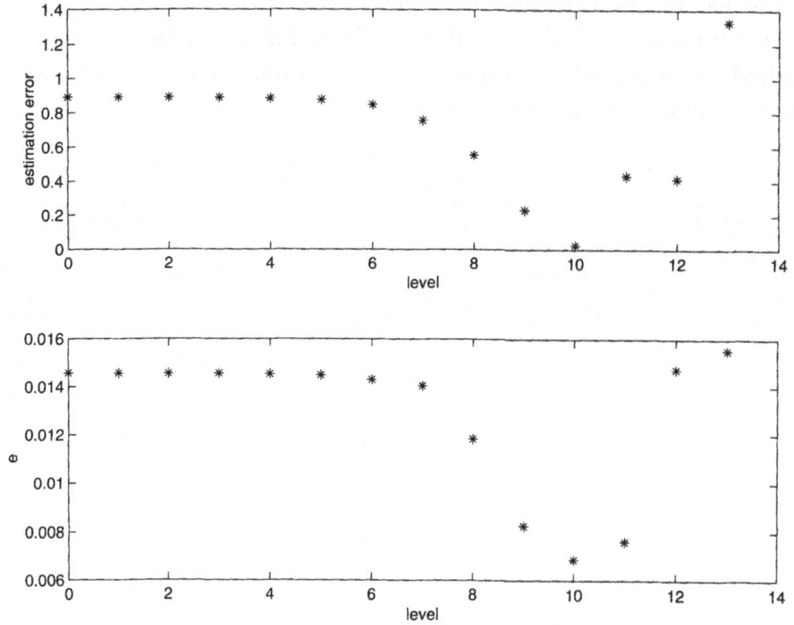

Fig. 3.15. Length Estimation Results of $\tilde{e}$ and $\left|\frac{\Delta L}{L}\right|$ for L=1.46m

The figures show the different estimation performances at different levels. The relationship between the estimation errors and the filtering levels can be found.

Figures 3.13, 3.14, 3.15 indicate that $\tilde{e}$ and the estimation error, delta L, have the same feature of changing with respect to the levels. The estimation reaches the minimum $\left|\frac{\Delta L}{L}\right| = 10.5\%, 7.9\%$ and 2.6% at level 11, 10 and 10, respectively. At the same level, the residual $\tilde{e}$ is also minimized. Thus, minimizing $\tilde{e}$, which can be computed on-line by the on-board computer, becomes the criterion for optimizing the estimation.

The figures also show that after the estimation level at which the estimation error takes a minimum value, the value of $\tilde{e}$ and the estimation error are increasing, due to lack of the normal frequency components of the true signal (serious distortion) at the further levels of low pass filtering. It also indicates that the true signal component of the measurement resides in certain bandwidth at low frequency range.

To estimate the kinematic length of a cart, the proposed method and traditional RLSM are used. The estimates by DWMI Algorithm, according to the proposed method, and the estimates by traditional RLSM without preprocessing the raw data are listed in Table 3.3. It can be seen that the estimation error by RLSM method is about $80\% - 90\%$, while the DWMI method can reduce the estimation error to about 10%. This is a significant improvement of estimation accuracy.

Table 3.3: Comparison of Length Estimation Results

Length (m)	*LS*		*DWMI*	
	$\hat{L}(m)$	*error*	$\hat{L}(m)$	*error*
0.93	0.0290	-96%	1.0278	10.5%
1.14	0.128	-89.3%	1.061	-7.0%
1.31	0.1213	-90%	1.415	7.9%
1.46	0.1577	-89%	1.50	2.6%

3.6 Conclusion

In this chapter, in order to solve the online model learning problem, a Discrete Wavelet based model Identification method has been proposed. The method provides a new criterion to optimize the parameter estimations in noisy environment by minimizing the least square residual. When the unknown noises generated by sensor measurements and numerical operations are uncorrelated, the least square residual is a monotonically increasing function of estimation error. Based on this, the estimation convergence theory is created and proved mathematically. This method offers significant advantages over the classical least square estimation methods in model identification for online estimation without prior statistical knowledge of measurement and operation noises. The experimental results show the improved estimation accuracy of the proposed method for identifying the mass and the length of a nonholonomic cart by interactive action in cart pushing,

Robotic manipulation has a wide range of applications in complex and dynamic environments. Many applications, including home care, search, rescue and so on, require the mobile manipulator to work in unstructured environments. Based on the method proposed in this chapter, the task model can be found by simple interactions between the mobile manipulator and the environment. This approach significantly improves the effectiveness of the operations.

References

1 N. Ali Akansu, J. T. Mark Smith, Subband and wavelet transforms: design and applications, Kluwer Academic Publishers, 1996.

2 Giordano A and Hsu MF (1985), Least square estimation with application to digital signal processing, A Wiley-Interscience Publication 1985.

3 L. Bushnell G., Tibury D. M., Sastry S. S(1995), `Steering three-input nonholonomic systems: The fire truck example', The International Journal of Robotics Research, pages 366-381, vol.14, No.4, 1995.

4 Choi A (1997), Real-Time fundamental frequency estimation by least-square fitting, *LIEEE Transactions on Speech and Audio Processing,* Vol.5, No. 2, pp 201-pp205, March, 1997.

5 Daubechies I(1992), 'Ten lectures on wavelets, Philadelphia, PA: SIAM 1992, Notes from the 1990 CBMS-NSF conference, Wavelets Applications, Lowell, MA, USA.
6 Desantis PM (1994) Path-tracking for a tracker-trailer-like robot, The International Journal of Robotics Research, pages 533-543. vol. 13, No. 5, 1994.
7 Polikar Robi, `The engineer's ultimate guide to wavelet analysis, the wavelettutorial", http://engineering.rowan.edu/~polikar/WAVELETS /WTtutorial.html.
8 Mohinder S. Grewal and Angus P. Andrews, (1993) *Kalman Filtering, theory and practice*, Prentice Hall Information and System Sciences Series, Thomas Kailath, Series Editor Englewood Cliffs, New Jersey, 1993.
9 Hsia T.C (1974), System Identification: Least Square Method, Lexington Books, 1974.
10 Isermann R (1982), Practical aspects of process identification, automatica, Vol, 16. pp. 575-587, 1982.
11 Kam M, Zhu X, Kalata P (1997), Sensor fustion for Mobile robot navigation, , Proceedings of the IEEE, pages 108-119, vol. 85, No. 1, 1997.
12 Li W and Slotine JJE (1987), `Parameter estimation strategies for robotic applications', A.S.M.E Winter Annual Meeting, 1987.
13 Samson C(1995), `Control of chained systems application to path following and time-varying point-stabilization of mobile robots', *IEEE Transactions on Automatic Control*, pages 64-77,vol. 40, No.1, 1995.
14 Sermann R and Baur U(1974), Two step process identification with correlation analysis and least squares parameter estimation, Transactions of ASME, Series G.J. of Dynamic Systems Measurement and Control, Vol.96, pp. 425-432, 1974.
15 Tan J and Xi N (2001), Unified model approach for planning and control of mobile manipulators, Proceedings of IEEE International Conference on Robotics and Automation, pages 3145-3152, Korea, May, 2001.
16 Tibury D, Murray R, Sastry SS, Trajectory generation for the n-trailer problem using goursat normal form, IEEE Transactions on Automatic Control, pages 802-819, vol. 40, No. 5, 1995.
17 Xi N, Tarn TJ and Bejczy, AK(1996), Intelligent planning and control for multirobot coordination: An event-based approach, IEEE Transactions on Robotics and Automation, pages 439-452, vol. 12, No. 3, 1996.

18 Yamamoto Y (1994), Control and coordination of locomotion and manipulation of a wheeled mobile manipulators, Ph. D Dissertation in University of Pennsylvania, August, 1994.
19 Zhuang H and Roth SZ(1993), A linear solution to the kinematic parameter identification of robot manipulators, IEEE Transactions on Robotics and Automation, Vol.9, No.2, 1993.

4 Continuous Reinforcement Learning Algorithm for Skills Learning in an Autonomous Mobile Robot

Mª Jesús López Boada[1], Ramón Barber[2], Verónica Egido[3], Miguel Ángel Salichs[2]

1. Mechanical Engineering Department. Carlos III University, Avd. de la Universidad, 30. 28911. Leganes. Madrid. Spain
mjboada@ing.uc3m.es
2. System Engineering and Automation Department, Carlos III University, Avd. de la Universidad, 30. 28911. Leganes. Madrid. Spain
{rbarber, salichs}@ing.uc3m.es
3. Computer Systems and Automation Department, European University of Madrid. 28670. Villaviciosa de Odón. Madrid, Spain.
veroeg@uem.es

4.1 Introduction

In the last years, one of the main challenges in robotics is to endow the robots with a grade of intelligence in order to allow them to extract information from the environment and use that knowledge to carry out their tasks safely. The intelligence allows the robots to improve their survival in the real world. Two main characteristics that every intelligent system must have are [1]:

1. Autonomy. Intelligent systems must be able to operate without the help of human being or other systems, and to have control over its own actions and internal state. Robots must have a wide variety of different behaviors to operate autonomously.
2. Adaptability. Intelligent systems must be able to learn to react to changes happening in the environment and on themselves in order to improve their behavior. Robots have to retain information about their personal experience to be able to learn.

A sign of intelligence is learning. Learning endows a mobile robot with a higher flexibility and allows it to adapt to changes occurring in the environment or in its internal state in order to improve its results. Learning is particularly difficult in robotics due to the following reasons [2] [3]:

M.J.L. Boada et al.: *Continuous Reinforcement Learning Algorithm for Skills Learning in an Autonomous Mobile Robot*, Studies in Computational Intelligence (SCI) **7**, 137–165 (2005)
www.springerlink.com

1. In most cases, the information provided by the sensors is incomplete and noisy.
2. Environment conditions can change.
3. Training data can not be available *off-line*. In this case, the robot has to move in its environment in order to acquire the necessary knowledge from its experience.
4. The learning algorithm has to achieve good results in a short period of time.

Despite these drawbacks, learning algorithms have been applied successfully in walking robots [4] [5], navigation [6] [7], tasks coordination [8], pattern recognition [9], etc.

According to the information received during the learning, learning methods can be classified as supervised and unsupervised [10]. In the supervised learning algorithms, there exists a *teacher* which provides the desired output for each input vector. These methods are very powerful because they work with a lot of information although they present the following drawbacks: the learning is performed *off-line* and it is necessary to know how the system has to behave.

In the unsupervised learning algorithms, there is not a *teacher* which appraises the suitable outputs for particular inputs. The reinforcement learning is included in these methods [11]. In this case, there exists a *critic* which provides more evaluative than instructional information. The idea lies in the system, explores the environment and observes the action results in order to achieve a learning results index. The main advantages are that there is no need for a complete knowledge of the system and the robot can continuously improve its performance while it is learning.

The more complex a task is performed by a robot, the slower the learning is, because the number of states increases so that it makes it difficult to find the best action. The task decomposition in simpler sub-tasks permits an improvement of the learning because each skill learns in a subset of possible states, so that the search space is reduced. The current tendency is to define basic robot behaviors, which are combined to execute more complex tasks [12] [13] [14].

In this work, we present a reinforcement learning algorithm using neural networks which allows a mobile robot to learn skills. The implemented neural network architecture works with continuous input and output spaces, has a good resistance to forget previously learned actions and learns quickly. Other advantages this algorithm presents are that on one hand, it is not necessary to estimate an expected reward because the robot receives a real continuous reinforcement each time it performs an action and, on the other hand, the robot learns *on-line*, so that the robot can adapt to changes produced in the environment. Finally, the learnt skills are combined to successfully perform a more complex skills called *Visual Approaching* and *Go To Goal Avoiding Obstacles*.

Section 2 describes a generic structure of an automatic skill. Automatic skills are the sensorial and motor capacities of the system. The skill's concept includes the basic and emergent behaviors' concepts of the behavior-based systems [15] [12]. Skills are the base of the robot control architecture AD proposed by R. Barber et al. [16]. This control architecture is inspired from the human being reasoning capacity and the actuation capacity and it is formed by two levels: Deliberative and Automatic. The Deliberative level is associated with the reflective processes and the Automatic level is associated to the automatic processes. Section 3 proposes three different methods for generating complex skills from simpler ones in the AD architecture. These methods are not exclusive, they can occur in the same skill. Section 4 gives an overview of the reinforcement learning and the main problems appeared in reinforcement learning systems. Section 5 shows a detailed description of the continuous reinforcement learning algorithm proposed. Section 6 presents the experimental results obtained from the learning of different automatic skills. Finally, in section 7, some conclusions based on the results presented in this work are provided.

4.2 Automatic Skills

Automatic skills are defined as the capacity of processing sensorial information and/or executing actions upon the robot's actuators [17]. Bonasso et al. [18] define skills as the robot's connection with the world. For Chatila et al. [19] skills are all built-in robot action and perception capacities. In the AD architecture skills are classified as perceptive and sensorimotor. Perceptive skills interpret the information perceived from the sensors, sensorimotor skills, or other perceptive skills. Sensorimotor skills perceive information from the sensors, perceptive skills or other sensorimotor skills and on the basis of that perform an action upon the actuators. All automatic skills have the following characteristics:

1. They can be activated by skills situated in the same level or in the higher level. A skill can only deactivate skills which it has activated previously.
2. Skills have to store their results in memory to be used by other skills.
3. A skill can generate different events and communicate with whom has requested to receive notification previously.

Fig. 4.1 shows the generic structure of a skill. It contains an active object, an event manager object and data objects. The active object is in charge of processing. When a skill is activated, it connects to data objects or to sensors' servers as required by the skill. Then, it processes the received input information, and finally, it stores the output results in its data objects. These objects contain different data structures depending on the type of stored data. When the skill is sensorimotor, it can connect to actuators' servers in order to send them movement commands.

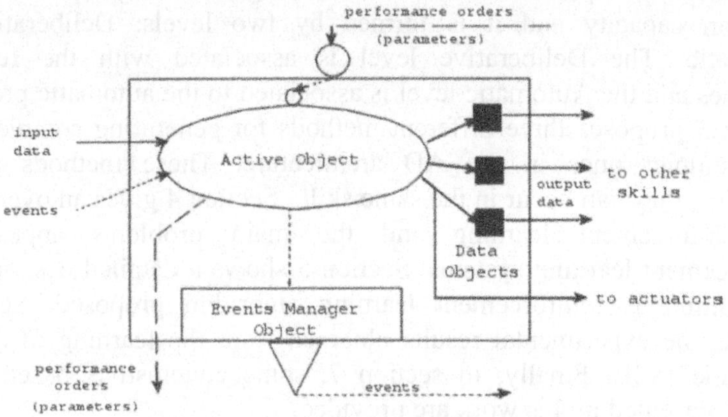

Fig. 4.1. Generic automatic skill's structure

Skills which can be activated are represented by a circle. There could be skills which are permanently active and in this case they are represented without circles. During the processing, the active object can generate events. For example, the sensorimotor skill called *Go To Goal* generates the event *GOAL_ REACHED* when the required task is achieved successfully. Events are sent to the event manager object, which is in charge of notifying skills of the produced event. Only the skills that they have previously registered on it will receive notification. During the activation of the skill, some parameters can be sent to the activated skill. For instance, the skill called *Go To Goal* receives as parameters the goal's position, the robot's maximum velocity and if the skill can send velocity commands to actuators directly or not.

4.3 Complex Skills Generation

Skills can be combined to obtain complex skills and these, in turn, can be recursively combined to form more complex skills. Owing to the modular characteristic of the skills, they can be used to build skills' hierarchies with higher abstraction levels. Skills are not organized *a priori*; they are, rather, used depending on the task being carrying out and on the state of the environment. The complex skill concept is similar to the emergent behavior concept of the behavior based systems [20].

The generation of complex skills from simpler ones presents the following main advantages:

1. Re-using of software. A skill can be used for different complex skills.
2. Reducing the programming complexity. The problem is divided into smaller and simpler problems.
3. Improving the learning rate. Each skill is learned in a subset of possible states, so that the search space is reduced.

In the literature, there exist different methods to generate new behaviors from simpler ones: direct, temporal and information flow based methods. In the first methods the emergent behavior's output is a combination of the simple behaviors' outputs. Within them, the competitive [12] and the cooperative methods [21] [22] can be found. In the temporal methods a sequencer is in charge of establishing the temporal dependencies among simple behaviors [23] [24]. In the information flow based methods the behaviors do not use the information perceived directly by the sensors. They receive information processed previously by other behaviors [25].

According to these ideas, we propose three different methods for generating complex skill from simple ones [17]:

1. *Sequencing* method. In the sequencing method the complex skill is formed by a sequencer which is in charge of deciding what skills have to be activated in each moment avoiding the simultaneous activation of other skills which act upon the same actuator (see Fig. 4.2).
2. *Output addition* method. In the output addition method the resulting movement commands are obtained by combining the movement commands of each skill (see Fig. 4.3). In this case, skills act upon the same actuator and are activated at the same time. Contrary to the previous method, simple skills do not connect to actuators directly. They have to store their results in the data objects in order to be used by the complex skill. When a skill is activated it does not know if it has to send the command to actuators or store its results in its data object. In order to solve this problem, one of the activation parameters sent to the skill determines if the skill has to connect to actuators or not.

3. *Data flow* method. In the data flow method, the complex skill is made up of skills which send information from one to the other as shown in Fig. 4.4. The difference from the above methods is that the complex skill does not have to be responsible for activating all skills. Simple skills activate skills from which they need their data.

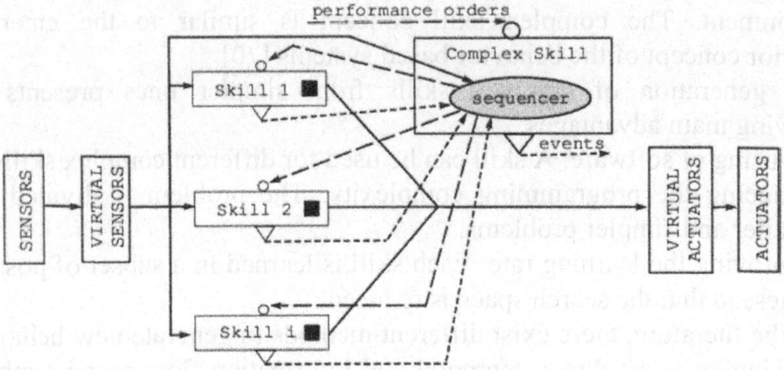

Fig. 4.2. Sequencing method

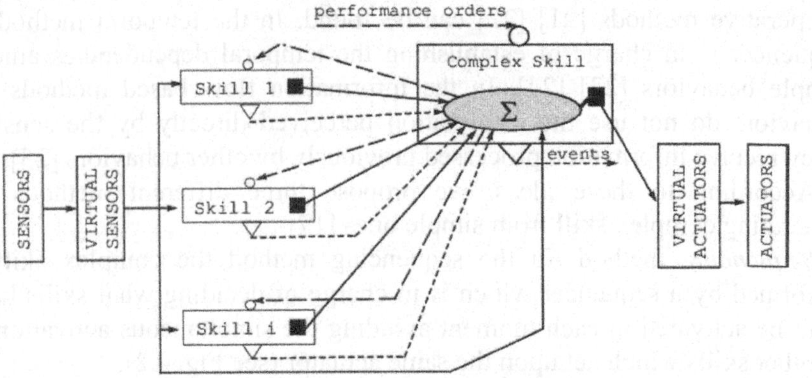

Fig. 4.3. Output addition method

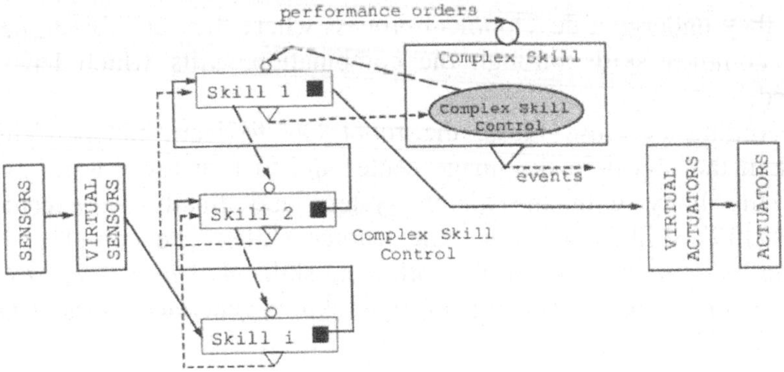

Fig. 4.4. Data flow method

Unlike other authors who only use one of the methods for generating emergent behaviors, the three proposed methods are not exclusive; they can occur in the same skill. A generic complex skill must have a structure which allows its generation by one or more of the methods described above (see Fig. 4.5).

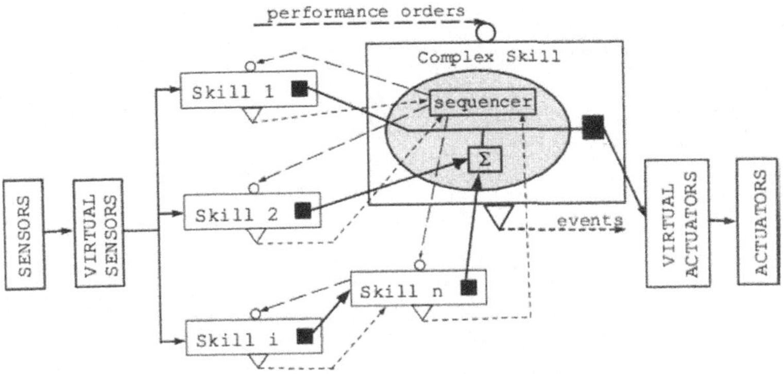

Fig. 4.5. Generic structure of a complex skill

4.3.1 Visual Approach Skill

Approaching a target means moving towards a stationary object [17][26]. In the process, the human performs to execute this skill using visual feedback is, first of all, to move his eyes and head to center the object in the image and then to align the body with the head while he is moving towards the target. Humans are not able to perform complex skill when they are born, they undergo a development process where they are able to perform more complex skills through the combination skills which have been learned.

According to these ideas, the robot has to learn independently to maintain the object in the image center and to turn towards the base to align the body with the vision system and finally to execute the approaching skill coordinating the learned skills. The complex skill is formed by a combination of the following skills *Watching*, *Object Center* and *Robot Orientation*, see Fig. 4.6. This skill is generated by the data flow method.

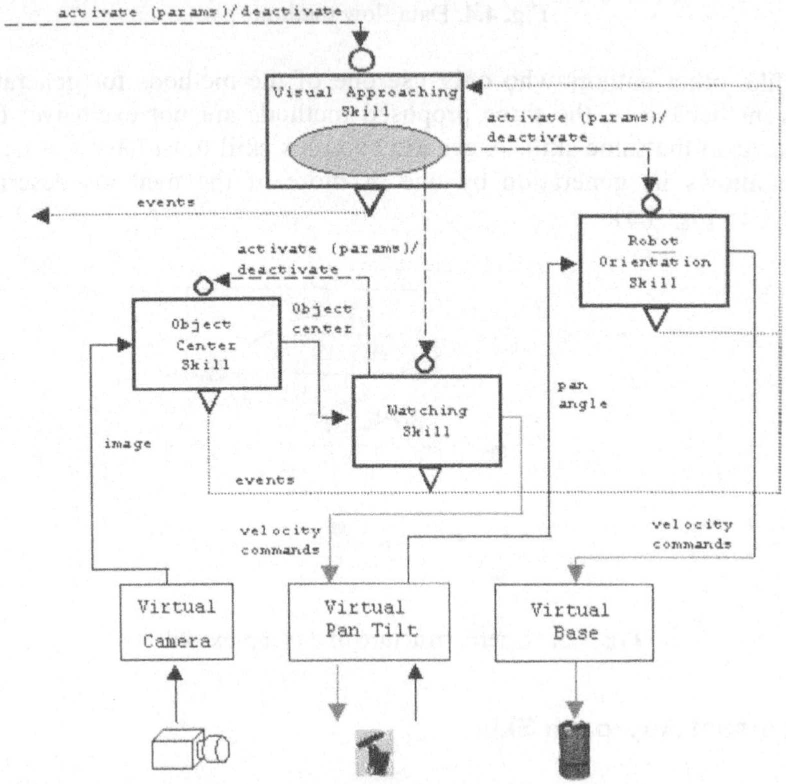

Fig. 4.6. *Visual Approaching* skill structure

Watching a target means keeping the eyes on it. The inputs, that the *Watching* skill receives are the object center coordinates in the image plane and the performed outputs are the pan tilt velocities. The information is not obtained from the camera sensor directly but it is obtained by the skill called *Object Center*. Object center means searching for an object on the image previously defined. The input is the image recorded with the camera and the output is the object center position on the image in pixels. If the object is not found, this skill sends the event *OBJECT_NOT_FOUND. Object Center* skill is perceptive because it does not produce any action upon the actuators but it only interprets the information obtained from the sensors. When the object is centered on the image, the skill *Watching* sends notification of the event *OBJECT_CENTERED.*

Orientating the robot means turning the robot's body to align it with the vision system. The turret is mounted on the robot so the angle formed by the robot body and the turret coincides with the turret angle. The input to the *Orientation* skill is the turret pan angle and the output is the robot angular velocity. The information about the angle is obtained from the encoder sensor placed on the pan tilt platform. When the turret is aligned with the robot body, this skill sends notification of the event *TURRET_ALIGNED.*

4.3.2 Go To Goal Avoiding Obstacles Skill

The skill called *Go To Goal Avoiding Obstacles* allows the robot to go towards a given goal without colliding with any obstacle [27]. It is formed by a sequencer which is in charge of sequencing different skills, see Fig. 4.7, such as *Go To Goal* and *Left* and *Right Following Contour.*

The *Go To Goal* skill estimates the velocity at which the robot has to move in order to go to the goal in a straight line without taking into account the obstacles in the environment. This skill generates the event *GOAL_ REACHED* when the required task is achieved successfully. The input that the skill receives is the robot's position obtained from the base's server.

The *Right* and *Left Following Contour* skills estimate the velocity by which the robot has to move in order to follow the contour of an obstacle placed on the right and left side respectively. The input received by the skills is the sonar readings.

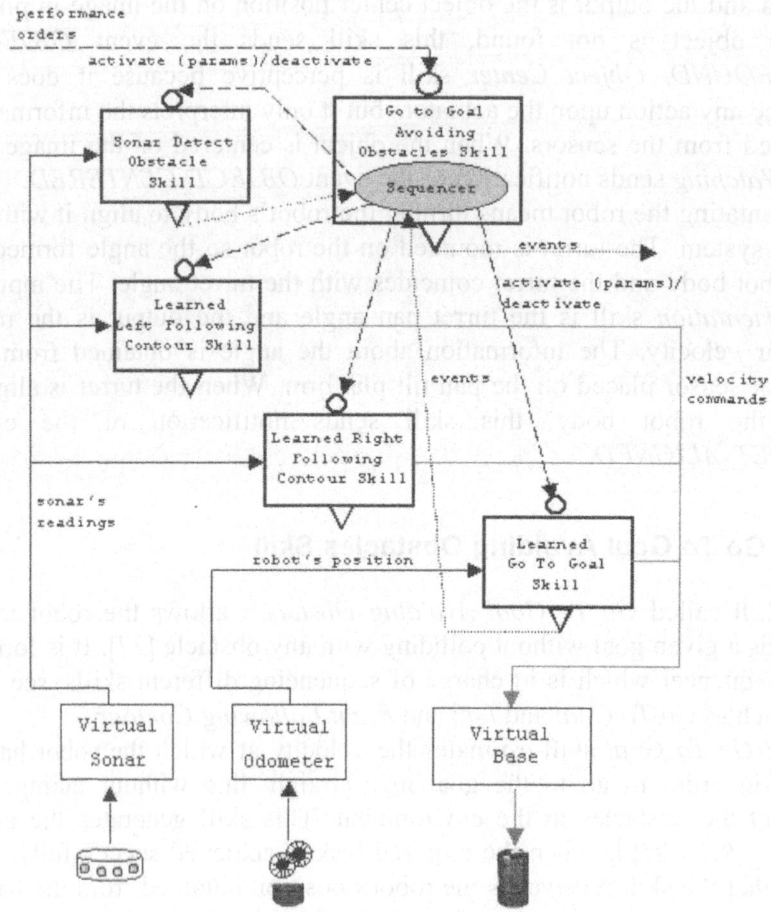

Fig. 4.7. Go to Goal Avoiding Obstacles skill structure

4.4 Reinforcement Learning

Reinforcement learning consists of mapping from situations to actions so as to maximize a scalar called reinforcement signal [11] [28]. It is a learning technique based on trial and error. A good performance action provides a reward, increasing the probability of recurrence. A bad performance action provides punishment, decreasing the probability. Reinforcement learning is used when there is not detailed information about the desired output. The system learns the correct mapping from situations to actions without *a priori* knowledge of its environment. Another advantage that the reinforcement learning presents is that the system is able to learn on-line, it does not require dedicated training and evaluation phases of learning, so that the system can dynamically adapt to changes produced in the environment.

A reinforcement learning system consists of an agent, the environment, a policy, a reward function, a value function, and, optionally, a model of the environment, see Fig. 4.8. The agent is a system that is embedded in an environment, and takes actions to change the state of the environment. The environment is the external system that an agent is *embedded* in, and can perceive and act on. The policy defines the learning agent's way of behaving at a given time. A policy is a mapping from perceived states of the environment to actions to be taken when in those states. In general, policies may be stochastic. The reward function defines the goal in a reinforcement learning problem. It maps perceived states (or state-action pairs) of the environment to a single number called reward or reinforcement signal, indicating the intrinsic desirability of the state. Whereas a reward function indicates what is good in an immediate sense, a value function specifies what is good in the long run. The value of a state is the total amount of reward an agent can expect to accumulate over the future starting from that state, and finally the model is used for planning, by which it means any way of deciding on a course of action by considering possible future situations before they are actually experienced.

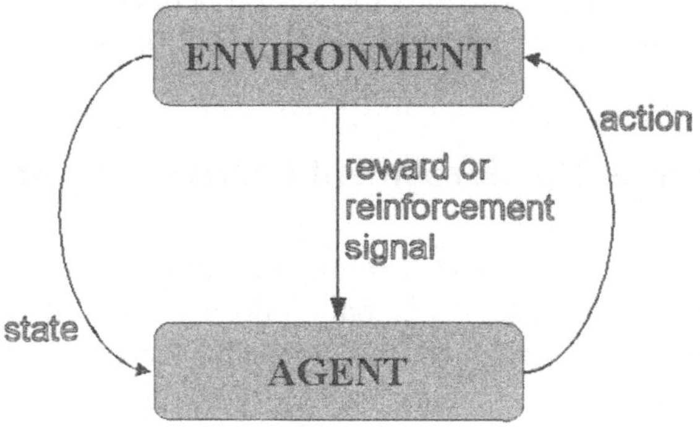

Fig. 4.8. Interaction among the elements of a reinforcement learning system

A reinforcement learning agent must explore the environment in order to acquire knowledge and to make better action selections in the future. On the other hand, the agent has to select that action which provides the better reward among actions which have been performed previously. The agent must perform a variety of actions and favor those that produce better re-wards. This problem is called tradeoff between exploration and exploitation. To solve this problem different authors combine new experience with old value functions to produce new and statistically improved value functions in different ways [29].

Reinforcement learning algorithms implies two problems [30]: temporal credit assignment problem and structural credit assignment or generalization problem. The temporal assignment problem appears due to the received reward or reinforcement signal may be delayed in time. The reinforcement signal informs about the success or failure of the goal after some sequence of actions have been performed. To cope with this problem, some reinforcement learning algorithms are based on estimating an expected reward or predicting future evaluations such as Temporal Differences TD(λ) [31]. Adaptive Heuristic Critic (AHC) [32] and Q'Learning [33] are included in these algorithms. The structural credit assignment problem arises when the learning system is formed by more than one component and the performed actions depend on several of them. In these cases, the received reinforcement signal has to be correctly assigned between the participating components. To cope with this problem, different methods have been proposed such as gradient methods, methods based on a minimum-change principle an based on a measure of worth of a network component [34] [35].

The reinforcement learning has been applied in different areas such as computer networks [36], game theory [37], power system control [38], road vehicle [39], traffic control [40], etc. One of the applications of the reinforcement learning in robotics focuses on behaviors' learning [41] [42] and behavior coordination's learning [43] [44] [45][46].

4.5 Continuous Reinforcement Learning Algorithm

In most of the reinforcement learning algorithms mentioned in previous section, the reinforcement signal only informs about if the system has crashed or if it has achieved the goal. In these cases, the external reinforcement signal is a binary scalar, typically (0, 1) (0 means bad performance and 1 means a good performance), and/or it is delayed in time. The success of a learning process depends on how the reinforcement signal is defined and when it is received by the control system. Later the system receives the reinforcement signal, the later it takes to learn.We propose a reinforcement learning algorithm which receives an external continuous reinforcement signal each time the system performs an action. This reinforcement is a continuous signal between 0 and 1.This value shows how well the system has performed the action. In this case, the system can compare the action result with the last action result performed in the same state, so it is not necessary to estimate an expected reward and this allows to increase the learning rate.

Most of these reinforcement learning algorithms work with discrete output and input spaces. However, some robotic applications requires to work with continuous spaces defined by continuous variables such as position, velocity, etc. One of the problems that appears working with continuous input spaces is how to cope the infinite number of the perceived states. A generalized method is to discretize the input space into bounded regions within each of which every input point is mapped to the same output [47] [48] [49].

The drawbacks of working with discrete output spaces are: some feasible solution could not take into account and the control is less smooth. When the space is discrete, the reinforcement learning is easy because the system has to choose an action among a finite set of actions, being this action which provides the best reward. If the output space is continuous, the problem is not so obvious because the number of possible actions is infinite. To solve this problem several authors use perturbed actions adding random noise to the proposed action [30] [50] [51].

In some cases, reinforcement learning algorithms use neural networks for their implementation because of their flexibility, noise robustness and adaptation capacity. Following, we describe the continuous reinforcement learning algorithm proposed for the learning of skills in an autonomous mobile robot. The implemented neural network architecture works with continuous input and output spaces and with real continuous reinforcement signal.

4.5.1 Neural Network Architecture

The neural network architecture proposed to implement the reinforcement learning algorithm is formed by two layers as is shown in Fig. 4.9. The input layer consists of *radial basis function* (RBF) nodes and is in charge to discretize the input space. The activation value for each node depends on the input vector proximity to the center of each node thus, if the activation level is 0 it means that the perceived situation is outside its receptive field. But it is 1, it means that the perceived situation corresponds to the node center.

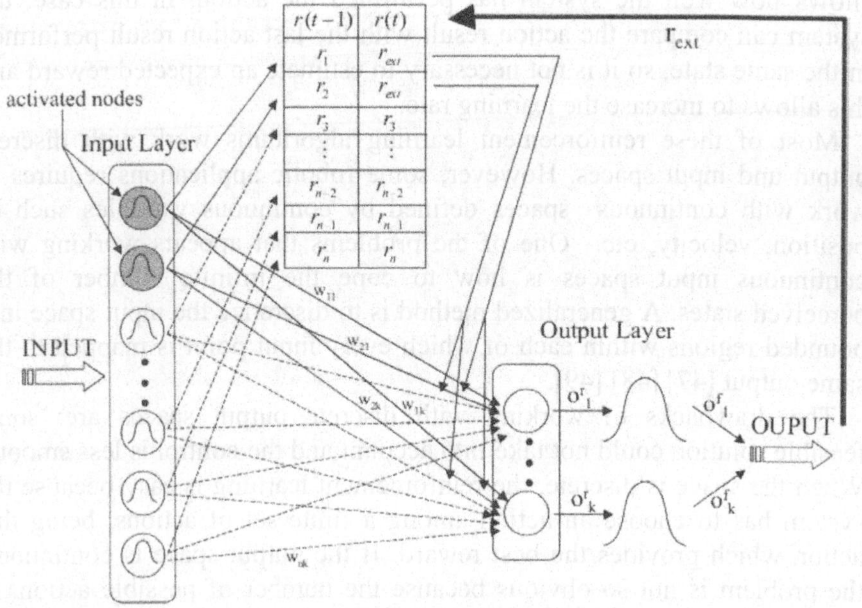

Fig. 4.9. Structure of the neural network architecture. Shaded RBF nodes of the input layer represent the activated ones for a perceived situation. Only the activated nodes will update its weights and reinforcement values

The output layer consists of linear stochastic units allowing the search for better responses in the action space. Each output unit represents an action. There exists a complete connectivity between the two layers.

4.5.1.1 Input Layer

The input space is divided into discrete, overlapping regions using RBF nodes. The activation value for each node is:

$$i_j = e^{\frac{\left\|\vec{i}-\vec{c}_j\right\|^2}{\sigma_{rbf}^2}}$$

where $\vec{i}$ is the input vector, $\vec{c}_j$ is the center of each node and σ_{rbf} the width of the activation function. Next, the obtained activation values are normalized:

$$i_j^n = \frac{i_j}{\sum_{k=1}^{n_n} i_k}$$

where n_n is the number of created nodes.

Nodes are created dynamically where they are necessary maintaining the network structure as small as possible. Each time a situation is presented to the network, the activation value for each node is calculated. If all values are lower than a threshold, $a_{\min}$, a new node is created. The center of this new node coincides with the input vector presented to the neural network, $\vec{c}_i = \vec{i}$. Connections weights, between the new node and the output layer, are initialised to randomly small values.

4.5.1.2 Output Layer

The output layer must find the best action for each situation. The recommended action is a weighted sum of the input layer given values:

$$o_k^r = \sum_{j=1}^{n_n} w_{jk} \cdot i_j^n, \qquad 1 \le k \le n_0$$

where n_0 is the number of output layer nodes. During the learning process, it is necessary to explore for the same situation all the possible actions to discover the best one. This is achieved adding noise to the recommended action. The real final action is obtained from a normal distribution centered in the recommended value and with variance:

$$o_k^f = N(o_k^r, \sigma)$$

As the system learns a suitable action for each situation, the value of σ is decreased. We state that the system can perform the same action for the learned situation.

To improve the results, the weights of the output layer are adapted according to the following equations:

$$w_{jk}(t+1) = w_{jk}(t) + \Delta w_{jk}(t)$$

$$\Delta w_{jk}(t) = \beta \cdot (r_{j'}(t) - r_{j'}(t-1)) \cdot \frac{\mu_{jk}(t)}{\sum_l \mu_{lk}(t)} \mid j' = \arg\max_j i_j^n$$

$$e_{jk}(t) = \frac{o_k^f - o_k^r}{\sigma} \cdot i_j^n$$

$$\mu_{jk}(t+1) = \nu \cdot \mu_{jk}(t) + (1-\nu) \cdot e_{jk}(t)$$

where β is the learning rate, μ_{jk} is the eligibility trace and e_{jk} is the eligibility of the weight w_{jk}, and ν is a value in the [0, 1] range. The weight eligibility measures how this weight influences in the action, and the eligibility trace allows rewarding or punishing not only the last action but the previous ones. Values of r_j associated with each weight are obtained from the expression:

$$r_j(t) = \begin{cases} r_{ext}(t) & \text{if } i_j'' \neq 0 \\ r_j(t-1) & \textit{otherwise} \end{cases}$$

where r_{ext} is the exterior reinforcement. Actions' results depend on the activated states, so that only the reinforcement values associated with these states will update.

4.6 Experimental Results

The experimental results have been carried out on a RWI-B21 mobile robot (see Fig. 4.10). It is equipped with different sensors such as sonars placed around it, a color CCD camera, a laser telemeter PLS from SICK which allow the robot to get information from the environment. On the other hand, the robot is endowed with different actuators which allow it to explore the environment such as the robot's base and pan tilt platform on which the CCD camera is mounted.

Fig. 4.10. B21 robot

The robot has to be capable of learning the simple skills such as Watching, Orientation, *Go To Goal* and *Right* and *Left Contour Following* and finally to execute the complex sensorimotor skills *Visual Approaching* and *Go To Goal Avoiding Obstacles* from the previously learnt skills. Skills are implemented in C++ Language using the CORBA interface definition language to communicate with other skills.

In the *Watching* skill, the robot must learn the mapping from the object center coordinates *(x,y)* to the turret velocity ($\dot{pan}-\dot{tilt}$). In our experiment, a cycle starts with the target on the image plane at an initial position (243,82) pixels, and ends when the target comes out of the image or when the target reaches the image center (0,0) pixels and stays there. The turret pan tilt movements are coupled so that a x-axis movement implies a y-axis movement and viceversa.

This makes the learning task difficult. The reinforcement signal that the robot receives when it performs this skill is:

$$r_{ext} = e^{-k\frac{error^2}{2}}$$

$$error = \sqrt{(x_{o_c} - x_{i_c})^2 + (y_{o_c} - y_{i_c})^2}$$

where x_{o_c} and y_{o_c} are the object center coordinates in the image plane, and x_{i_c} and y_{i_c} are the image center coordinates.

Fig. 4.11 shows the robot performance while learning the watching skill. The plots represent the X-Y object coordinates on the image plane. As seen in the figure, the robot is improving its performance while its learning. In the first cycles, the target comes out of the image in a few learning steps, while in cycle 6 robot is able to center the target on the image rapidly.

The learning parameters values are $\beta = 0.01$, $\mu = 0.3$, $\sigma_{rbf} = 0.2$ and $a_{min} = 0.2$. Fig. 4.12 shows how the robot is able to learn to center the object on the image from different initial positions. The turret does not describe a linear movement by the fact that the pan-tilt axis are coupled. The number of created nodes, taking into account all the possible situations which can be presented to the neural net, is 40.

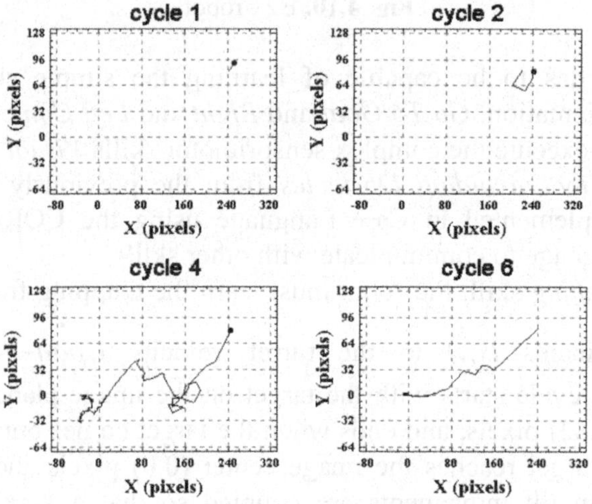

Fig. 4.11. Learning results of the skill *Watching* (I)

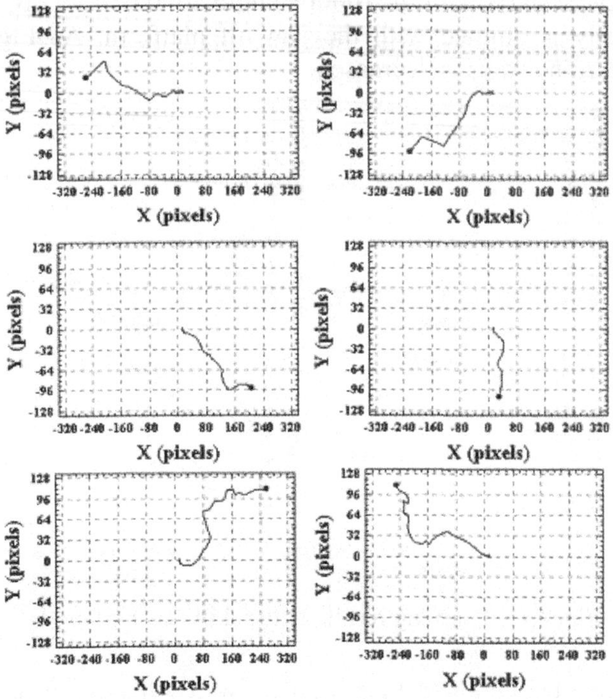

Fig. 4.12. Learning results of the skill *Watching* (II)

Once the robot has achieved a good level of performance in the *Watching* skill, it learns the *Orientation* skill. In this case, the robot must learn the mapping from the turret pan angle to the robot angular velocity ($\dot{\theta}$). To align the robot's body with the turret, maintaining the target in the center image, the robot has to turn an angle. Because the turret is mounted on the robot's body, the target is displaced on the image. The learned *Watching* skill obliges the turret to turn to center the object so the robot's body-turret angle decreases. The reinforcement signal that the robot receives when it performs this skill is:

$$r_{ext} = e^{-k\frac{error^2}{2}}$$

$$error = angle_{turret\_robot}$$

where $angle_{turret\_robot}$ is the angle formed by the robot body and the turret.

The experimental results for this skill are shown in Fig. 4.13. The plots represent the robot's angle as a function of the number of learning steps. In this case, a cycle starts with the robot's angle at –0.61 radians, and it ends when the body is aligned with the pan tilt platform. As Fig. 4.13, the number of learning steps is decreased.

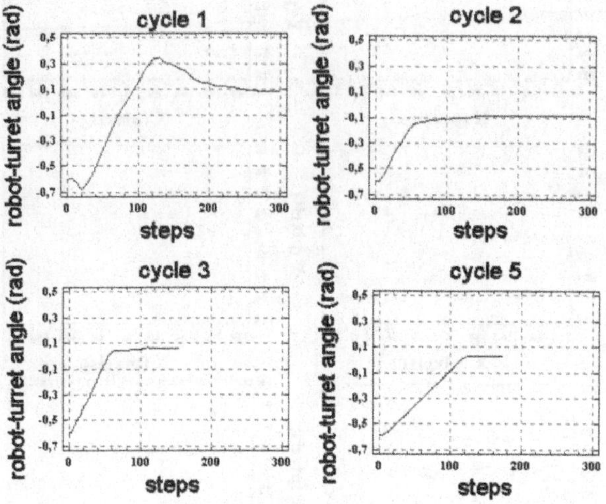

Fig. 4.13. Learning results of the skill *Orientation* (I)

The learning parameters values are $\beta = 1.0$, $\mu = 0.3$, $\sigma_{rbf} = 0.1$ and $a_{min} = 0.2$. Fig. 4.14 shows how the robot is able to align its body with the turret from different initial positions. Its behavior is different depending on the orientation sign. This is due to two reasons: on one hand, during the learning, a noise is added so that the robot can explore different situations, and on the other hand, the turret axis are coupled. The number of created nodes, taking into account all the possible situations which can be presented to the neural net, is 22.

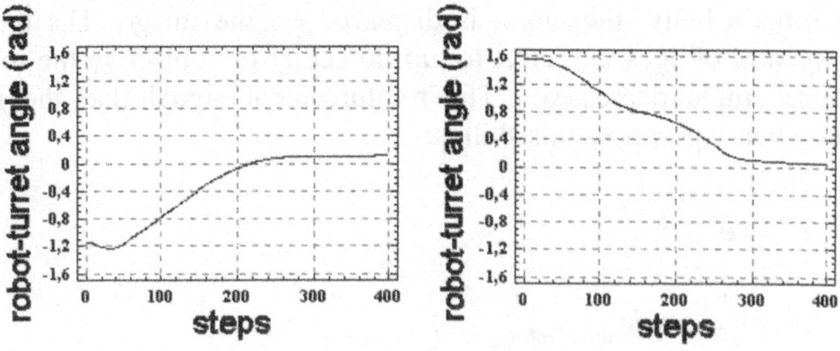

Fig. 4.14. Learning results of the skill *Orientation (*II)

Once the robot has learned the above skills, the robot is able to perform the *Approaching* skill by coordinating them. Fig. 4.15 shows the experimental results obtained from the performing of this complex skill. This experiment consists of the robot going towards a goal which is a visual target. First of all, the robot moves the turret to center the target on the image and then the robot moves towards the target.

In the *Go To Goal* skill, the robot must learn the mapping from the distance between the robot and the goal (*dist*) and the angle formed by them (α), see figure 16, to angular and linear velocities. The reinforcement signal that the robot receives when it performs this skill is:

$$r_{ext} = e^{-k_1 \cdot dist} \cdot e^{-k_2 \cdot \alpha \cdot e^{-k_3 (1-dist)}}$$

The shorter the distance between the robot and the goal the greater the reinforcement is. The reinforcement becomes maximum when the robot reaches the goal.

Fig. 4.15. Experimental results obtained from the performing of complex skill *Visual Approaching*

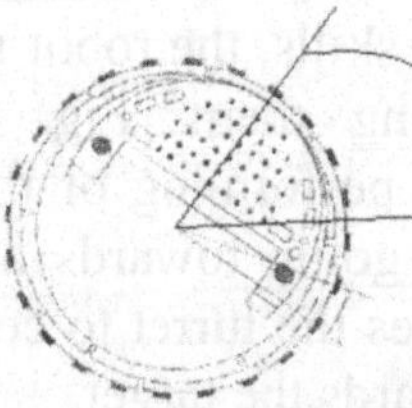

Fig. 4.16. Input variables for the learning of the skill *Go To Goal*

Fig. 4.17 shows the robot performance once it has learnt the skill. The robot is capable of going towards the goal placed 4.5 meters in front of it with different initial orientations and with a minimum of oscillations. The maximum translation velocity by which the robot can move is 30 cm/s. As the robot learns, the σ value is decreased in order to reduce the noise and achieve the execution of the same actions for the same input. The parameters used for the learning of the skill called *Go To Goal* are $\beta = 0.1$, $\mu = 0.2$, $\sigma_{rbf} = 0.1$ and $a_{min} = 0.1$. The number of created nodes, taking into account all the possible situations which can be presented to the neural net, is 20.

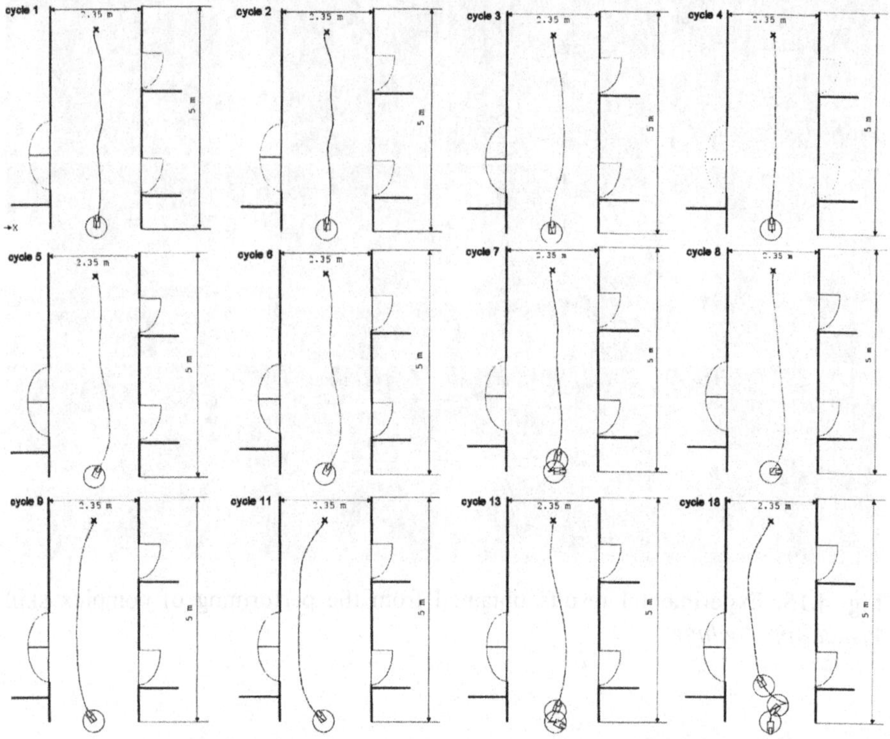

Fig. 4.17. Learning results of the skill *Go To Goal*

In the *Left* and *Right Contour Following* skills, the robot must learn the mapping from the sensor which provides the minimum distance to the obstacle (*minSensor*) and the minimum distance (*minDist*), see Fig. 4.18, to angular and linear velocities.

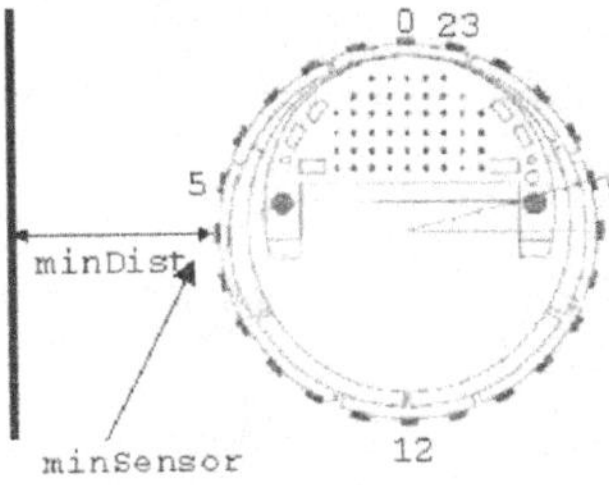

Fig. 4.18. Input variables for the learning of the skill *Left Contour Following*

The reinforcement that the robot receives when it performs this skill is :

$$r_{ext} = e^{-k_3 \cdot \left| \frac{distMin - distLim}{distLim} \right|} \cdot e^{-k_4 \cdot angSensMin}$$

where *distLim* is the distance to which the robot has to follow the contour and *minSensAng* is the sonar sensor angle which provides the minimum distance. The *minSensAng* is calculated so that the value 0 corresponds to sensor number 6 when the skill is *Left Following Contour* and 18 when the skill is *Right Following Contour*. These values correspond when the robot is parallel to the wall. The reinforcement is maximum when the distance between the robot and the robot coincides with *distLim* and the robot is parallel to the wall.

Fig. 4.19 shows the robot performance once it has learnt the simple skill *Left Contour Following*. The robot is capable of following the contour of a wall at 0.65 meters. The last two graphics show the results obtained when the robot has to go around obstacles of 30 and 108.5 cm wide. The parameters used for learning the *Left Contour Following* skill are $\beta = 0.1$, $\mu = 0.2$, $\sigma_{rbf} = 0.1$ and $a_{min} = 0.1$. The number of created nodes, taking into account all the possible situations which can be presented to the neural net, is 11.

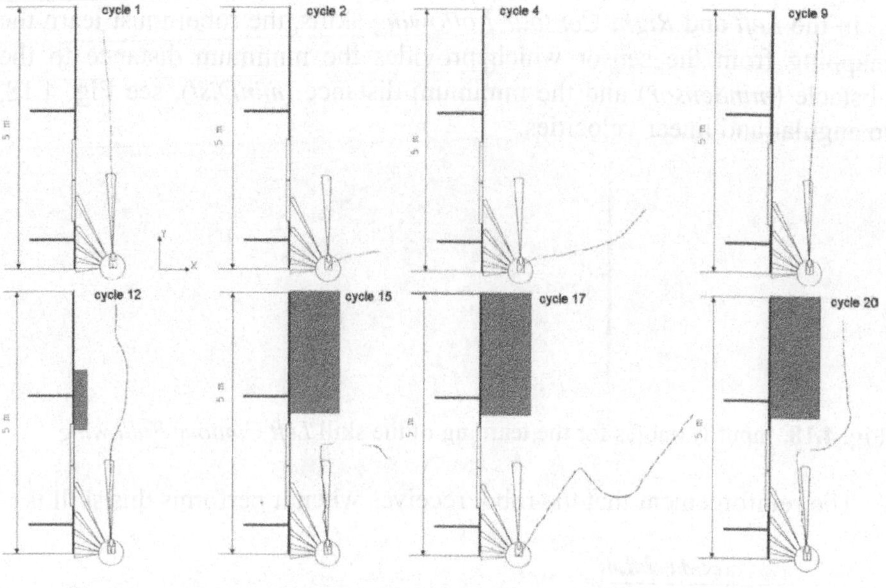

Fig. 4.19. Learning results of the skill *Left Contour Following*

The learnt simple skills can be combined to obtain the complex skill called *Go To Goal Avoiding Obstacles*. Fig. 4.20 shows the experimental results obtained during complex skill execution. The robot has to go towards a goal situated at (8, 0.1) meters. When an obstacle is detected, the robot is able to avoid it and keep going to the goal once the obstacle is behind the robot.

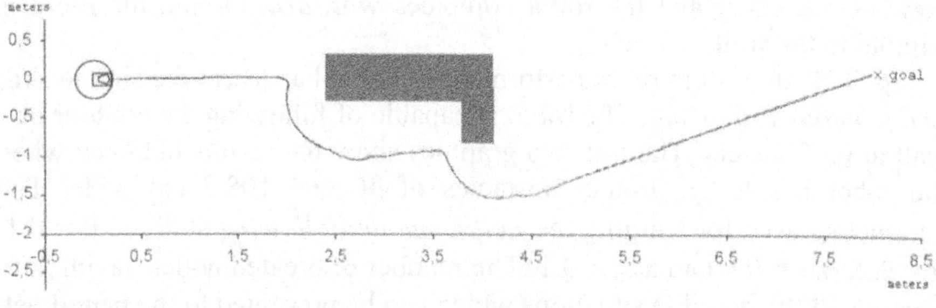

Fig. 4.20. Experimental results obtained during the execution of the complex skill called *Go To Goal Avoiding Obstacles*

4.7 Conclusions

We propose a reinforcement learning algorithm which allows a mobile robot to learn simple skills. The implemented neural network architecture works with continuous input and output spaces, has a good resistance to forget previously learned actions and learns quickly. Nodes of the input layer are allocated dynamically. Situations where the robot has explored are only taken into account so that the input space is reduced. Other advantages this algorithm presents are that on one hand, it is not necessary to estimate an expected reward because the robot receives a real continuous reinforcement each time it performs an action and , on the other hand, the robot learns *on-line*, so that the robot can adapt to changes produced in the environment.

This work also presents a generic structure definition to implement perceptive and sensorimotor skills. All skills have the same characteristics: they can be activated by other skills from the same level or from a higher level, the output data are stored in data objects in order to be used by other skills, and skills notify events to other skills about its performance. Skills can be combining to generate complex ones through three different methods called *sequencing*, *output addition* and *data flow*. Unlike other authors who only use one of the methods for generating emergent behaviors, the three proposed methods are not exclusive; they can occur in the same skill.

The proposed reinforcement learning algorithm has been tested in an autonomous mobile robot in order to learn simple skills showing a good results. Finally the learnt simple skills are combined to successfully perform a more complex skills called *Visual Approaching and Go To Goal Avoiding Obstacles*.

Acknowledgements

The authors would like to acknowledge that papers with brief versions of the work presented in this chapter have been published in conference proceedings for the *IEEE International Conference on Industrial Electronics Society* [27] and the Journal *Robotics and Autonomous Systems* [17]. The authors also thank Ms. **Cristina Castejon from Carlos III University** for her invaluable help and comments when proof-reading the chapter text.

References

1 R. C. Arkin, Behavior-Based Robotics, The MIT Press, 1998.
2 S. Mahadevan, "Machine learning for robots: A comparison of different paradigms", in *Workshop on Towards Real Autonomy. IEEE/RSJ International Conference on Intelligent Robots and Systems (IROS'96)*, 1996.
3 M. J. Mataric, "Behavior-based robotics as a tool for synthesis of artificial behavior and analysis of natural behavior", *Cognitive Science*, vol. 2, pp. 82–87, 1998.
4 Y. Kusano and K. Tsutsumi, "Hopping height control of an active suspension type leg module based on reinforcement learning and a neural network", in *Proceedings IEEE/RSJ International Conference on Intelligent Robots and Systems*, 2002, vol. 3, pp. 2672–2677.
5 D. Zhou and K. H. Low, "Combined use of ground learning model and active compliance to the motion control of walking robotic legs", in *Proceedings 2001 ICRA. IEEE International Conference on Robotics and Automation*, 2001, vol. 3, pp. 3159–3164.
6 C Ye, N. H. C. Yung, and D.Wang, "A fuzzy controller with supervised learning assisted reinforcement learning algorithm for obstacle avoidance", *IEEE Transactions on Systems, Man and Cybernetics, Part B (Cybernetics)*, vol. 33, no. 1, pp. 17–27, 2003.
7 T. Belker, M. Beetz, and A. B. Cremers, "Learning action models for the improved execution of navigation plans", *Robotics and Autonomous Systems*, vol. 38, no. 3-4, pp. 137–148, 2002.
8 Z. Kalmar, C. Szepesvari, and A. Lorincz, "Module-based reinforcement learning: Experiments with a real robot", *Autonomous Robots*, vol. 5, pp. 273–295, 1998.
9 M. Mata, J. M. Armingol, A. de la Escalera, and M. A. Salichs, "A visual landmark recognition system for topological navigation of mobile robots", *Proceedings of the 2001 IEEE International Conference on Robotics and Automation*, pp. 1124,–1129, 2001.
10 T. Mitchell, *Machine Learning*, McGraw Hill, 1997.
11 L. P. Kaelbling, M. L. Littman, and A. W. Moore, "Reinforcement learning: A survey", *Artificial Intelligent Research*, , no. 4, pp. 237–285, 1996.
12 R. A. Brooks, "A robust layered control system for a mobile robot", *IEEE Journal of Robotics and Automation, RA-2(1)*, pp. 14–23, 1986.

13 M. Becker, E. Kefalea, E. Maël, C. Von Der Malsburg, M. Pagel, J. Triesch, J. C. Vorbruggen, R. P. Wurtz, and S. Zadel, "Gripsee: A gesture-controlled robot for object perception and manipulation", in *Autonomous Robots*, 1999, vol. 6, pp. 203–221.

14 Y. Hasegawa and T. Fukuda, "Learning method for hierarchical behavior controller", in *Proceedings of the 1999 IEEE. International Conference on Robotics and Automation*, 1999, pp. 2799–2804.

15 M. J. Mataric, *Interaction and Intelligent Behavior*, PhD thesis, Massachusetts Institute of Technology, May 1994.

16 R. Barber and M. A. Salichs, "A new human based architecture for intelligent autonomous robots", in *The Fourth IFAC Symposium on Intelligent Autonomous Vehicles, IAV 01*, 2001, pp. 85–90.

17 M. J. L. Boada, R. Barber, and M. A. Salichs, "Visual approach skill for a mobile robot using learning and fusion of simple skills", *Robotics and Autonomous Systems.*, vol. 38, pp. p157–170, March 2002.

18 R. P. Bonasso, J. Firby, E. Gat, D. Kortenkamp, D. P. Miller, and M. G. Slack, "Experiences of robotics and automation, RA-2", *Journal of Experimental Theory of Artificial Intelligence*, vol. 9, pp. 237–256, 1997.

19 R. Alami, R. Chatila, S. Fleury, M. Ghallab, and F. Ingrand, "An architecture for autonomy", *The International Journal of Robotics Research*, vol. 17, no. 4, pp. 315–337, 1998.

20 M. J. Mataric, "Learning to behave socially", in *From Animals to Animats: International Conference on Simulation of Adaptive Behavior*, 1994, pp. 453–462.

21 D. Gachet, M. A. Salichs, L. Moreno, and J. R. Pimentel, "Learning emergent tasks for an autonomous mobile robot", in *Proceedings of the IEEE/RSJ/GI. International Conference on Intelligent Robots and Systems. Advanced Robotic System and the Real World.*, 1994, pp. 290–297.

22 O. Khatib, "Real-time obstacle avoidance for manipulators and mobile robots", in *Proceedings of the IEEE International Conference on Robotics and Automation*, 1985, pp. 500–505.

23 R. J. Firby, "The rap language manual", Tech. Rep., University of Chicago, 1995.

24 D. Schreckenghost, P. Bonasso, D. Kortenkamp, and D. Ryan, "Three Tier architecture for controlling space life support systems", in *IEEE International Joint Symposium on Intelligence and Systems*, 1998, pp. 195–201.

25 R. C. Arkin, "Motor schema-based mobile robot navigation", *International Journal of Robotics Research*, vol. 8, no. 4, pp. 92–112, 1989.
26 P. N. Prokopowicz, M. J. Swain, and R. E. Kahn, "Task and environment-sensitive tracking", in *Proceedings of the Workshop on Visual Behaviors*, 1994, pp. 73–78.
27 M. J. L. Boada, R. Barber, V. Egido and M. A. Salichs, "Continuous Reinforcement Learning Algorithm for Skills Learning in an Autonomous Mobile Robot", in *Proceedings of the IEEE International Conference on Industrial Electronics Society*, 2002.
28 R. S. Sutton and A. G. Barto, *Reinforcement Learning: An Introduction*, The MIT Press, 1998.
29 S. Singh, T. Jaakkola, and C. Szepesvari M. L. Littman, "Convergence results for single-step on-policy reinforcement-learning algorithms", *Machine Learning*, vol. 38, no. 3, pp. 287–308, 2000.
30 V. Gullapalli, *Reinforcement learning and its application to control*, PhD thesis, Institute of Technology and Science. University of Massachusetts, 1992.
31 R. S. Sutton, "Learning to predict by the method of temporal differences", *Machine Learning*, vol. 3, no. 1, pp. 9–44, 1988.
32 A. G. Barto, R. S. Sutton, and C. W. Anderson, "Neurolike elements that can solve difficult learning control problems", in *IEEE Transactions on Systems, Man and Cybernetics*, 1983, vol. 13, pp. 835–846.
33 C. J. C. H. Watkins and P. Dayan, "Technical note: Q'Learning", *Machine Learning*, vol. 8, pp. 279–292, 1992.
34 D. Rumelhart, G. Hinton, and R. Williams, "Learning representations by backpropagation errors", *Nature*, vol. 323, pp. 533–536, 1986.
35 C. W. Anderson, "Strategy learning with multi-layer connectionist representations", in *Proceedings of the Fourth International Workshop on Machine Learning*, 1987, pp. 103–114.
36 T. C.-K. Hui and C.-K. Tham, "Adaptive provisioning of differentiated services networks based on reinforcement learning", *IEEE Transactions on Systems, Man and Cybernetics, Part C*, vol. 33, no. 4, pp. 492–501, 2003.
37 W. A. Wright, "Learning multi-agent strategies in multi-stage collaborative games", in *IDEAL*. 2002, vol. 2412 on *Lecture Notes in Computer Science*, pp. 255–260, Springer.

38 T. P. I. Ahamed, P. S. N. Rao, and P. S. Satry, "A reinforcement learning approach to automatic generation control", *Electric Power Systems Research*, vol. 63, no. 1, pp. 9–26, 2002.
39 N. Krodel and K.-D. Kuhner, "Pattern matching as the nucleus for either autonomous driving or driver assistance systems", in *Proceedings of the IEEE Intelligent Vehicle Symposium*, 2002, pp. 135–140.
40 M. C. Choy, D. Srinivasan, and Ruey Long Cheu, "Cooperative, hybrid agent architecture for real-time traffic signal control", *IEEE Transactions on systems, man, and cybernetics: PART A.*, vol. 33, no. 5, pp. 597–607, 2003.
41 R. A. Grupen and J. A. Coelho, "Acquiring state from control dynamics to learn grasping policies for robot hands", *Advanced Robotics*, vol. 16, no. 5, pp. 427–443, 2002.
42 G. Hailu, "Symbolic structures in numeric reinforcement for learning optimum robot trajectory", *Robotics and Autonomous Systems*, vol. 37, no. 1, pp. 53–68, 2001.
43 P. Maes and R. A. Brooks, "Learning to coordinate behaviors", in *Proceedings, AAAI-90*, 1990, pp. 796–802.
44 L. P. Kaelbling, *Learning in Embedded Systems*, PhD thesis, Standford University, 1990.
45 S. Mahadevan and J. Connell, "Automatic programming of behavior-based robots using reinforcement learning", in *Proceedings of the AAAI-91*, 1991, pp. 8–14.
46 G. J. Laurent and E. Piat, "Learning mixed behaviors with parallel q-learning", in *Proceedings IEEE/RSJ International Conference on Intelligent Robots and Systems*, 2002, vol. 1, pp. 1002–1007.
47 I. O. Bucak and M. A. Zohdy, "Application of reinforcement learning to dextereous robot control", in *Proceedings of the 1998 American Control Conference. ACC'98*, USA, 1998, vol. 3, pp. 1405–1409.
48 D. F. Hougen, M. Gini, and J. Slagle, "Rapid unsupervised connectionist learning for backing a robot with two trailers", in *IEEE International Conference on Robotics and Automation.*, 1997, pp. 2950–2955.
49 F. Fernandez and D. Borrajo, *VQQL. Applying Vector Quantization to Reinforcement Learning*, pp. 292–303, Lecture Notes in Computer Science. 2000.
50 S. Yamada, M. Nakashima, and S. Shiono, "Reinforcement learning to train a cooperative network with both discrete A. J andcontinuous output neurons", *IEEE. Transactions on Neural Network*, vol. 9, no. 6, pp. 1502–1508, November 1998.
51 A.J. Smith, "Applications of the self-organizing map to reinforcement learning", *Neural Network*, vol. 15, no. 8-9, pp. 1107–1124, 2002.

5 Efficient Incorporation of Optical Flow into Visual Motion Estimation in Tracking

Gozde Unal[1], Anthony Yezzi[2], Hamid Krim[3]

1. Intelligent Vision and Reasoning, Siemens Corporate Research, Princeton NJ 08540 USA
gozde.unal@siemens.com
2. School of Electrical Engineering, Georgia Institute of Technology, Atlanta, GA 30332, USA
ayezzi@ece.gatech.edu
3. Electrical and Computer Engineering, North Carolina State University, Raleigh NC 27695 USA
ahk@eos.ncsu.edu

5.1 Introduction

Recent developments in digital technology have increased acquisition of digital video data, which in turn have led to more applications in video processing. Video sequences provide additional information about how scenes and objects change over time when compared to still images. The problem of tracking moving objects remains of great research interest in computer vision on account of various applications in video surveillance, monitoring, robotics, and video coding. For instance, MPEG-4 video standard introduced video object plane concept, and a decomposition of sequences into object planes with different motion parameters [1]. Video surveillance systems are needed in traffic and highway monitoring, in law enforcement and security applications by banks, stores, and parking lots. Algorithms for extracting and tracking over time moving objects in a video sequence are hence of importance.

Tracking methods may be classified into two categories [2]: (i) **Motion-based approaches**, which use motion segmentation of temporal image sequences by grouping moving regions over time, and by estimating their motion models [3–6]. This region tracking, not being object-based is not well-adapted to the cases where prior shape knowledge of the moving object is provided. (ii) **Model-based approaches** exploit some model structure to combat generally noisy conditions in the scene. Objects are usually tracked using a template of the 3D object such as 3D models in [7–11].

G. Unal et al.: *Efficient Incorporation of Optical Flow into Visual Motion Estimation in Tracking*, Studies in Computational Intelligence (SCI) **7**, 167–202 (2005)
www.springerlink.com

Usage of this high level semantic information yields robust algorithms at a high computational cost.

Another classification of object tracking methods due to [2] is based on the type of information that the tracking algorithm uses. Along these lines, tracking methods which exploit either boundary-based information or region-based information have been proposed:

Boundary-based methods use the boundary information along the object's contour, and are flexible because usually no object shape model and no motion model are required. Methods using snake models such as [12–14], employ parameterized snakes (such as B-splines), and constrain the motion by assuming certain motion models, e.g. rigid, or affine. In [12], a contour's placement in a subsequent frame is predicted by an iterative registration process where rigid objects and rigid motion are assumed. In another tracking method with snakes [15], the motion estimation step is skipped, and the snake position from any given image frame is carried to the next frame. Other methods employ geodesic active contour models [16], which also assumes rigid motion and rigid objects, and [2] for tracking of object contours.

Region-based methods such as [3–5, 17, 18] segment a temporal image sequence into regions with different motions. Regions segmented from each frame by a motion segmentation technique are matched to estimate motion parameters [19]. They usually employ parametric motion models, and they are computationally more demanding than boundary-based tracking methods because of the cost of matching regions.

Another tracking method, referred to as Geodesic Active Regions [20], incorporates both boundary based and region-based approaches. An affine motion model is assumed in this technique, and successive different estimation steps involved increases its computational load. In feature-based trackers, one usually seeks similar features in subsequent frames. For instance in [21], the features in subsequent frames are matched by a deformation of a current feature image onto the next following feature image, and a level set methodology is used to carry out this approach. One of the advantages of our technique is that of avoiding to have to match features, e.g. boundary contours, in a given image frame to those on successive ones. There are also approaches to object tracking which use posterior density estimation techniques [22, 23]. These algorithms maintain high computational costs, which we mean to avoid in this study.

Our goal is to build on the existing achievements, and the corresponding insight to develop a simple and efficient boundary-based tracking algorithm well adapted to polygonal objects. This is in effect an extension of our evolution models which use region-based data distributions to capture polygonal object boundaries.

5.1.1 Motion Estimation

Motion of objects in 3D real world scene are projected onto 2D image plane, and this projected motion is referred to as "apparent motion", "2D image motion", or sometimes as "optical flow", which is to be estimated. In a time-varying image sequence, I(x,y,t) : [0,a] × [0,b] × [0,T] → R^+, image motion may be described by a 2-D vector field **V**(x,y,t), which specifies the direction and speed of the moving target at each point (x,y) and time t. The measurement of visual motion is equivalent to computing V(x,y,t) from I(x,y,t) [24]. Estimating the velocity field remains an important research topic in light of its ubiquitous presence in many applications and as reflected by the wealth of previously proposed techniques.

The most popular group of motion estimation techniques are referred to as **differential techniques**, and solve an optical flow equation which states that intensity or brightness of an image remains constant with time. They use spatial and temporal derivatives of the image sequence in a gradient search, and sometimes referred to as **gradient-based techniques**. The basic assumption that a point in the 3D shape, when projected onto the 2D image plane, has a constant intensity over time, may be formulated as (**x** =(x,y)),

$$I(\mathbf{x},t) \approx I(\mathbf{x}+\delta\mathbf{x}, t+\delta t)$$

where δx is the displacement of the local image region at (**x** ,t) after time δt . A first-order Taylor series expansion on the right-hand-side yields

$$I(\mathbf{x},t) = I(\mathbf{x},t) + \nabla \mathrm{I} \cdot \delta\mathbf{x} + I_t \delta t + O^2$$

where O^2 denotes second and higher order derivatives. Dividing both sides of the equation by δt , and neglecting O^2, the **optical flow constraint equation** is obtained as

$$\frac{\partial \mathrm{I}}{\partial t} + \nabla \mathrm{I} \cdot \mathbf{V} = 0 \qquad (1)$$

This constraint is, however, not sufficient to solve for both components of V (**x** ,t)=(u(**x** ,t),v(**x** ,t)), and additional constraints on the velocity field are required to address the ill-posed nature of the problem.

The direction of motion of an object boundary B monitored through a small aperture A (small with respect to the moving unit) (see Figure 5.1) can not be determined uniquely (known as the **aperture problem**). Experimentally, it can be observed that when viewing the moving edge B through aperture A, it is not possible to determine whether the edge has moved towards the direction c or direction d. The observation of the moving edge only allows for the detection and hence computation of the velocity component normal to the edge (vector towards n in Figure 5.1), with the tangential component remaining undetectable. Uniquely determining the velocity field hence requires more than a single measurement, and it necessitates a combination stage using the local measurements [25]. This in turn means that computing the velocity field involves regularizing constraints such as its smoothness and other variants.

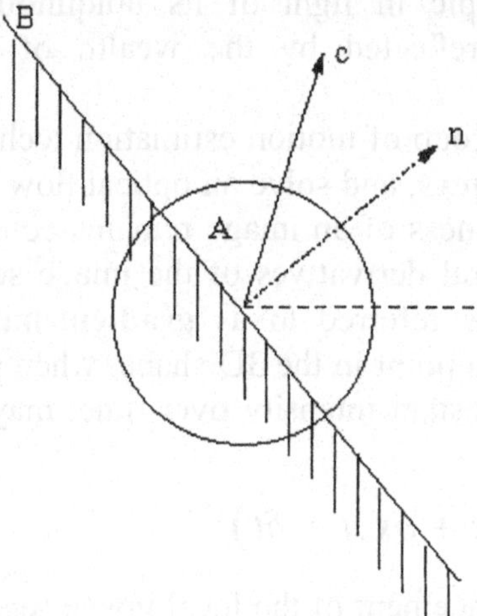

Fig. 5.1. The aperture problem: when viewing the moving edge B through aperture A, it is not possible to determine whether the edge has moved towards the direction c or direction d

Horn and Schunck, in their pioneering work [26], combined the optical flow constraint with a global smoothness constraint on the velocity field to define an energy functional whose minimization

$$\arg\min_{u,v} = \int_{\Omega} [(\nabla I \cdot \mathbf{V} + I_t)^2 + \lambda^2 (\| \nabla u \|^2 + \| \nabla u \|^2)] d\mathbf{x}$$

can be carried out by solving its gradient descent equations. A variation on this theme, would adopt an L1 norm smoothness constraint, (in contrast to

Horn-Schunck's L2 norm), on the velocity components, and was given in [27]. Lucas and Kanadc, in contrast to Horn and Schunck's regularization based on post-smoothing, minimized a pre-smoothed optical constraint

$$\int_R W^2(\mathbf{x})[(\nabla I(\mathbf{x},t)\cdot \mathbf{V} + I_t(\mathbf{x},t))]^2 d\mathbf{x}$$

where W(**x**) denotes a window function that gives more weight to constraints near the center of the neighborhood R[28].

Imposing the regularizing smoothness constraint on the velocity over the whole image leads to over-smoothed motion estimates at the discontinuity regions such as occlusion boundaries and edges. Attempts to reduce the smoothing effects along steep edge gradients included modifications such as incorporation of an oriented smoothness constraint by [29], or a directional smoothness constraint in a multi-resolution framework by [30]. Hildreth [24] proposed imposing the smoothness constraint on the velocity field only along contours extracted from time-varying images. One advantage of imposing smoothness constraint on the velocity field is that it allows for the analysis of general classes of motion, i.e., it can account for the projected motion of 3D objects that move freely in space, and deform over time [24].

Spatio-temporal energy-based methods make use of energy concentration in 3D spatio-temporal frequency domain. A translating 2D image pattern transformed to the Fourier domain shows that its velocity is a function of its spatio-temporal frequency [31]. A family of Gabor filters which simultaneously provide spatio-temporal and frequency localization, were used to estimate velocity components from the image sequences [32, 33].

Correlation-based methods estimate motion by correlating or by matching features such as edges, or blocks of pixels between two consecutive frames [34], either as block matching in spatial domain, or phase correlation in the frequency domain. Similarly, in another classification of motion estimation techniques, **token-matching schemes**, first identify features such as edges, lines, blobs or regions, and then measure motion by matching these features over time, and detecting their changing positions [25]. There are also **model-based approaches** to motion estimation, and they use certain motion models. Much work has been done in motion estimation, and the interested reader is referred to [31, 34–36] for a more compulsive literature.

5.1.2 Kalman Filtering Approach to Tracking

Another popular approach to tracking is based on Kalman filtering theory. The dynamical snake model of Terzopoulos and Szeliski [37] introduces a time-varying snake which moves until its kinetic energy is dissipated. The potential function of the snake on the other hand represents image forces, and a general framework for a sequential estimation of contour dynamics is presented. The state space framework is indeed well adapted to tracking not only for sequentially processing time varying data but also for increasing robustness against noise. The dynamic snake model of [37] along with a motion control term are expressed as the system equations whereas the optical flow constraint and the potential field are expressed as the measurement equations by Peterfreund [38]. The state estimation is performed by Kalman filtering. An analogy can be formed here since a state prediction step which uses the new information of the most current measurement is essential to our technique.

A generic dynamical system can be written as

$$\begin{aligned} \dot{\mathbf{P}}(\mathbf{t}) &= \mathbf{F}(\mathbf{P}(\mathbf{t})) + \mathbf{V}(\mathbf{t}) \\ \mathbf{Y} &= \mathbf{H}(\mathbf{P}(\mathbf{t})) + \mathbf{W}(\mathbf{t}), \end{aligned} \tag{2}$$

where **P** is the state vector (here the coordinates of a set of vertices of a polygon), **F** and **H** are the nonlinear vector functions describing the system dynamics and the output respectively, **V** and **W** are noise processes, and **Y** represents the output of the system. Since only the output **Y** of the system is accessible by measurement, one of the most fundamental steps in model based feedback control is to infer the complete state **P** of the system by observing its output **Y** over time. There is a rich literature dealing with the problem of state observation. The general idea [39] is to simulate the system (2) using a sufficiently close approximation of the dynamical system, and to account for noise effects, model uncertainties, and measurement errors by augmenting the system simulation by an output error term designed to push the states of the simulated system towards the states of the actual system. The observer equations can then be written as

$$\dot{\mathbf{P}}(\mathbf{t}) = \mathbf{F}(\mathbf{P}(\mathbf{t})) + \mathbf{L}(\mathbf{t})(\mathbf{Y}(\mathbf{t}) - \hat{\mathbf{H}}(\hat{\mathbf{P}}(\mathbf{t}))), \tag{3}$$

where L(t) is the error feedback gain, determining the error dynamics of the system. It is immediately clear, that the art in designing such an observer is in choosing the "right" gain matrix L(t). One of the most influential ways in designing this gain is the Kalman filter [40]. Here L(t) is usually called the Kalman gain matrix K and is designed so to minimize the mean square estimation error (the error between simulated and measured output) based on the known or estimated statistical properties of the noise processes **V** (t) and **W** (t) which are assumed to be Gaussian. Note, that for a general, nonlinear system as given by Equation (2) an extended Kalman filter is required. In visual tracking we deal with a sampled continuous reality, i.e. objects being tracked move continuously, but we are only able to observe the objects at specific times (e.g. depending on the frame rate of a camera). Thus, we will not have measurements Y at every time instant t; they will be sampled. This requires a slightly different observer framework, which can deal with an underlying continuous dynamics *and* sampled measurements. For the Kalman filter this amounts to using the continuous-discrete extended Kalman filter given by the state estimate propagation equation

$$\dot{\mathbf{P}}(t) = \hat{\mathbf{F}}(\hat{\mathbf{P}}(t)) \tag{4}$$

and the state estimate update equation

$$\hat{\mathbf{P}}_k(+) = \hat{\mathbf{P}}_k(-) + \mathbf{K}_k(\mathbf{Y}_k - \mathbf{H}_k(\hat{\mathbf{P}}_k(-))), \tag{5}$$

where + denotes values after the update step, − values obtained from Equation (4) and k is the sampling index. We assume that P contains the (x,y) coordinates of the vertices of the active polygon. We note that Equations (4) and (5) then correspond to a two step approach to tracking: (i) state propagation and (ii) state update.

In our approach, given a time-varying image sequence, and assuming boundary contours of an object are initially outlined, step (i) is a **prediction step**, which predicts the position of a polygon at time step k based on its position and the optical flow field along the contour at time step k − 1. This is like a state update step. Step (ii) refines the position obtained by step (i) through a spatial segmentation, referred to as a **correction step**, which is like a state propagation step. Past information is only conveyed by means of the location of the vertices and the motion is assumed to be piecewise constant from frame to frame.

5.1.3 Strategy

Given the vast literature on optical flow, we first give an explanation and implementation of previous work on its use on visual tracking, to acknowledge what has already been done, and to fairly compare our results and show the benefits of novelties of our contribution. Our contribution, rather than the idea of adding a prediction step to active contour based visual tracking using optical flow with appropriate regularizers, is computation and utilization of an optical flow based prediction step directly through the parameters of an active polygon model for tracking. This automatically gives a regularization effect connected with the structure of the polygonal model itself due to the integration of measurements along polygon edges and avoiding the need for adding ad-hoc regularizing terms to the optical flow computations.

Our proposed tracking approach may somewhat be viewed as model-based because we will fully exploit a polygonal approximation model of objects to be tracked. The polygonal model is, however, inherently part of an ordinary differential equation model we developed in [41]. More specifically, and with minimal assumption on the shape or boundaries of the target object, an initialized generic active polygon on an image, yields a flexible approximation model of an object. The tracking algorithm is hence an adaptation of this model and is inspired by evolution models which use region-based data distributions to capture polygonal object boundaries [41]. A fast numerical approximation of an optimization of a newly introduced information measure first yields a set of coupled ODEs, which in turn, define a flow of polygon vertices to enclose a desired object.

To better contrast existing continuous contour tracking methods to those based on polygonal models, we will describe the two approaches in this sequel. As will be demonstrated, the polygonal approach presents several advantages over continuous contours in video tracking. The latter case consists of having each sample point on the contour be moved with a velocity which ensures the preservation of curve integrity. Under noisy conditions, however, the velocity field estimation usually requires regularization upon its typical initialization as the component normal to the direction of the moving target boundaries, as shown in Figure 5.2. The polygonal approximation of a target on the other hand, greatly simplifies the prediction step by only requiring a velocity field at the vertices as illustrated in Figure 5.2. The reduced number of vertices provided by the polygonal approximation is clearly well adapted to man-made objects and appealing in its simple and fast implementation and efficiency in its rejection of undesired regions.

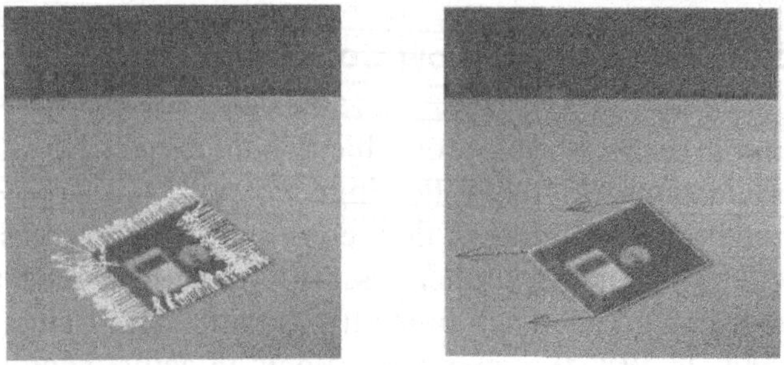

Fig. 5.2. Velocity vectors perpendicular to local direction of boundaries of an object which is translating horizontally towards left. Right: Velocity vectors at vertices of the polygonal boundary

The chapter is organized as follows. In the next section, we present a continuous contour tracker, with an additional smoothness constraint. In Section 5.3, we present a polygonal tracker and compare it to the continuous tracker. We provide simulation results and conclusions in Section 5.4.

5.2 Tracking with Active Contours

Evolution of curves is a widely used technique in various applications of image processing such as filtering, smoothing, segmentation, tracking, registration, to name a few. Curve evolutions consist of propagating a curve via partial differential equations (PDEs). Denote a family of curves by $\mathbf{C}(p, t) = (\mathbf{X}(p, t'), \mathbf{Y}(p, t'))$, a mapping from $R \times [0, T'] \rightarrow R^2$, where p is a parameter along the curve, and t parameterizes the family of curves. This curve may serve to optimize an energy functional over a region R, and thereby serve to capture contours of given objects in an image with the following [41, 42]

$$\mathbf{E(C)} = \iint_{\mathbf{R}} \mathbf{f}\ \mathbf{dxdy} = \oint_{\mathbf{C}=\partial\mathbf{R}} < \mathbf{F}, \mathbf{N} > \mathbf{ds}, \tag{6}$$

where $\mathbf{N}$ denotes the outward unit normal to $\mathbf{C}$ (the boundary of R), ds the Euclidean arclength element, and where $\mathbf{F} = (F_1, F_2)$ is chosen so that $\nabla \cdot \mathbf{F} = \mathbf{f}$. Towards optimizing this functional, it may be shown [42] that a gradient flow for $\mathbf{C}$ with respect to E may be written as

$$\frac{\partial \mathbf{C}}{\partial t'} = \mathbf{fN}, \tag{7}$$

where t' denotes the evolution time variable for the differential equation.

5.2.1 Tracker with Optical Flow Constraint

Image features such as edges or object boundaries are often used in tracking applications. In the following, we will similarly exploit such features in addition to an optical flow constraint which serves to predict a velocity field along object boundaries. This in turn is used to move the object contour in a given image frame I(x ,t) to the next frame I(x ,t + 1). If a 2-D vector field **V**(x ,t) is computed along an active contour, the curve may be moved with a speed V in time according to

$$\frac{\partial \mathbf{C}(p,t)}{\partial t} = \mathbf{V}(p,t),$$

This is effectively equivalent to

$$\frac{\partial \mathbf{C}(p,p)}{\partial p} = (\mathbf{V}(p,p)\cdot \mathbf{N}(p,p))\mathbf{N}(p,p),$$

as it is well known that a re-parameterization of a general curve evolution equation is always possible, and in this case yields an evolution along the normal direction to the curve [43]. The velocity field at each point on the contour at time t by **V** (x) may hence be represented in terms of parameter p as **V** (p)= $v^{\perp}$(p)**N** (p) + v^{T}(p)**T** (p), with **T** (p) and **N** (p) respectively denoting unit vectors in the tangential and normal directions to an edge (Figure 5.3).

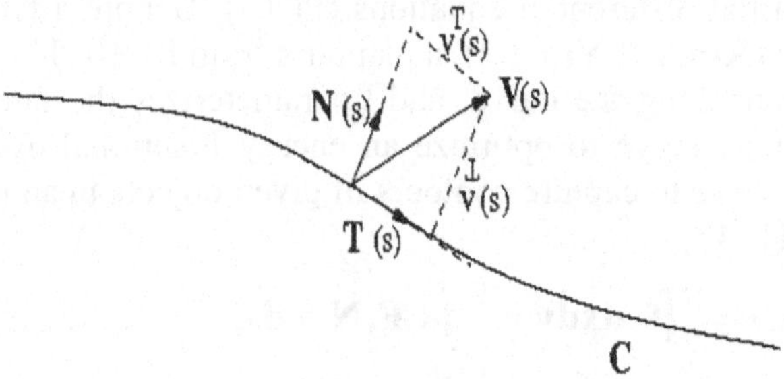

Fig. 5.3. 2-D velocity field along a contour

Using Eq.(1), we may proceed to compute the estimate of the orthogonal component $v^{\perp}$. Using a set of local measurements derived from the time-varying image I(x ,t) and brightness constraints, would indeed yield

$$v^{\perp}(x,y) = -\frac{I_t}{\| \nabla I \|}, \tag{8}$$

This provides the magnitude of the velocity field in the direction orthogonal to the local edge structure which may in turn be used to write a curve evolution equation which preserves a consistency between two consecutive frames,

$$\frac{\partial \mathbf{C}(p,t)}{\partial t} = \mathbf{v}^{\perp}(p,t)\mathbf{N}(p,t), \quad 0 \leq t \leq 1, \tag{9}$$

An efficient method for implementation of curve evolutions, due to Osher and Sethian [44], is the so-called, level set method. The parameterized curve **C** (p, t) is embedded into a surface, which is called a level set function Φ(x, y, t) : $R^2 \times [0, T] \rightarrow R$, as one of its level sets. This leads to an evolution equation for Φ, which amounts to evolving C in Eq. (7), and written as

$$\frac{\partial \Phi}{\partial t} = -f \| \nabla \Phi \|. \tag{10}$$

The prediction of the new location of the active contour on the next image frame of the image sequence can hence be obtained as the solution of the following PDE

$$\frac{\partial \Phi}{\partial t} = -v^{\perp} \| \nabla \Phi \|, \quad 0 \leq t \leq 1. \tag{11}$$

In the implementation, a narrowband technique which solves the PDE only in a band around the zero level set is utilized [45]. Here, $v^{\perp}$ is computed on the zero level set and extended to other levels of the narrowband. Most active contour models require some regularization to preserve the integrity of the curve during evolution, and a widely used form of the regularization is the arc length penalty. Then the evolution for **the prediction step** takes the form,

$$\frac{\partial \Phi}{\partial t} = \alpha \, \kappa - v^{\perp} \| \nabla \Phi \|, \quad 0 \leq t \leq 1, \tag{12}$$

where κ(x, y, t) is the curvature of the level set function Φ(x, y, t), and $\alpha \geq 0 \in R$ is a weight determining the desired amount of regularization.

Upon predicting the curve at the next image frame, a correction/propagation step is usually required in order to refine the position of the contour on the new image frame. One typically exploits region-based active contour models to update the contour or the level set function. These models assume that the image consists of a finite number of regions that are characterized by a pre-determined set of features or statistics such as means, and variances. These region characteristics are in turn used in the construction of an energy functional of the curve which aims at maximizing a divergence measure among the regions. One simple and convenient choice of a region based characteristic is the mean intensity of regions inside and outside a curve [46, 47], which leads the image force f in Eq.(10) to take the form

$$f(x, y) = 2(u - v)(I(x, y) - (u + v)/2), \tag{13}$$

where u and v respectively represent the mean intensity inside and outside the curve. Region descriptors based on information-theoretic measures or higher order statistics of regions may also be employed for increasing the robustness against noise and textural variations in an image [41]. **The correction step** is hence carried out by

$$\frac{\partial \Phi}{\partial t'} = \alpha' \kappa - f \,\|\nabla \Phi\|, \quad 0 \le t' \le T' \tag{14}$$

on the next image frame I(x, y, t + 1). Here, $\alpha' \geq 0 \in R$ is included as a very small weight to help preserve the continuity of the curve evolution, and T' is an approximate steady-state reaching time for this PDE.

To clearly show the necessity of the prediction step in Eq. (12) in lieu of a correction step alone, we show in the next example a video sequence of two marine animals. In this clear scene, a curve evolution is carried out on the first frame so that the boundaries of the two animals are outlined at the outset. Several images from this sequence shown in Figure 5.4 demonstrate the tracking performance with and without prediction respectively in (rows 3 and 4) and (rows 1 and 2). This example clearly shows that the prediction step is crucial to a sustained tracking of the target, as a loss of target tracking results rather quickly without prediction. Note that the continuous model's "losing track" is due to the fact that region based active contours are usually based on non-convex energies, with many local minima, which may sometimes drive a continuous curve into a single point, usually due to the regularizing smoothness terms.

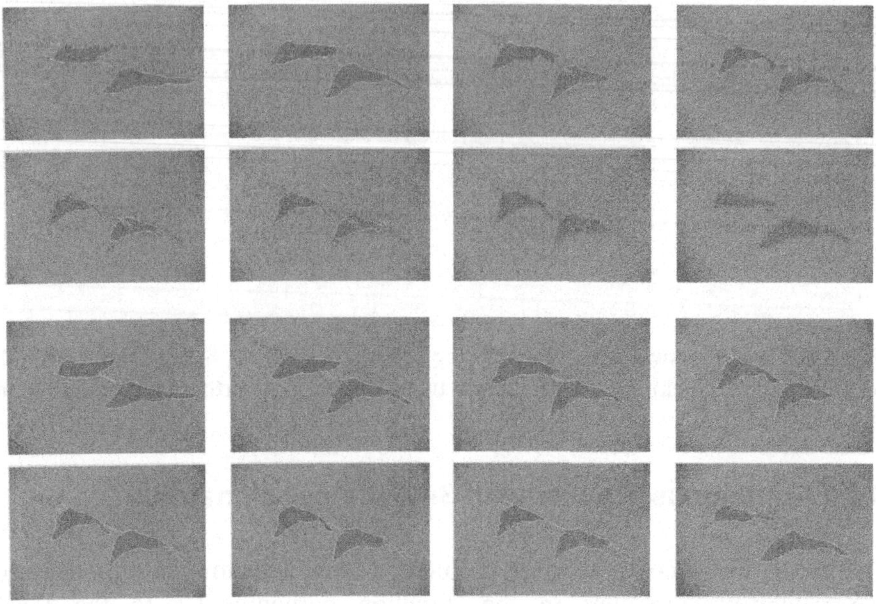

Fig. 5.4. Two rays are swimming gently in the sea (Frames 1, 10, 15, 20, 22, 23, 24, 69 are shown left-right top-bottom). Rows 1 and 2: Tracking without prediction. Rows 3 and 4: Tracking with prediction using optical flow orthogonal component

In the noisy scene of Figure 5.5 (e.g. corrupted with Gaussian noise), we show a sequence of frames for which a prediction step with an optical flow-based normal velocity, may lead to a failed tracking on account to the excessive noise. Unreliable estimates from the image at the prediction stage are the result of the noise. At the correction stage, on the other hand, the weight of the regularizer, i.e. the arc length penalty, requires a significant increase. This in turn leads to rounding and shrinkage effects around the target object boundaries. This is tantamount to saying that the joint application of prediction and correction cannot guarantee an assured tracking under noisy conditions as may be seen in Figure 5.5. One may indeed see that the active contour loses track of the rays after some time. This is a strong indication that additional steps have to be taken into account in reducing the effect of noise. This may be in the form of regularization of the velocity field used in the prediction step.

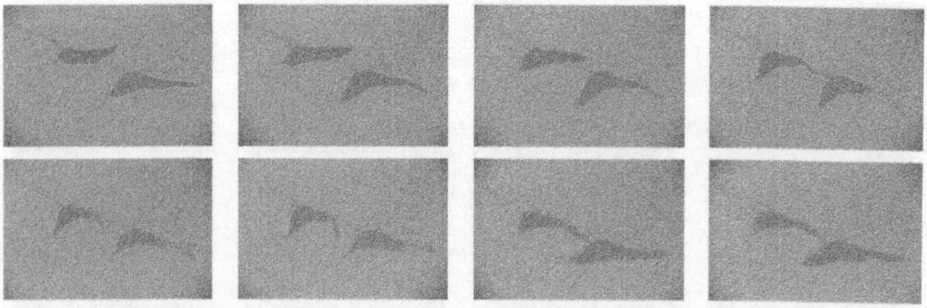

Fig. 5.5. Two rays-swimming video noisy version (Frames 1, 8, 13, 20, 28, 36, 60, 63 are shown). Tracking with prediction using optical flow orthogonal component

5.2.2 Continuous Tracker with Smoothness Constraint

Due to the well-known aperture problem, a local detector can only capture the velocity component in the direction perpendicular to the local orientation of an edge. Additional constraints are hence required to compute the correct velocity field. A smoothness constraint, introduced in [24] relies on the physical assumption that surfaces are generally smooth, and generate a smoothly varying velocity field when they move. Still, there are infinitely many solutions. A single solution may be obtained by finding a smooth velocity field that exhibits the least amount of variation among the set of velocity fields that satisfy the constraints derived from the changing image. The smoothness of the velocity field along a contour can be introduced by a familiar approach such as $\oint_{\mathbf{C}} \left\| \frac{\partial \mathbf{V}}{\partial s} \right\|^2 ds$. Image constraints may be satisfied by minimizing the difference between the measurements $v^{\perp}$ and the projection of the velocity field **V** onto the normal direction to the contour, i.e. **N** . The overall energy functional thus defined by Hildreth [24] is given by

$$E(\mathbf{V}) = \oint_{\mathbf{C}} \left\| \frac{\partial \mathbf{V}}{\partial s} \right\|^2 ds + \beta \oint_{\mathbf{C}} \left[\mathbf{V} \cdot \mathbf{N} - v^{\perp} \right]^2 ds \tag{15}$$

where β is a weighting factor that expresses the confidence in the measured velocity constraints. The estimate of the velocity field V may be obtained by way of minimizing this energy. This is in turn carried out by seeking a steady state solution of a PDE corresponding to the Euler Lagrange equations of the functional. In light of our implementation of the active contour model via a level set method, the target object's contour is implicitly represented as the zero level set of the higher dimensional embedding function Φ. The solution for the velocity field $\mathbf{V}$, defined over an implicit contour embedded in Φ, is obtained with additional constraints such as derivatives that depend on $\mathbf{V}$ which are intrinsic to the curve (a different case where data defined on a surface embedded into a 3D level set function is given in [48]). Following the construction in [48], the smoothness constraint of the velocity field, i.e. the first term in Eq. (15), corresponds to the Dirichlet integral with the intrinsic gradient, and using the fact that the embedding function Φ is chosen as a signed distance function, the gradient descent of this energy can be obtained as

$$\frac{\partial \mathbf{V}}{\partial t'} = \mathbf{V}_{ss} - \beta \left(\mathbf{V} \cdot \frac{\nabla \Phi}{\| \nabla \Phi \|} - v^{\perp} \right) \frac{\nabla \Phi}{\| \nabla \Phi \|}, \quad 0 \le t' \le T'. \qquad (16)$$

Also by construction, the extension of the data defined on the curve C over the narrowband satisfies, $\nabla \mathbf{V} \cdot \nabla \Phi = 0$, which helped lead to Eq. (16) (here the gradient operator ∇ also acts on each component of V separately). This PDE can be solved with an initial condition taken as the $v^{\perp} \mathbf{N}$, to provide estimates for full velocity vector $\mathbf{V}$ at each point on the contour, indeed at each point of the narrowband.

A blowup of a simple object subjected to a translational motion from a video sequence is shown in Figure 5.6 with a velocity vector at each sample point on the active contour moving from one frame to the next. The initial normal velocities are shown on the left, and the final velocity field is obtained as a steady state solution of the PDE in (16) and is shown on the right. It can be observed that the correct velocity on the boundary points, is closely approximated by the solution depicted on the right. Note that the zero initial normal speeds over the top and bottom edges of the object have been corrected to nonzero tangential speeds as expected.

The noisy video sequence of two-rays-swimming shown in the previous section, is also tested with the same evolution technique, replacing the direct normal speed measurements $v^{\perp}$ by the projected component of the estimated velocity field, which is $\mathbf{V} \cdot \mathbf{N}$ as explained earlier. It is observed in Figure 5.7 that the tracking performance is, unsurprisingly, improved upon utilizing Hildreth's method, and the tracker kept a better lock on objects. This validates the adoption of a smoothness constraint on the velocity field. The noise presence, however, heavily penalizes the length of the tracking contours has to be significantly high, which in turn, leads to severe roundedness in the last few frames. If we furthermore consider its heavy computational load, we realize that the continuous tracker with its Hildreth-based smoothness constraint is highly impractical.

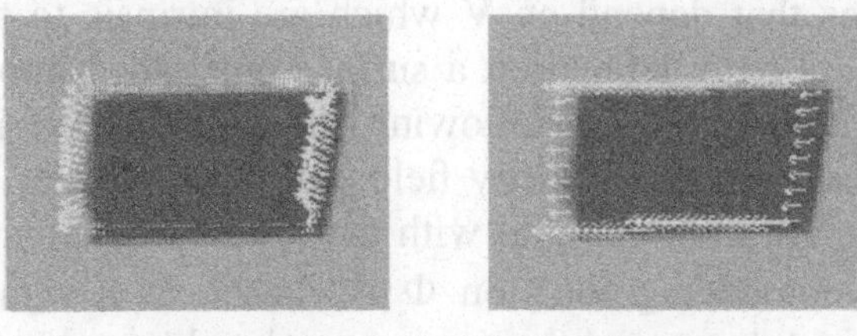

Fig. 5.6. Velocity normal to local direction of boundaries of an object which is translating horizontally as shown on the left, and the velocity field computed from (16) is given on the right (with β=0.1, a time step of 0.24 and number of iterations=400)

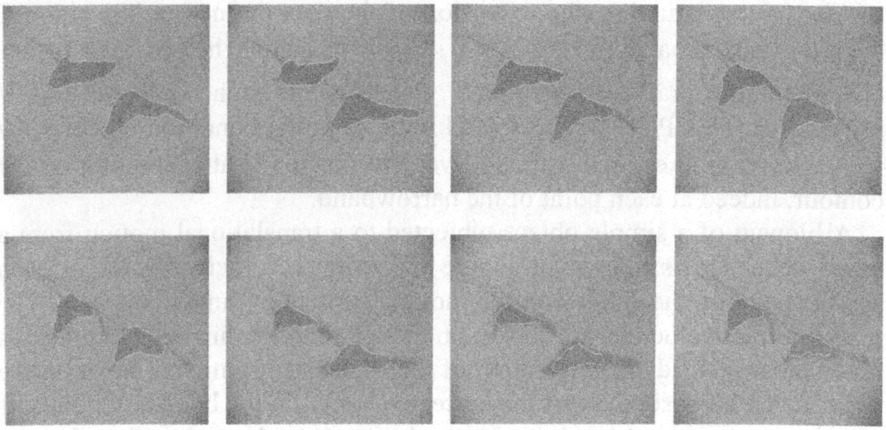

Fig. 5.7. Two rays swimming video noisy version (Frames 1, 8, 13, 20, 28, 36, 60, 63 are shown). Tracking with prediction using optical flow computed via Eq. (16)

In an attempt to address these problems and to better consider issues related to speed, we next propose a polygonal tracker nearly an order of magnitude faster than the most effective continuous tracker introduced in the previous sections. The advantage of our proposed technique is made clear by the resulting tracking speeds of various approaches displayed in Figure 5.8. It is readily observed that the smoothness constraint on the velocity field of a continuous tracker significantly increases the computation time of the algorithm, and that a more robust performance is achievable.

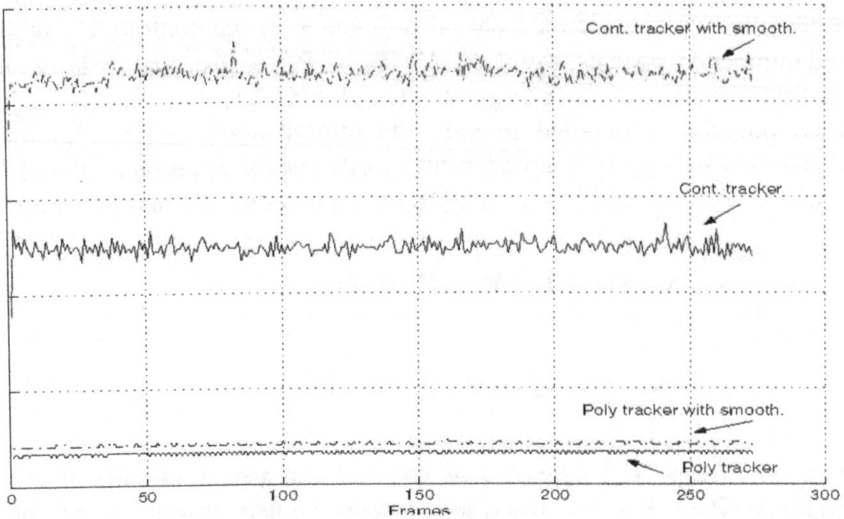

Fig. 5.8. Speed comparisons among different trackers introduced in this study. From top to bottom plots depicted are: continuous tracker with smoothness constraint; continuous tracker; polygonal tracker with smoothness constraint; polygonal tracker

5.3 Polygonal Tracker

The goal of this section is the development of a simple and efficient boundary-based tracking algorithm well adapted to polygonal objects. The idea is built on the insights gained from both the continuous tracker model and the polygon evolution model introduced in [41]. The latter provides sufficient structure to capture an object, resulting in a coarse yet a descriptive representation of a target. Its enhanced robustness in segmentation applications of noisy and/or textural regions, and its fast implementation secured by a reduced number of degrees of freedom, put this model at a great advantage. Its suitability to tracking problems and its amenability to Kalman Filter-inspired prediction and correction steps make it an all around good choice as we elaborate next.

5.3.1 Velocity Estimation at Vertices

We presented in [41] gradient flows which could move polygon vertices so that an image domain be parsed into meaningfully different regions.

Specifically, we considered a closed polygon **P** as the contour **C** , with a fixed number of vertices, say $n \in N$, $\{\mathbf{P}_1, \ldots, \mathbf{P}_n\} = \{(xi, yi), i = 1, \ldots n\}$. The first variation of an energy functional E(**C**) in Eq. (6) for such a closed polygon is detailed in [41]. Its minimization yields a gradient descent flow by a set of coupled ordinary differential equations (ODEs) for the whole polygon, and hence an ODE for each vertex $\mathbf{P}_k$, and given by

$$\frac{\partial P_k}{\partial t'} = \mathbf{N}_{1,k} \int_0^1 p \; f(L(p, \mathbf{P}_{k-1}, \mathbf{P}_k)) \; dp + \mathbf{N}_{2,k} \int_0^1 p \; f(L(p, \mathbf{P}_k, \mathbf{P}_{k+1})) \; dp, \tag{17}$$

where $\mathbf{N}_{1,k}$ (resp. $\mathbf{N}_{2,k}$) denotes the outward unit normal of edge ($P_{k-1} - P_k$) (resp. ($P_k - P_{k+1}$)), and L parameterizes a line between P_{k-1} and P_k or P_k and P_{k+1}. We note the similarity between this polygonal evolution equation which may simply be written in the form

$$\frac{\partial \mathbf{P}_k}{\partial t'} = \tilde{\mathbf{f}}_1 \mathbf{N}_{k,k-1} + \tilde{\mathbf{f}}_2 \mathbf{N}_{k+1,k},$$

and the curve evolution model given in Eq. (7), and recall that each of f_1 and f_2 corresponds to an integrated f on both neighboring edges of vertex $\mathbf{P}_k$. Whereas each point of the curve in the continuous model moves as a single entity driven by a functional f of local as well as global quantities, each polygon edge in the proposed approach moves as a single unit moved along by its end vertices. The latter motion is in turn driven by information gleaned from two neighboring edges via f. In addition to the pertinent information captured by the functional f, its integration along edges provides an enhanced and needed immunity to noise and textural variability. This clear advantage over the continuous tracker, highlights the added gain from a reduced number of well separated vertices and its distinction from snake-based models.

The integrated spatial image information along adjacent edges of a vertex $\mathbf{P}_k$ may also be used to determine the speed and direction of a vertex on a single image, as well as to estimate its velocity field on an active polygon laid on a time-varying image sequence. The estimated velocity vector at each vertex $\mathbf{P}_k$ using the two adjacent edges is schematically illustrated in Figure 5.9.

orthogonal direction to the local edge structure. Instantaneous measurements are unfortunately insufficient to determine the motion, and an averaged information is

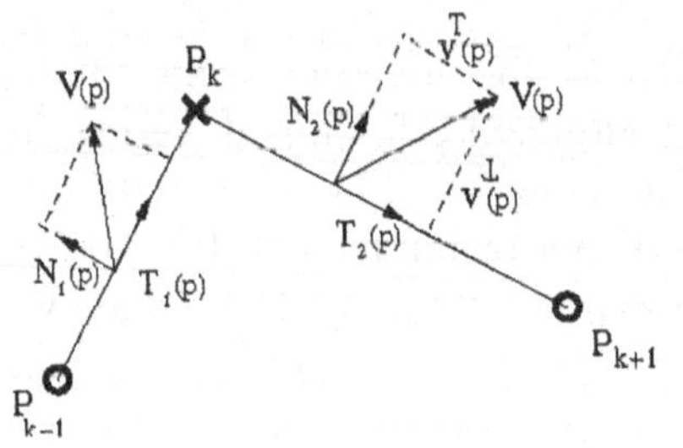

Fig. 5.9. 2-D velocity field along two neighbor edges of a polygon vertex

The velocity field **V** (x_r, y) at each point of an edge may be represented as **V** $(p) = v^{\perp}(p)\mathbf{N}_i(p) + v^{T}(p)\mathbf{T}_i(p)$, where $\mathbf{T}_i(p)$ and $\mathbf{N}_i(p)$ are unit vectors in the tangential and normal directions of edge i. Once an active polygon locks onto a target object, the unit direction vectors **N** and may readily be determined. A set of local measurements $v^{\perp}$ (Eq. (8)) obtained from the optical flow constraint yield the magnitude of a velocity field in an shown to be critical for an improved point velocity estimation. To that end, we utilize a joint contribution from two edges of a vertex to infer its resultant motion. Specifically, we address the sensitivity of the normal velocity measurements to noise by their weighted integration along neighboring edges of a vertex of interest. This leads to our prediction equation of vertex velocity,

$$\frac{\partial \mathbf{P}_k}{\partial t} = \mathbf{V}_k = \mathbf{u}^{\perp}_{1,k} \int_0^1 p \;\; v^{\perp}(L(p, \mathbf{P}_{k-1}, \mathbf{P}_k)) \; dp \\ + \mathbf{u}^{\perp}_{2,k} \int_0^1 p \;\; v^{\perp}(L(p, \mathbf{P}_k, \mathbf{P}_{k+1})) \; dp, \tag{18}$$

for k =1, ...n. To introduce further robustness and to achieve more reliable estimates in the course of computing $v^{\perp}$, we may make use of smoother spatial derivatives (larger neighborhoods).

To fully exploit the vertices of the underlying polygon, our tracking procedure is initialized by delineating target boundaries by either region-based active polygon segmentation or manually. The prediction step of the velocity vector is carried out in Eq. (18), which in turn determines the locations of the polygon vertices at the next time instance on I(x, y, t + 1). In a discrete setting, the ODE simply corresponds to

$$\mathbf{P}_k(t+1) = \mathbf{P}_k(t) + \mathbf{V}_k(t) \tag{19}$$

if the time step in the discretization is chosen as 1.

The correction step of the tracking seeks to minimize the deviation between current measurement/estimate of vertex location and predicted vertex location, by applying Eq. (17). Since both the prediction as well as the correction stages of our technique call for a polygonal delineation of a target contour, a global regularizing technique we introduced in great detail in [41] is required to provide stability. Specifically, it makes use of the notion of an electrostatic field among the polygon edges as a means of self-repulsion. This global regularizer technique provides an evolution without degeneracies and preserves the topology of the evolving polygon as a simple shape. The polygon-based segmentation/approximation of a target assumes an adequate choice of the initial number of vertices. Should this prior knowledge be lacking, we have developed in [41] a procedure which automatically adapts this number by periodic additions/deletions of new/redundant vertices as the case may be. In some of the examples given below, this adaptive varying number of vertices approach is lumped together with the correction step and will be pointed out in due course.

One may experimentally show that the velocity estimation step (prediction) of the polygonal tracker indeed improves performance. The following sequence in Figure 5.10 shows a black fish swimming among a school of other fish. Tracking which uses only the spatial polygonal segmentation with an adaptive number of vertices, (i.e., just carries the active polygon from one image frame onto the consecutive one after a number of spatial segmentation iterations), may lose track of the black fish. In particular, as one notes in Fig. 5.10 a partial occlusion of the black fish leads to a track loss (frame marked by LOST). The active polygon may be re-initialized after the occlusion scene (frame marked by RE-INITIALIZED), but to no avail as another track loss follows as soon as the fish turns around (second frame marked by LOST).

On the other hand and as may be observed in Figure 5.11, the polygonal tracker with the prediction step could follow the black fish under rougher visibility conditions such as partial occlusions and small visibility area when the fish is making a turn around itself. A successful tracking continues for all 350 frames of the sequence. This example demonstrates that the tracking performance is improved with the addition of the optical flow estimation step, which, as described earlier, merely entails the integration of the normal optical flow field along the polygon adjacent edges to yield a motion estimate of a vertex.

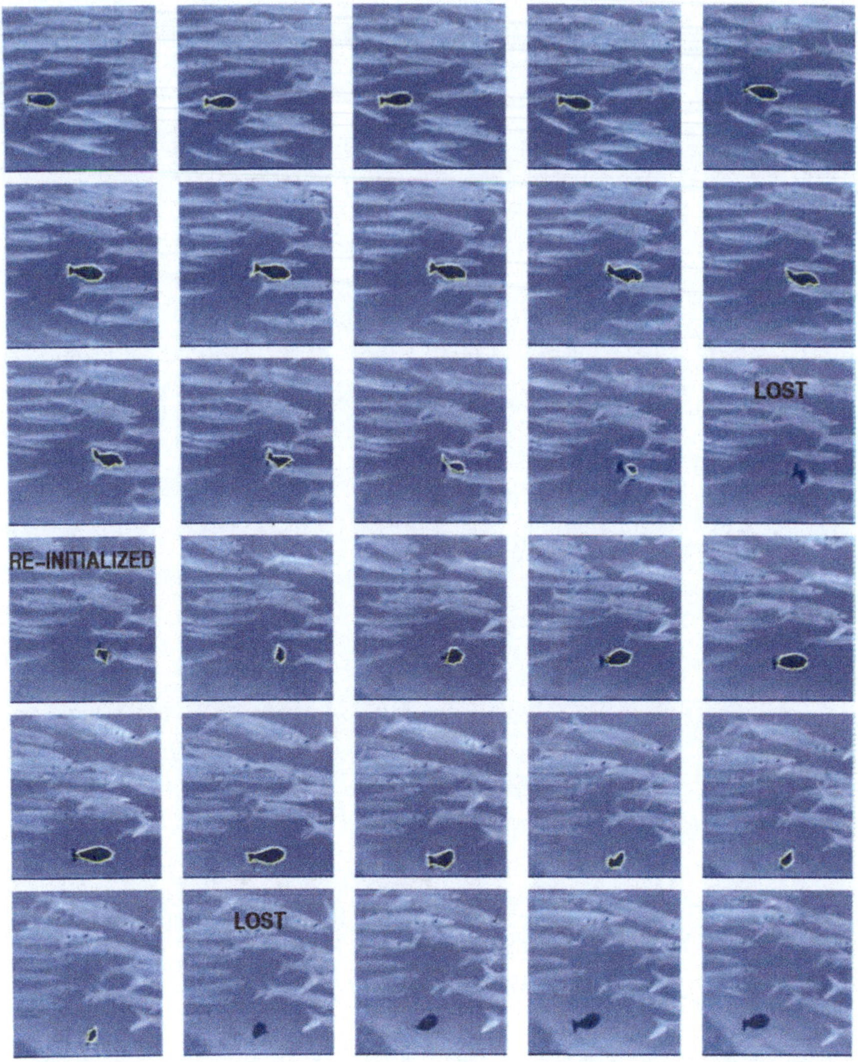

Fig. 5.10. A black fish swims among a school of other fish. Polygonal tracker with only the correction stage may lose track of the black fish when it is partly occluded by other fish, or turning backwards

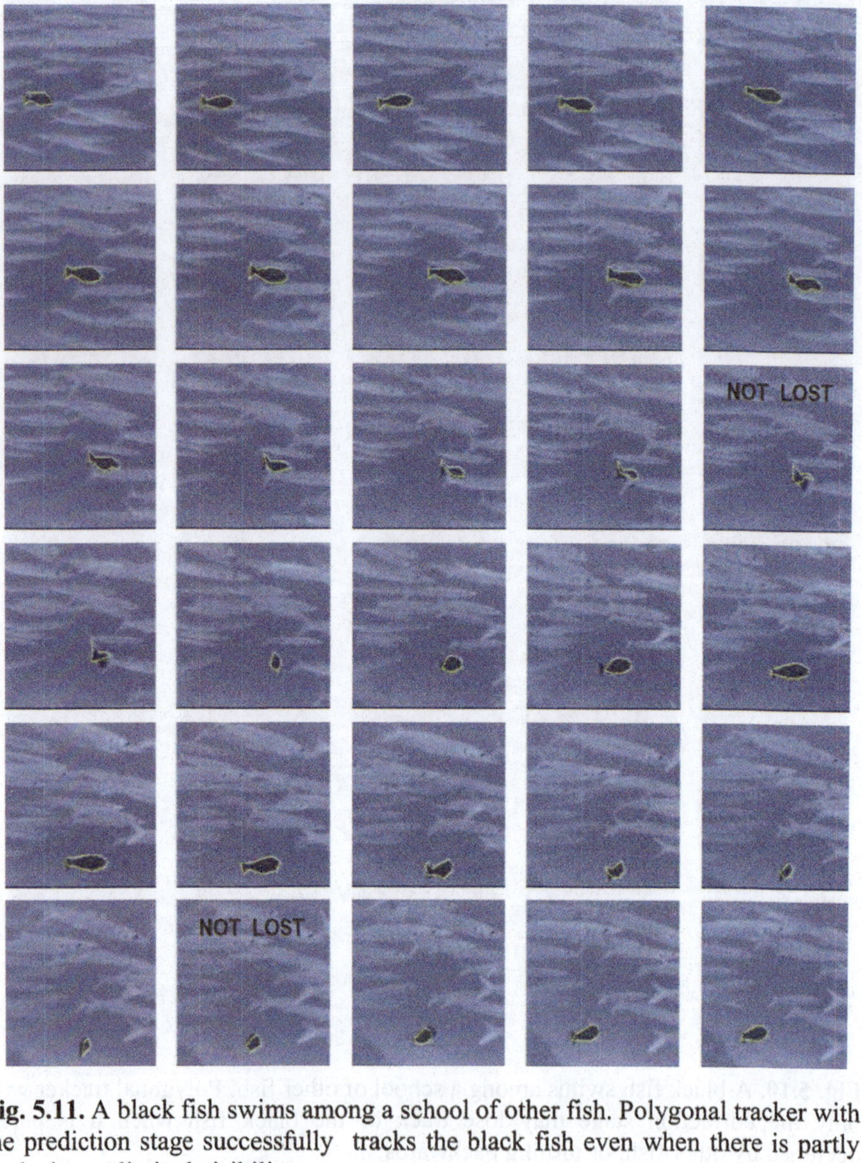

Fig. 5.11. A black fish swims among a school of other fish. Polygonal tracker with the prediction stage successfully tracks the black fish even when there is partly occlusion or limited visibility

5.3.2 Polygonal Tracker With Smoothness Constraint

A smoothness constraint may also be directly incorporated into the polygonal framework, with in fact much less effort than required by the continuous framework in Section 2.2. In the prediction stage, an initial vector of normal optical flow could be computed all along the polygon over a sparse sampling on edges between vertices. A minimization of the continuous energy functional (15) is subsequently carried out by directly discretizing it, and taking its derivatives with respect to the x and y velocity field components. This leads to a linear system of equations which can be solved by a mathematical programming technique, e.g. the conjugate gradients as suggested in [24]. We have carried out this numerical minimization in order to obtain the complete velocity field $\mathbf{V}$ along all polygon edges. For visualizing the effect of the smoothness constraint on the optical flow, a snapshot from a simple object in translational motion is shown in Figure 5.12 where the first picture in a row depicts the normal optical flow component $v^{\perp}\,\mathbf{N}$ initialized over the polygon. In this figure, the first row corresponds to a clean sequence whereas the second row corresponds to the noisy version of the former. The velocity at a vertex may be computed by integrating according to Eq. (18), and shown in the second picture in a row. The complete velocity $\mathbf{V}$ obtained as a result of the minimization of the discrete energy functional is shown in the third picture. It is observed that the estimated velocity field is smooth, and satisfies the image constraints, and very closely approximates the true velocity. This result could be used in the active polygon framework by integrating the velocity field along the neighbor edge pair of each vertex $\mathbf{P}_k$ for yet additional improvement on the estimate $\mathbf{V}_k$

$$\mathbf{V}_k = \int_0^1 p\ V(L(p, \mathbf{P}_{k-1}, \mathbf{P}_k))\ dp + \int_0^1 p\ V(L(p, \mathbf{P}_k, \mathbf{P}_{k+1}))\ dp, \quad k = 1, \ldots, n \tag{20}$$

as demonstrated on the right in Fig. 5.12 for n = 4. The active polygon can now be moved directly with Eq. (19) onto the consecutive image frame. The correction step follows the prediction step to continue the process.

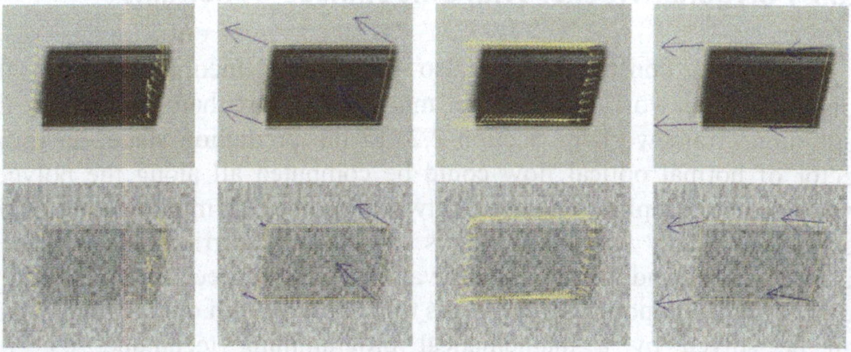

Fig. 5.12. An object is translating horizontally. Row 1:clean version. Row 2: noisy version. (left-right) Picture 1: Velocity normal to local direction of boundaries.; 2: The overall integrated velocity at the vertices from picture 1; 3: Velocity field computed through minimization of (15) with conjugate gradients technique; 4; The overall integrated velocity field at the vertices

5.4 Discussions and Results

In this section, we substantiate our proposed approach by a detailed discussion contrasting it to existing approaches, followed by numerical experiments.

5.4.1 Comparison between the Continuous and the Polygonal Approaches

A comparison between the continuous and the polygonal approaches may be made on the basis of the following:

If the true velocity field **V** were to be *exactly* computed , the polygonal model would move the vertices of the polygon directly with the full velocity onto the next frame by $\frac{\partial \mathbf{C}}{\partial t} = \mathbf{V}$ with no need for update. Such information could not, however, be so readily used by a continuous tracker, as its update would require a solution to a PDE $\frac{\partial \Phi}{\partial t} = (\mathbf{V} \cdot \mathbf{N})\mathbf{N}$ (by level set method). The zero-level set curve motion, as a solution to the PDE, only depends on the normal component of the velocity vector, and is hence unable to account for the complete direction of the velocity. Moreover, additive noise in continuous contours causes irregular displacements of contour points, break-ups and others. The well-separated vertex locations of the polygonal model, on the other hand takes full advantage of the complete optical flow field to avoid such problems.

The polygonal approach owes its robustness to an averaging of information gathered at all pixels along edges adjacent to a moving vertex; in contrast to a pixelwise information in the continuous model. The noisy video sequence of two-rays-swimming constitutes a good case study to unveil the pros and cons of both approaches. The continuous tracker via the level set implementation autonomously handles topological changes, and conveniently takes care of multiply connected regions, here the two swimming animals. Adapting the polygonal model to allow topology changes may be done by observing the magnitudes of its self-repulsion forces (which kicks in when polygonal edges are about to cross each other). This term can communicate to us when and where a topological change should occur. For our intended applications we do not pursue this approach. Handling multiple targets is easier than handling topology changes though, because the models we developed can be extended to multiple polygons which evolve separately with coupled ODEs. Snapshots from the noisy two-rays-swimming sequence illustrate the polygonal tracker (here for sake of example, two animals could be separately tracked and the results are overlaid) in Figure 5.13. The ability of the continuous tracker to automatically handle topological changes, is overshadowed by its sensitivity to noise which is likely to cause breakdown making this property less pronounced. The prediction and correction steps with a statistical filtering perspective, improve the robustness of the polygonal approach. As already seen in Figures 5.5 and 5.7 shrinkage and rounding effects may be very severe in the presence of a significant amount of noise in the scene due to necessary large regularization in continuous tracking. This is in contrast to the electrostatic forces used in conjunction with the polygonal model as well as the latter's resilience to external textural variability. We also note here that the region-based descriptor f used in the update step is the same in both the continuous and polygonal tracker examples shown, and is as given in Eq.(13).

The lower number of degrees of freedom present in moving a polygon makes leaking through background regions more unlikely than for a continuous curve being easily attracted towards unwanted regions. The following example illustrates this in Figure 5.14, which shows a fish swimming in a rocky sea terrain. As the background bears similar region characteristics as the fish, the continuous tracker with its ease in split and merge encloses unrelated regions other than the target fish in the update step. The polygonal contour, in this case, follows the fish by preserving the topology of its boundaries. This is also an illustration for handling topological changes automatically may be either an advantage or a disadvantage.

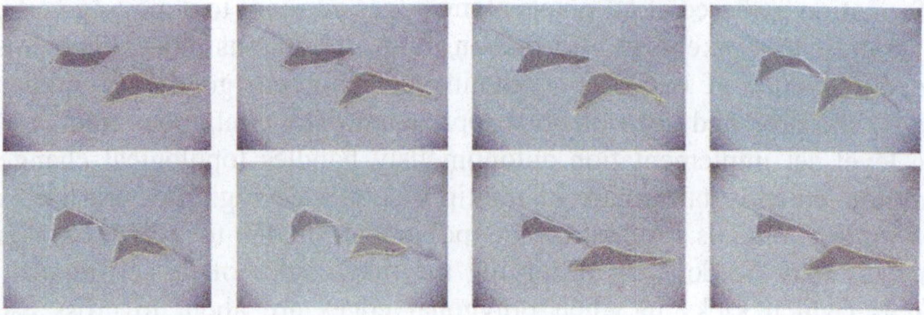

Fig. 5.13. Two-rays-swimming video noisy version (Frames 1, 8, 13, 20, 28, 36, 60, 63 are shown). Tracking via active polygons with prediction using optical flow normal component

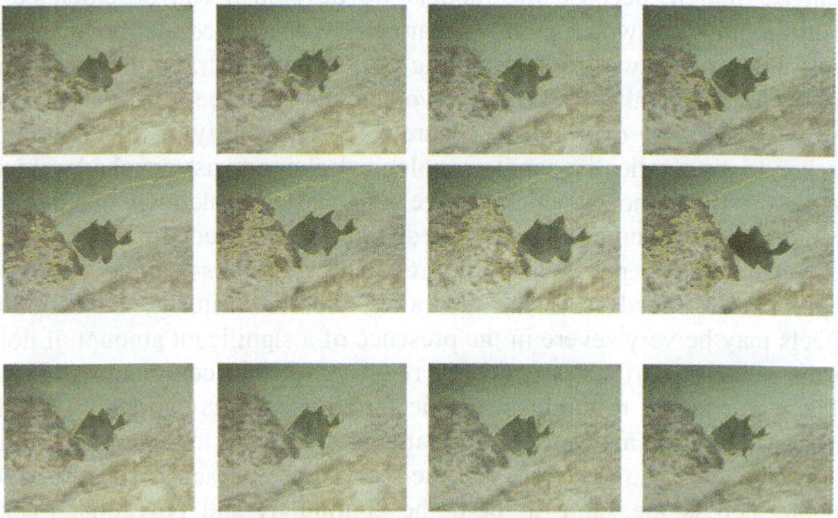

Fig. 5.14. A swimming fish in a rocky terrain in the sea (Frames 1, 10, 20, 30, 40, 70, 110, 143 are shown left-right-top-bottom). Rows 1 and 2: Continuous tracker fails to track the fish. Rows 3 and 4: Polygonal tracker successfully tracks the fish

The speed performance of the polygonal tracker is superior to that of the continuous tracker. A comparison is given in Fig. 5.8, where the plots depict the computation speed versus frames for both the polygonal and the continuous models. The polygonal tracker with or without the smoothness constraint is approximately 8 times faster than thc continuous model with or without the smoothness constraint.

The proposed polygonal tracker is intrinsically more regular by a natural regularizer term which keeps polygonal edges from crossing each other, and only kicks in significantly when such a pathology is close to occuring.

5.4.2 Experimental Results

Figure 5.15 illustrates tracking in snapshots from a video sequence of a person walking in a parking lot . The insertion of a prediction step in the tracking methodology is to speed up the computations by helping the active polygon to glide onto a new image frame in the sequence and smoothly adapt to displaced object's boundaries. The temporal resolution of the given sequence is quite high, and the scene changes from frame to frame are minimal. Nonetheless, when we plot the speeds of the polygonal tracker with and without the velocity prediction as depicted in Figure 5.16 (left), we observe that the former is faster, confirming the expected benefit of the prediction step. To verify this effect for a sequence with lower temporal resolution, we decimated the sequence by six in time, and plotted the speeds in Fig. 5.16 (right). When the temporal resolution of the sequence is decreased, the processing time for each frame increases as expected for both tracking methods. Even though our velocity prediction scheme gives rough estimates, however, the tracking polygon is mapped to a position which is closer to the new object position in the new scene or frame. This is reflected in the given speed plots where the polygonal tracker without the prediction step, takes longer to flow the polygon towards the desired object boundaries.

Fig. 5.15. A walking person (Frames shown L-R-top-bottom) is tracked by the polygonal tracker

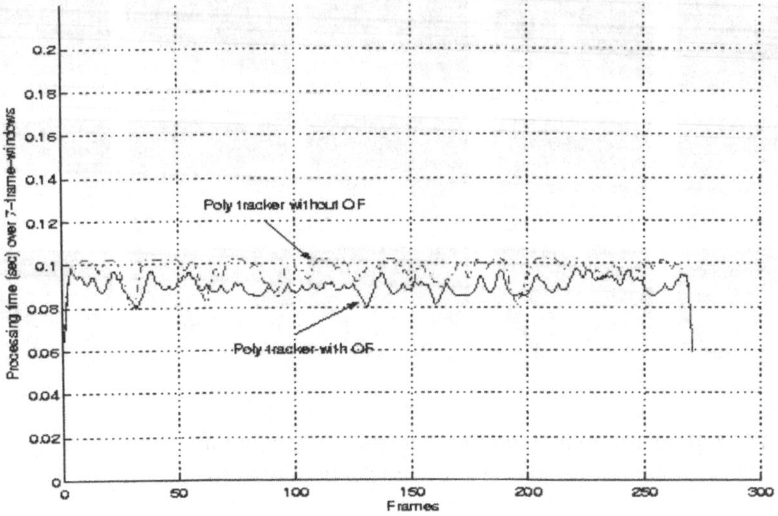

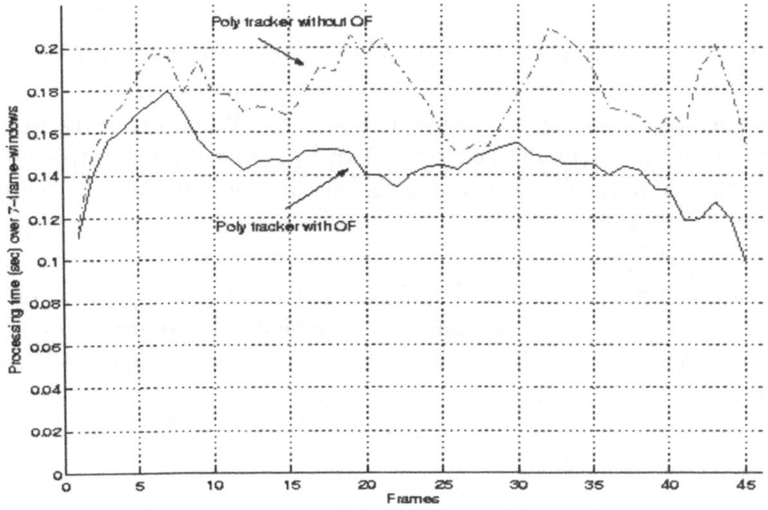

Fig. 5.16. Processing time (averaged over 7-window-frames) vs frames: for the original sequence (*left*), for the sequence subsampled by 6 in time (*right*)

The polygonal tracker with its ability to utilize various region-based descriptors could be used for tracking textured objects on textured backgrounds. A specific choice based on an information-theoretic measure [41]

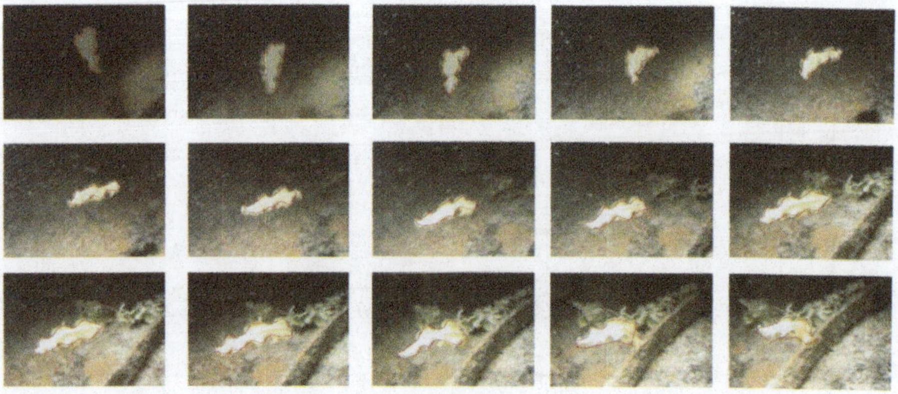

Fig. 5.17. A flatworm in a textured sea terrain (15 frames are shown left-right top-bottom). Polygonal tracker successfully tracks the flatworm

whose approximation uses high order moments of the data distributions leads the image based integrand f in Eq.(17) to take the form $f = \sum_{j=1}^{m}(u_j - v_j)((G_j(I) - u_j) + (G_j(I) - v_j))$ with functions G chosen for instance as $G_1(\xi)=\xi e^{-\xi^2/2}$ and $G_2(\xi) = e^{-\xi^2/2}$. When the correction step of our method involves the descriptor f just given with a adaptive number of vertices, a flatworm swimming in the bottom of the sea could be captured through the highly textured sequence by the polygonal tracker in Fig. 5.17. The speed plots in Fig. 5.18 depict the speeds for the tracker with and without prediction. The figure on the right is for the original sequence (whose plot is given on the left) which is temporally subsampled by two. Varying the number of vertices to account for shape variations of the worm slows down the tracking in general. However, the tracker with prediction still performs faster than the tracker without prediction as expected. The difference in speeds becomes more pronounced in the subsampled sequence on the left. Similarly, a clownfish on a host anemone shown in Figure 5.19, could be tracked in a highly textured scene. The continuous trackers we have introduced in this study do not provide a continuous tracking in either of these examples, and they split, leak to background regions, and lose track of the target completely.

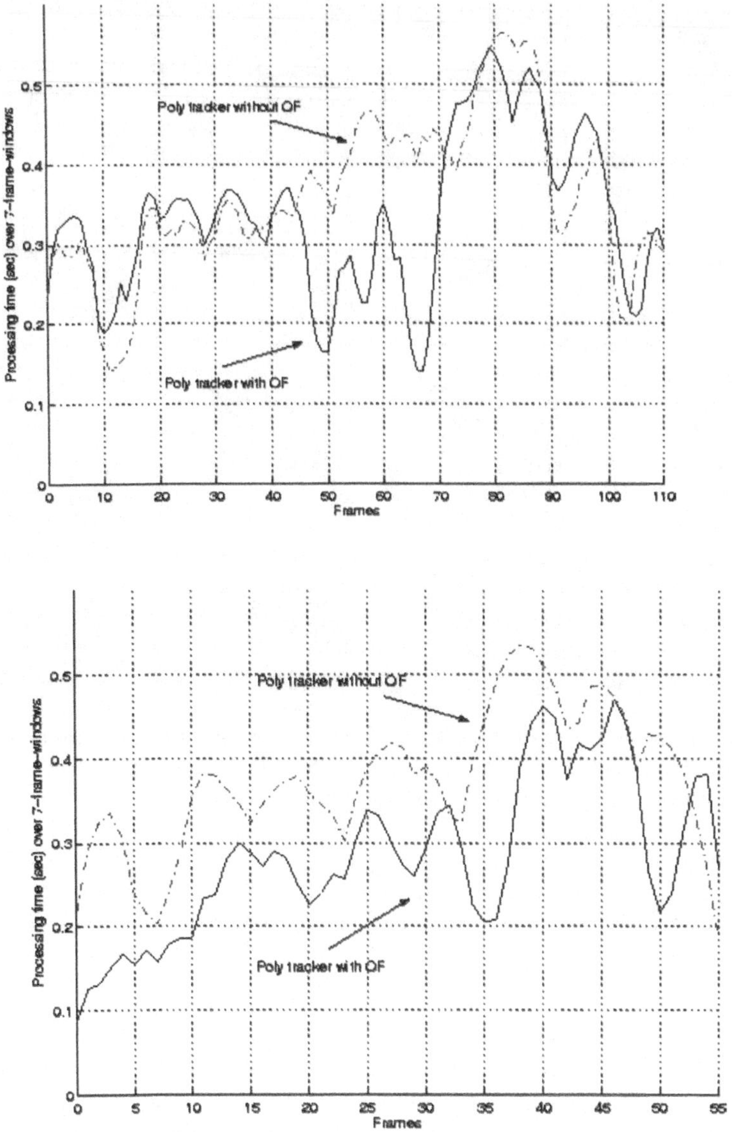

Fig. 5.18. Processing time (averaged over 7-window-frames) vs frames: for the original sequence (*left*), for the sequence subsampled by 2 in time (*right*)

Fig. 5.19. A clownfish with its textured body swims in its host anemone (Frames 1, 13, 39,59, 64, 67, 71, 74, 78, 81, 85, 95, 105, 120, 150, 155 are shown *left-right top-bottom*). Polygonal tracker successfully tracks the fish

5.5 Conclusions

In this chapter, we have presented a simple but efficient approach to object tracking combining active contours framework with the optical-flow based motion estimation. Both curve evolution and polygon evolution models are utilized to carry out the tracking. The ODE model obtained in the polygonal tracker, can act on vertices of a polygon for their intra-frame as well as inter-frame motion estimation according to region-based characteristics as well as the optical-flow field's known properties. The latter is easily estimated from a well-known image brightness constraint. We have demonstrated by way of example and discussion that our proposed tracking approach effectively and efficiently moves vertices through integrated local information with a resulting superior performance.

We note moreover that no prior shape model assumptions on targets are made, since any shape may be approximated by a polygon. While the topology-change property provided by continuous contours in the level-set framework is not attained, this limitation may be an advantage if the target region stays simply connected. We also note that there are no assumptions, such as a static camera which is widely employed in the literature by other object tracking methods utilizing also a motion detection step. A motion detection step can also be added to this framework to make the algorithm more unsupervised in detecting motion in the scene, or the presence of multiple moving targets in the scene.

References

1. C. Kim and J. N. Hwang, "Fast and automatic video object segmentation and tracking for content based applications," IEEE Trans. Circuits and Systems on Video Technology, vol. 12, no. 2, pp. 122–129, 2002.
2. N. Paragios and R. Deriche, "Geodesic active contours and level sets for the detection and tracking of moving objects," IEEE Trans. Pattern Analysis, and Machine Intelligence, vol. 22, no. 3, pp. 266–280, 2000.
3. F. G. Meyer and P. Bouthemy, "Region-based tracking using affine motion models in long image sequences," Computer Vision, Graphics, and Image Processing, vol. 60, no. 2, pp. 119–140, 1994.
4. B. Bascle and R. Deriche, "Region tracking through image sequences," in Proc. Int. Conf. on Computer Vision, 1995, pp. 302–307.
5. J. Wang and E. Adelson, "Representing moving images with layers," IEEE Trans. Image Process., vol. 3, no. 5, pp. 625–638, 1994.
6. T.J. Broida and R. Chellappa, "Estimation of object motion parameters from noisy images," IEEE Trans. Pattern Analysis, and Machine Intelligence, vol. 8, no. 1, pp. 90–99, 1986.
7. D. Koller, K. Daniilidis, and H. H. Nagel, "Model-based object tracking in monocular image sequences of road traffic scenes," Int. J. Computer Vision, vol. 10, no. 3, pp. 257–281, 1993.
8. J. Regh and T. Kanade, "Model-based tracking of self-occluding articulated objects," in Proc. IEEE Conf. on Computer Vision and Pattern Recognition, 1995, pp. 612–617.
9. D. Gavrial and L. Davis, "3-d model-based tracking of humans in action: A multi-view approach," in Proc. IEEE Conf. on Computer Vision and Pattern Recognition, 1996, pp. 73–80.

10. D. Lowe, "Robust model based motion tracking through the integration of search and estimation," Int. J. Computer Vision, vol. 8, no. 2, pp. 113–122, 1992.
11. E. Marchand, P. Bouthemy, F. Chaumette, and V. Moreau, "Robust real-time visual tracking using a 2D-3D model-based approach,," in Proc. Int. Conf. on Computer Vision, 1999, pp. 262–268.
12. M. O. Berger, "How to track efficiently piecewise curved contours with a view to reconstructing 3D objects,," in Proc. Int. Conf. on Pattern Recognition, 1994, pp. 32–36.
13. M. Isard and A. Blake, "Contour tracking by stochastic propagation of conditional density,," in Proc. European Conf. Computer Vision, 1996, pp. 343–356.
14. Y. Fu, A. T. Erdem, and A. M. Tekalp, "Tracking visible boundary of objects using occlusion adaptive motion snake," IEEE Trans. Image Process., vol. 9, no. 12, pp. 2051–2060, 2000.
15. F. Leymarie and M. Levine,"Tracking deformable objects in the plane using an active contour model," IEEE Trans. Pattern Analysis, and Machine Intelligence, vol. 15, no. 6, pp. 617–634, 1993.
16. V. Caselles and B. Coll, "Snakes in movement," SIAM Journal on Numerical Analysis, vol. 33, no. 12, pp. 2445–2456, 1996.
17. J. Badenas, J. M. Sanchiz, and F. Pla, "Motion-based segmentation and region tracking in image sequences," Pattern Recognition, vol. 34, pp. 661–670, 2001.
18. F. Marques and V. Vilaplana, "Face segmentation and tracking based on connected operators and partition projection," Pattern Recognition, vol. 35, pp. 601–614, 2002.
19. J. Badenas, J.M. Sanchiz, and F. Pla, "Using temporal integration for tracking regions in traffic monitoring sequences," in Proc. Int. Conf. on Pattern Recognition, 2000, pp. 1125–1128.
20. N. Paragios and R. Deriche, "Geodesic active regions for motion estimation and tracking," Tech. Report INRIA 1999.
21. M. Bertalmio, G. Sapiro, and G. Randall, "Morphing active contours," IEEE Trans. Pattern Analysis, and Machine Intelligence, vol. 22, no. 7, pp. 733–737, 2000.
22. A. Blake and M. Isard, Active Contours, Springer Verlag, London, Great Britain, 1998.
23. B. Li and R. Chellappa, "A generic approach to simultaneous tracking and verification in video," IEEE Trans. Image Process., vol. 11, no. 5, pp. 530–544, 2002.
24. E. C. Hildreth, "Computations underlying the measurement of visual motion," AI, vol. 23, pp. 309–354, 1984.

25. S. Ullman, "Analysis of visual motion by biological and computer systems," IEEE Computer, vol. 14, no. 8, pp. 57–69, 1981.
26. B. K. P. Horn and B. G. Schunck, "Determining optical flow," AI, vol. 17, pp. 185–203, 1981.
27. A. Kumar, A. R. Tannenbaum, and G. J. Balas, "Optical flow: A curve evolution approach," IEEE Trans. Image Process., vol. 5, no. 4, pp. 598–610, 1996.
28. B. D. Lucas and T. Kanade, "An iterative image registration technique with an application to stereo vision," Proc. Imaging Understanding Workshop, pp. 121–130, 1981.
29. H. H. Nagel and W. Enkelmann, "An investigation of smoothness constraints for the estimation of displacement vector fields from image sequences," IEEE Trans. Pattern Analysis, and Machine Intelligence, vol. 8, no. 5, pp. 565–593, 1986.
30. S. V. Fogel, "The estimation of velocity vector fields from time-varying image sequences," CVGIP: Image Understanding, vol. 53, no. 3, pp. 253–287, 1991.
31. S. S. Beauchemin and J. L. Barron, "The computation of optical flow," ACM Computing Surveys, vol. 27, no. 3, pp. 433–467, 1995.
32. D. J. Heeger, "Optical flow using spatiotemporal filters," IJCV, vol. 1, pp. 279–302, 1988.
33. D. J. Fleet and A. D. Jepson, "Computation of component image velocity from local phase information," Int. J. Computer Vision, vol. 5, no. 1, pp. 77–104, 1990.
34. A. M. Tekalp, Digital Video Processing, Prentice Hall, 1995.
35. M.I. Sezan and R.L. Lagendijk (eds.), Motion Analysis and Image Sequence Processing, Norwell, MA: Kluwer, 1993.
36. W. E. Snyder (Ed.), "Computer analysis of time varying images, special issue," IEEE Computer, vol. 14, no. 8, pp. 7–69, 1981.
37. D. Terzopoulos and R. Szeliski, Active Vision, chapter Tracking with Kalman Snakes, pp. 3–20, MIT Press, 1992.
38. N. Peterfreund, "Robust tracking of position and velocity with Kalman snakes," IEEE Trans. Pattern Analysis, and Machine Intelligence, vol. 21, no. 6, pp. 564–569, 1999.
39. D. G. Luenberger, "An introduction to observers," IEEE Transactions on Automatic Control, vol. 16, no. 6, pp. 596–602, 1971.
40. A. Gelb, Ed., Applied Optimal Estimation, MIT Press, 1974.
41. G. Unal, A. Yezzi, and H. Krim, "Information-theoretic active polygons for unsupervised texture segmentation," May-June 2005, IJCV.

42. S. Zhu and A. Yuille, "Region competition: Unifying snakes, region growing, and Bayes/MDL for multiband image segmentation," ," IEEE Trans. Pattern Analysis, and Machine Intelligence, vol. 18, no. 9, pp. 884–900, 1996.
43. B.B. Kimia, A. Tannenbaum, and S. Zucker, "Shapes, shocks, and deformations I," Int. J. Computer Vision, vol. 31, pp. 189–224, 1995.
44. S. Osher and J.A. Sethian, "Fronts propagating with curvature dependent speed: Algorithms based on the Hamilton-Jacobi formulation," J. Computational Physics, vol. 49, pp. 12–49, 1988.
45. D. Peng, B. Merriman, S. Osher, H-K. Zhao, and M. Kang, "A PDE-based fast local level set method," J. Computational Physics, vol. 255, pp. 410–438, 1999.
46. T.F. Chan and L.A. Vese, "An active contour model without edges," in Int. Conf. Scale-Space Theories in Computer Vision, 1999, pp. 141–151.
47. A. Yezzi, A. Tsai, and A. Willsky, "A fully global approach to image segmentation via coupled curve evolution equations," J. Vis. Commun. Image Representation, vol. 13, pp. 195–216, 2002.
48. M. Bertalmio, L.T Cheng, S. Osher, and G. Sapiro, "Variational problems and partial differential equations on implicit surfaces," J. Computational Physics, vol. 174, no. 2, pp. 759–780, 2001.

6 3-D Modeling of Real-World Objects Using Range and Intensity Images

Johnny Park[1], Guilherme N. DeSouza[2]

1. School of Electrical and Computer Engineering, Purdue University, West Lafayette, Indiana, U.S.A.
   jpark@purdue.edu
2. School of Electrical, Electronic & Computer Engineering, The University of Western Australia, Australia
   gdesouza@ee.uwa.edu.au

6.1 Introduction

In the last few decades, constructing accurate three-dimensional models of real-world objects has drawn much attention from many industrial and research groups. Earlier, the 3D models were used primarily in robotics and computer vision applications such as bin picking and object recognition. The models for such applications only require salient geometric features of the objects so that the objects can be recognized and the pose determined. Therefore, it is unnecessary in these applications for the models to faithfully capture every detail on the object surface. More recently, however, there has been considerable interest in the construction of 3D models for applications where the focus is more on visualization of the object by humans. This interest is fueled by the recent technological advances in range sensors, and the rapid increase of computing power that now enables a computer to represent an object surface by millions of polygons which allows such representations to be visualized interactively in real-time. Obviously, to take advantage of these technological advances, the 3D models constructed must capture to the maximum extent possible of the shape and surface-texture information of real-world objects. By real-world objects, we mean objects that may present self-occlusion with respect to the sensory devices; objects with shiny surfaces that may create mirror-like (specular) effects; objects that may absorb light and therefore not be completely perceived by the vision system; and other types of optically uncooperative objects. Construction of such photo-realistic 3D models of real-world objects is the main focus of this chapter. In general, the construction of such 3D models entails four main steps:

1. **Acquisition of geometric data**:
   First, a range sensor must be used to acquire the geometric shape of the exterior of the object. Objects of complex shape may require a large number of range images viewed from different directions so that all of

J. Park and G.N. DeSouza: *3-D Modeling of Real-World Objects Using Range and Intensity Images*, Studies in Computational Intelligence (SCI) **7**, 203–264 (2005)
www.springerlink.com

the surface detail is captured, although it is very difficult to capture the entire surface if the object contains significant protrusions.

2. **Registration**:
The second step in the construction is the registration of the multiple range images. Since each view of the object that is acquired is recorded in its own coordinate frame, we must register the multiple range images into a common coordinate system called the world frame.

3. **Integration**:
The registered range images taken from adjacent viewpoints will typically contain overlapping surfaces with common features in the areas of overlap. This third step consists of integrating the registered range images into a single connected surface model; this process first takes advantage of the overlapping portions to determine how the different range images fit together and then eliminates the redundancies in the overlap areas.

4. **Acquisition of reflection data**:
In order to provide a photo-realistic visualization, the final step acquires the reflectance properties of the object surface, and this information is added to the geometric model.

Each of these steps will be described in separate sections of this chapter.

6.2 Acquisition of Geometric Data

The first step in 3D object modeling is to acquire the geometric shape of the exterior of the object. Since acquiring geometric data of an object is a very common problem in computer vision, various techniques have been developed over the years for different applications.

6.2.1 Techniques of Acquiring 3D Data

The techniques described in this section are not intended to be exhaustive; we will mention briefly only the prominent approaches. In general, methods of acquiring 3D data can be divided into passive sensing methods and active sensing methods.

Passive Sensing Methods

The passive sensing methods extract 3D positions of object points by using images with ambient light source. Two of the well-known passive sensing methods are Shape-From-Shading (SFS) and stereo vision. The Shape-From-Shading method uses a single image of an object. The main idea of this method derives from the fact that one of the cues the human visual system uses to infer the shape of a 3D object is its shading information. Using the variation in brightness of an object, the SFS method recovers the 3D shape of an object. There are three major drawbacks of this method: First, the shadow areas of an object cannot be recovered reliably since they do not provide enough intensity information. Second, the method assumes that the entire surface of an object has uniform reflectance property, thus the method cannot be applied to general objects. Third, the method is very sensitive to noise since the computation of surface gradients is involved.

The stereo vision method uses two or more images of an object from different viewpoints. Given the image coordinates of the same object point in two or more images, the stereo vision method extracts the 3D coordinate of that object point. A fundamental limitation of this method is the fact that finding the correspondence between images is extremely difficult.

The passive sensing methods require very simple hardware, but usually these methods do not generate dense and accurate 3D data compare to the active sensing methods.

Active Sensing Methods

The active sensing methods can be divided into two categories: contact and non-contact methods. Coordinate Measuring Machine (CMM) is a prime example of the contact methods. CMMs consist of probe sensors which provide 3D measurements by touching the surface of an object. Although CMMs generate very accurate and fine measurements, they are very expensive and slow. Also, the types of objects that can be used by CMMs are limited since physical contact is required.

The non-contact methods project their own energy source to an object, then observe either the transmitted or the reflected energy. The computed tomography (CT), also known as the computed axial tomography (CAT), is one of the techniques that records the transmitted energy. It uses X-ray beams at various angles to create cross-sectional images of an object. Since the computed tomography provides the internal structure of an object, the method is widely used in medical applications.

The active stereo uses the same idea of the passive sensing stereo method, but a light pattern is projected onto an object to solve the difficulty of finding corresponding points between two (or more) camera images.

The laser radar system, also known as LADAR, LIDAR, or optical radar, uses the information of emitted and received laser beam to compute the depth. There are mainly two methods that are widely used: (1) using amplitude modulated continuous wave (AM-CW) laser, and (2) using laser pulses.

The first method emits AM-CW laser onto a scene, and receives the laser that was reflected by a point in the scene. The system computes the phase difference between the emitted and the received laser beam. Then, the depth of the point can be computed since the phase difference is directly proportional to depth. The second method emits a laser pulse, and computes the interval between the emitted and the received time of the pulse. The time interval, well known as *time-of-flight*, is then used to compute the depth given by $t = 2z/c$ where t is time-of-flight, z is depth, and c is speed of light. The laser radar systems are well suited for applications requiring medium-range sensing from 10 to 200 meters.

The structured-light methods project a light pattern onto a scene, then use a camera to observe how the pattern is illuminated on the object surface. Broadly speaking, the structured-light methods can be divided into scanning and non-scanning methods. The scanning methods consist of a moving stage and a laser plane, so either the laser plane scans the object or the object moves through the laser plane. A sequence of images is taken while scanning. Then, by detecting illuminated points in the images, 3D positions of corresponding object points are computed by the equations of camera calibration. The non-scanning methods project a spatially or temporally varying light pattern onto an object. An appropriate decoding of the reflected pattern is then used to compute the 3D coordinates of an object.

The system that acquired all the 3D data presented in this chapter falls into a category of a scanning structured-light method using a single laser plane. From now on, such a system will be referred to as a structured-light scanner.

6.2.2 Structured-Light Scanner

Structured-light scanners have been used in manifold applications since the technique was introduced about two decades ago. They are especially suitable for applications in 3D object modeling for two main reasons: First, they acquire dense and accurate 3D data compared to passive sensing methods. Second, they require relatively simple hardware compared to laser radar systems.

In what follows, we will describe the basic concept of structured-light scanner and all the data that can be typically acquired and derived from this kind of sensor.

A Typical System

A sketch of a typical structured-light scanner is shown in Figure 6.1. The system consists of four main parts: linear stage, rotary stage, laser projector, and camera. The linear stage moves along the X axis and the rotary stage mounted on top of the linear stage rotates about the Z axis where XYZ are

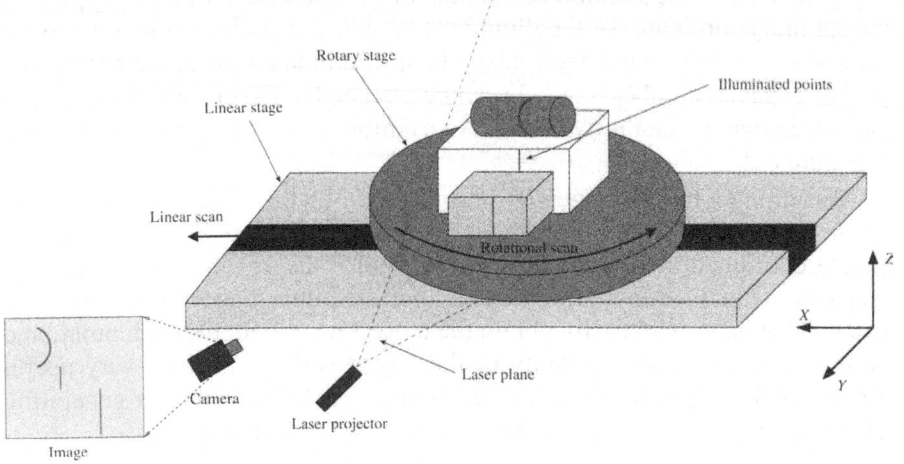

Fig. 6.1: A typical structured-light scanner

the three principle axes of the reference coordinate system. A laser plane parallel to the YZ plane is projected onto the objects. The intersection of the laser plane and the objects creates a stripe of illuminated points on the surface of the objects. The camera captures the scene, and the illuminated points in that image are extracted. Given the image coordinates of the extracted illuminated points and the positions of the linear and rotary stages, the corresponding 3D coordinates with respect to the reference coordinate system can be computed by the equations of camera calibration; we will describe the process of camera calibration shortly. Such process only acquires a set of 3D coordinates of the points that are illuminated by the laser plane. In order to capture the entire scene, the system either translates or rotates the objects through the laser plane while the camera takes the sequence of images. Note that it is possible to have the objects stationary, and move the sensors (laser projector and camera) to sweep the entire scene.

Acquiring Data: Range Image

The sequence of images taken by the camera during a scan can be stored in a more compact data structure called *range image*, also known as *range map*, *range data*, *depth map*, or *depth image*. A range image is a set of distance measurements arranged in a $m \times n$ grid. Typically, for the case of structured-light scanner, m is the number of horizontal scan lines (rows) of camera image, and n is the total number of images (i.e., number of stripes) in the sequence. We can also represent a range image in a parametric form $r(i, j)$ where r is the column coordinate of the illuminated point at the ith row in the jth image. Sometimes, the computed 3D coordinate (x, y, z) is stored instead of the column coordinate of the illuminated point. Typically, the column coordinates of the illuminated points are computed in a sub-pixel accuracy as will be described next. If an illuminated point cannot be detected, a special number (e.g., -1) can be assigned to the corresponding entry indicating that no data is available. An example of a range image is depicted in Figure 6.2.

Assuming a range image $r(i, j)$ is acquired by the system shown in Figure 6.1, i is related mainly to the coordinates along the Z axis of the reference coordinate system, j the X axis, and r the Y axis. Since a range image is maintained in a grid, the neighborhood information is directly provided. That is, we can easily obtain the closest neighbors for each point, and even detect spatial discontinuity of the object surface. This is very useful especially for computing normal directions of each data point, or generating triangular mesh; the discussion of these topics will follow shortly.

Computing Center of Illuminated Points

In order to create the range images as described above, we must collect one (the center) of the illuminated points in each row as the representative of that row. Assuming the calibrations of both the camera and the positioning stages are perfect, the accuracy of computing 3D coordinates of object points primarily depends on locating the true center of these illuminated points. A typical intensity distribution around the illuminated points is shown in Figure 6.3.

Ideally only the light source (e.g., laser plane) should cause the illumination, and the intensity curve around the illuminated points should be Gaussian. However, we need to be aware that the illumination may be affected by many different factors such as: CCD camera error (e.g., noise and quantization error); laser speckle; blurring effect of laser; mutual-reflections of object surface; varying reflectance properties of object surface; high curvature on object surface; partial occlusions with respect to camera or laser plane; etc. Although eliminating all these sources of error is unlikely, it is important to use an algorithm that will best estimate the true center of the illuminated points.

Here we introduce three algorithms: (1) center of mass, (2) Blais and Rioux algorithm, and (3) Gaussian approximation. Let $I(i)$ be the intensity value at i coordinate, and let p be the coordinate with peak intensity. Then, each algorithm computes the center c as follows:

1. **Center of mass**: This algorithm solves the location of the center by computing weighted average. The size of kernel n should be set such

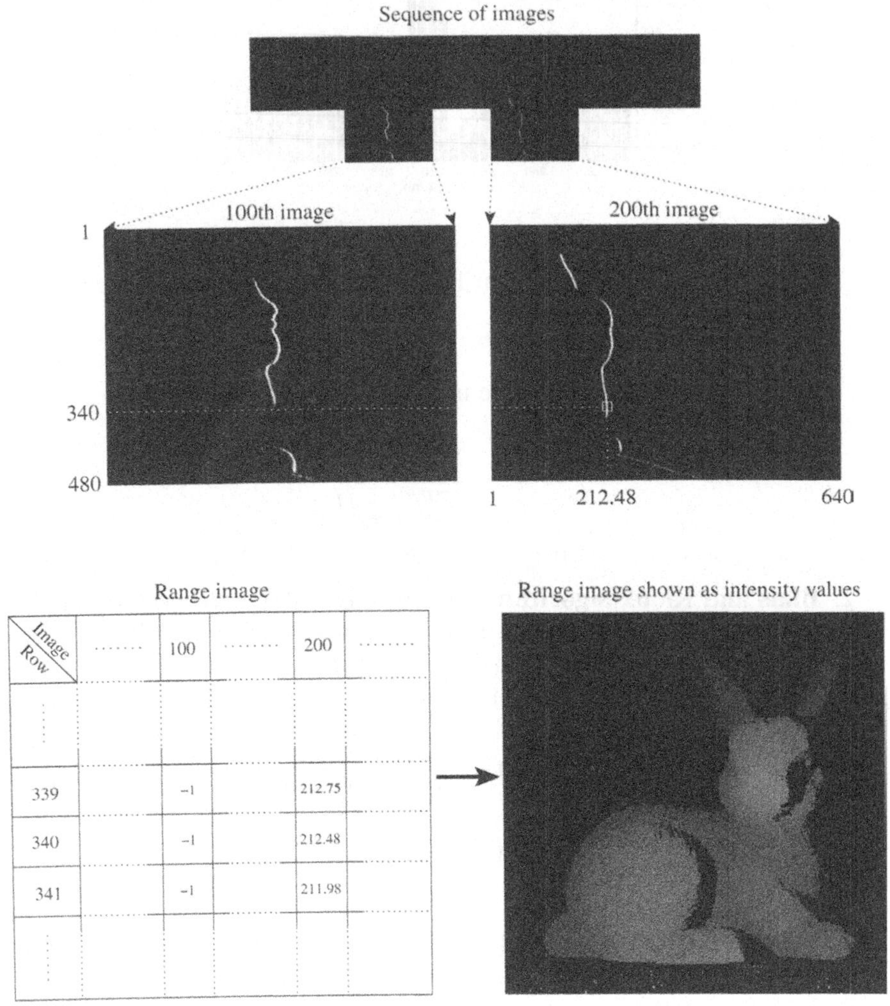

Fig. 6.2: Converting a sequence of images into a range image

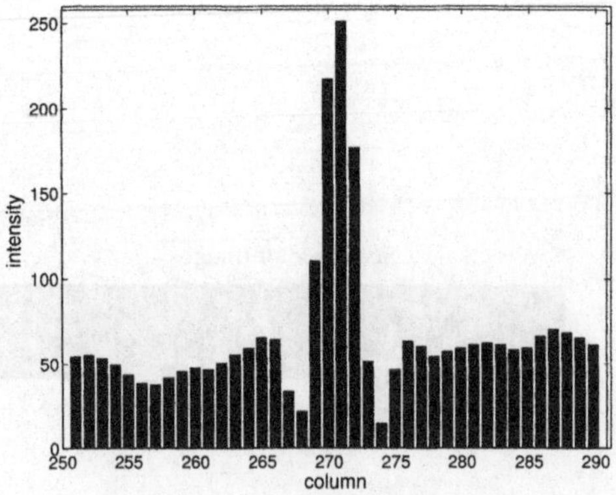

Fig. 6.3: Typical intensity distribution around illuminated points

that all illuminated points are included.

$$c = \frac{\sum_{i=p-n}^{p+n} iI(i)}{\sum_{i=p-n}^{p+n} I(i)}$$

2. **Blais and Rioux algorithm** [9]: This algorithm uses a finite impulse response filter to differentiate the signal and to eliminate the high frequency noise. The zero crossing of the derivative is linearly interpolated to solve the location of the center.

$$c = p + \frac{h(p)}{h(p) - h(p+1)}$$

where $h(i) = I(i-2) + I(i-1) - I(i+1) - I(i+2)$.

3. **Gaussian approximation** [55]: This algorithm fits a Gaussian profile to three contiguous intensities around the peak.

$$c = p - \frac{1}{2}\frac{\ln(I(p+1)) - \ln(I(p-1))}{\ln(I(p-1)) - 2\ln(I(p)) + \ln(I(p+1))}$$

After testing all three methods, one would notice that the center of mass method produces the most reliable results for different objects with varying reflection properties. Thus, all experimental results shown in this chapter were obtained using the center of mass method.

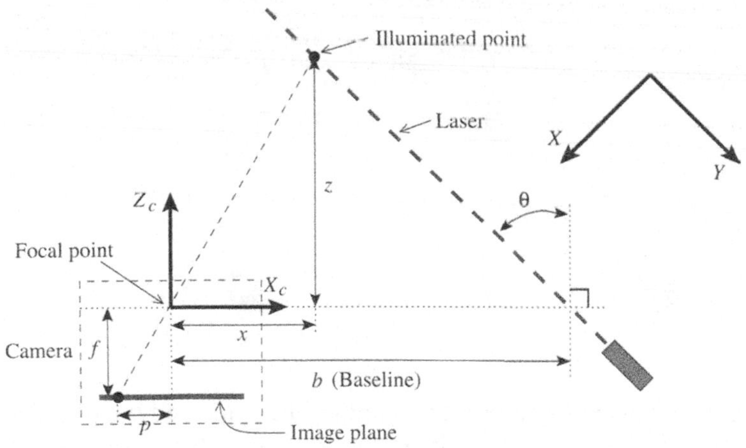

Fig. 6.4: Optical triangulation

Optical Triangulation

Once the range image is complete, we must now calculate the 3D structure of the scanned object. The measurement of the depth of an object using a structured-light scanner is based on optical triangulation. The basic principles of optical triangulation are depicted in Figure 6.4. X_c and Z_c are two of the three principle axes of the camera coordinate system, f is the focal length of the camera, p is the image coordinate of the illuminated point, and b (baseline) is the distance between the focal point and the laser along the X_c axis . Notice that the figure corresponds to the top view of the structured-light scanner in Figure 6.1.

Using the notations in Figure 6.4, the following equation can be obtained by the properties of similar triangles:

$$\frac{z}{f} = \frac{b}{p + f \tan\theta} \tag{1}$$

Then, the z coordinate of the illuminated point with respect to the camera coordinate system is directly given by

$$z = \frac{fb}{p + f \tan\theta} \tag{2}$$

Given the z coordinate, the x coordinate can be computed as

$$x = b - z \tan\theta \tag{3}$$

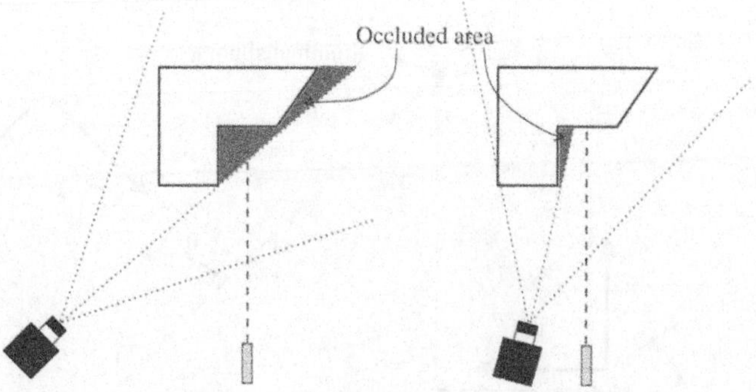

Fig. 6.5: Tradeoff between the length of baseline and the occlusion.
As the length of baseline increases, a better accuracy in the measurement can be achieved, but the occluded area due to shadow effect becomes larger, and vice versa

The error of z measurement can be obtained by differentiating Eq. (2):

$$\Delta z = \frac{fb}{(p + f\tan\theta)^2}\Delta p + \frac{fb\,(f\sec^2\theta)}{(p + f\tan\theta)^2}\Delta\theta \tag{4}$$

where Δp and $\Delta\theta$ are the measurement errors of p and θ respectively. Substituting the square of Eq. (2), we now have

$$\Delta z = \frac{z^2}{fb}\Delta p + \frac{z^2\sec^2\theta}{b}\Delta\theta \tag{5}$$

This equation indicates that the error of the z measurement is directly proportional to the square of z, but inversely proportional to the focal length f and the baseline b. Therefore, increasing the baseline implies a better accuracy in the measurement. Unfortunately, the length of baseline is limited by the hardware structure of the system, and there is a tradeoff between the length of baseline and the sensor occlusions – as the length of baseline increases, a better accuracy in the measurement can be achieved, but the occluded area due to shadow effect becomes larger, and vice versa. A pictorial illustration of this tradeoff is shown in Figure 6.5.

Computing 3D World Coordinates

The coordinates of illuminated points computed by the equations of optical triangulation are with respect to the camera coordinate system. Thus, an additional transformation matrix containing the extrinsic parameters of the camera (i.e., a rotation matrix and a translation vector) that transforms the camera coordinate system to the reference coordinate system needs to be

(a)

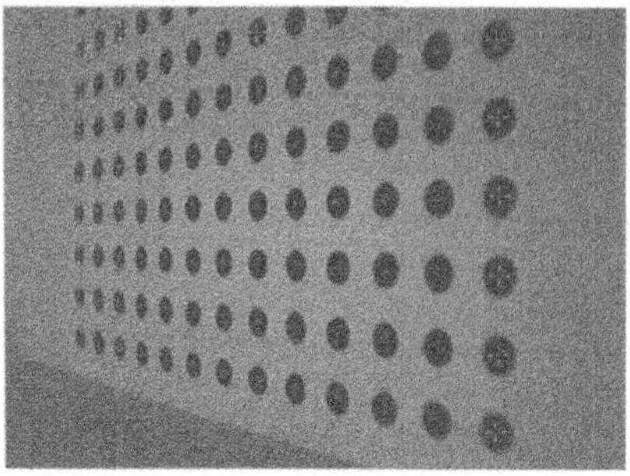

(b)

Fig. 6.6: Calibration pattern
(a): A calibration pattern is placed in such a way that the pattern surface is parallel to the laser plane (i.e., YZ plane), and the middle column of the pattern (i.e., 7th column) coincides the Z axis of the reference coordinate system. (b): Image taken from the camera. Crosses indicate extracted centers of circle patterns

found. However, one can formulate a single transformation matrix that contains the optical triangulation parameters and the camera calibration parameters all together. In fact, the main reason we derived the optical triangulation equations is to show that the uncertainty of depth measurement is related to the square of the depth, focal length of the camera, and the baseline.

The transformation matrix for computing 3D coordinates with respect to the reference coordinate system can be obtained as follows. Suppose we have n data points with known reference coordinates and the corresponding image coordinates. Such points can be obtained by using a calibration pattern placed in a known location, for example, the pattern surface is parallel to the laser plane and the middle column of the pattern coincides the Z axis (See Figure 6.6).

Let the reference coordinate of the ith data point be denoted by (x_i, y_i, z_i), and the corresponding image coordinate be denoted by (u_i, v_i). We want to solve a matrix $\mathbf{T}$ that transforms the image coordinates to the reference coordinates. It is well known that the homogeneous coordinate system must be used for linearization of 2D to 3D transformation. Thus, we can formulate the transformation as

$$\mathbf{T}_{4\times 3} \begin{bmatrix} u_i \\ v_i \\ 1 \end{bmatrix} = \begin{bmatrix} \overline{x_i} \\ \overline{y_i} \\ \overline{z_i} \\ \rho \end{bmatrix} \quad (6)$$

or

$$\begin{bmatrix} t_{11} & t_{12} & t_{13} \\ t_{21} & t_{22} & t_{23} \\ t_{31} & t_{32} & t_{33} \\ t_{41} & t_{42} & t_{43} \end{bmatrix} \begin{bmatrix} u_i \\ v_i \\ 1 \end{bmatrix} = \begin{bmatrix} \overline{x_i} \\ \overline{y_i} \\ \overline{z_i} \\ \rho \end{bmatrix} \tag{7}$$

where

$$\begin{bmatrix} x_i \\ y_i \\ z_i \end{bmatrix} = \begin{bmatrix} \overline{x_i} \,/\, \rho \\ \overline{y_i} \,/\, \rho \\ \overline{z_i} \,/\, \rho \end{bmatrix} \tag{8}$$

We use the free variable ρ to account for the non-uniqueness of the homogeneous coordinate expressions (i.e., scale factor). Carrying our the first row and the fourth row of Eq. (7), we have

$$\begin{aligned} \overline{x_1} &= t_{11}u_1 + t_{12}v_1 + t_{13} \\ \overline{x_2} &= t_{11}u_2 + t_{12}v_2 + t_{13} \\ &\vdots \\ \overline{x_n} &= t_{11}u_n + t_{12}v_n + t_{13} \end{aligned} \tag{9}$$

and

$$\begin{aligned} \rho &= t_{41}u_1 + t_{42}v_1 + t_{43} \\ \rho &= t_{41}u_2 + t_{42}v_2 + t_{43} \\ &\vdots \\ \rho &= t_{41}u_n + t_{42}v_n + t_{43} \end{aligned} \tag{10}$$

By combining these two sets of equations, and by setting $\overline{x_i} - \rho x_i = 0$, we obtain

$$\begin{aligned} t_{11}u_1 + t_{12}v_1 + t_{13} - t_{41}u_1x_1 - t_{42}v_1x_1 - t_{43}x_1 &= 0 \\ t_{11}u_2 + t_{12}v_2 + t_{13} - t_{41}u_2x_2 - t_{42}v_2x_2 - t_{43}x_2 &= 0 \\ &\vdots \\ t_{11}u_n + t_{12}v_n + t_{13} - t_{41}u_nx_n - t_{42}v_nx_n - t_{43}x_n &= 0 \end{aligned} \tag{11}$$

Since we have a free variable ρ, we can set $t_{43} = 1$ which will appropriately scale the rest of the variables in the matrix **M**. Carrying out the same procedure that produced Eq. (11) for y_i and z_i, and rearranging all the equations into a matrix form, we obtain

$$\begin{bmatrix} u_1 & v_1 & 1 & 0 & 0 & 0 & 0 & 0 & 0 & -u_1x_1 & -v_1x_1 \\ u_2 & v_2 & 1 & 0 & 0 & 0 & 0 & 0 & 0 & -u_2x_2 & -v_2x_2 \\ \vdots & \vdots & \vdots & \vdots & \vdots & \vdots & \vdots & \vdots & \vdots & \vdots & \vdots \\ u_n & v_n & 1 & 0 & 0 & 0 & 0 & 0 & 0 & -u_nx_n & -v_nx_n \\ 0 & 0 & 0 & u_1 & v_1 & 1 & 0 & 0 & 0 & -u_1y_1 & -v_1y_1 \\ 0 & 0 & 0 & u_2 & v_2 & 1 & 0 & 0 & 0 & -u_2y_2 & -v_2y_2 \\ \vdots & \vdots & \vdots & \vdots & \vdots & \vdots & \vdots & \vdots & \vdots & \vdots & \vdots \\ 0 & 0 & 0 & u_n & v_n & 1 & 0 & 0 & 0 & -u_ny_n & -v_ny_n \\ 0 & 0 & 0 & 0 & 0 & 0 & u_1 & v_1 & 1 & -u_1z_1 & -v_1z_1 \\ 0 & 0 & 0 & 0 & 0 & 0 & u_2 & v_2 & 1 & -u_2z_2 & -v_2z_2 \\ \vdots & \vdots & \vdots & \vdots & \vdots & \vdots & \vdots & \vdots & \vdots & \vdots & \vdots \\ 0 & 0 & 0 & 0 & 0 & 0 & u_n & v_n & 1 & -u_nz_n & -v_nz_n \end{bmatrix} \begin{bmatrix} t_{11} \\ t_{12} \\ t_{13} \\ t_{21} \\ t_{22} \\ t_{23} \\ t_{31} \\ t_{32} \\ t_{33} \\ t_{41} \\ t_{42} \end{bmatrix} = \begin{bmatrix} x_1 \\ x_2 \\ \vdots \\ x_n \\ y_1 \\ y_2 \\ \vdots \\ y_n \\ z_1 \\ z_2 \\ \vdots \\ z_n \end{bmatrix} \tag{12}$$

If we rewrite Eq. (12) as $\mathbf{Ax} = \mathbf{b}$, then our problem is to solve for $\mathbf{x}$. We can form the normal equations and find the linear least squares solution by solving $(\mathbf{A}^T\mathbf{A})\mathbf{x} = \mathbf{A}^T\mathbf{b}$. The resulting solution $\mathbf{x}$ forms the transformation matrix $\mathbf{T}$. Note that Eq. (12) contains $3n$ equations and 11 unknowns, therefore the minimum number of data points needed to solve this equation is 4.

Given the matrix $\mathbf{T}$, we can now compute 3D coordinates for each entry of a range image. Let $\mathbf{p}(i,j)$ represent the 3D coordinates (x,y,z) of a range image entry $r(i,j)$ with respect to the reference coordinate system; recall that $r(i,j)$ is the column coordinate of the illuminated point at the ith row in the jth image. Using Eq. (6), we have

$$\begin{bmatrix} \overline{x} \\ \overline{y} \\ \overline{z} \\ \rho \end{bmatrix} = \mathbf{T} \begin{bmatrix} i \\ r(i,j) \\ 1 \end{bmatrix} \tag{13}$$

and the corresponding 3D coordinate is computed by

$$\mathbf{p}(i,j) = \begin{bmatrix} \overline{x} \,/\, \rho \\ \overline{y} \,/\, \rho \\ \overline{z} \,/\, \rho \end{bmatrix} + \begin{bmatrix} x_0 + (j-1)\Delta x \\ 0 \\ 0 \end{bmatrix} \tag{14}$$

where x_0 is the x coordinate of the laser plane at the beginning of the scan, and Δx is the distance that the linear slide moved along the X axis between two consecutive images.

The transformation matrix $\mathbf{T}$ computed by Eq. (12) is based on the assumption that the camera image plane is perfectly planar, and that all the data points are linearly projected onto the image plane through an infinitely small focal point. This assumption, often called as *pin-hole camera model*, generally works well when using cameras with normal lenses and small calibration error is acceptable. However, when using cameras with wide-angle lenses or large aperture, and a very accurate calibration is required, this assumption may not be appropriate. In order to improve the accuracy of camera calibration, two types of camera lens distortions are commonly accounted for: radial distortion and decentering distortion. Radial distortion is due to flawed radial curvature curve of the lens elements, and it causes inward or outward perturbations of image points. Decentering distortion is caused by non-collinearity of the optical centers of lens elements. The effect of the radial distortion is generally much more severe than that of the decentering distortion.

In order to account for the lens distortions, a simple transformation matrix can no longer be used; we need to find both the intrinsic and extrinsic parameters of the camera as well as the distortion parameters. A widely accepted calibration method is Tsai's method, and we refer the readers to [56, 34] for the description of the method.

Computing Normal Vectors

Surface normal vectors are important to the determination of the shape of an object, therefore it is necessary to estimate them reliably. Given the 3D coordinate $\mathbf{p}(i, j)$ of the range image entry $r(i, j)$, its normal vector $\mathbf{n}(i, j)$ can be computed by

$$\mathbf{n}(i, j) = \frac{\frac{\partial \mathbf{p}}{\partial i} \times \frac{\partial \mathbf{p}}{\partial j}}{\left\| \frac{\partial \mathbf{p}}{\partial i} \times \frac{\partial \mathbf{p}}{\partial j} \right\|} \tag{15}$$

where $\times$ is a cross product. The partial derivatives can be computed by finite difference operators. This approach, however, is very sensitive to noise due to the differentiation operations. Some researchers have tried to overcome the noise problem by smoothing the data, but it causes distortions to the data especially near sharp edges or high curvature regions.

An alternative approach computes the normal direction of the plane that best fits some neighbors of the point in question. In general, a small window (e.g., 3×3, or 5×5) centered at the point is used to obtain the neighboring points, and the PCA (Principal Component Analysis) for computing the normal of the best fitting plane.

Suppose we want to compute the normal vector $\mathbf{n}(i, j)$ of the point $\mathbf{p}(i, j)$ using a $n \times n$ window. The center of mass $\mathbf{m}$ of the neighboring points is computed by

$$\mathbf{m} = \frac{1}{n^2} \sum_{r=i-a}^{i+a} \sum_{c=j-a}^{j+a} \mathbf{p}(r,c) \tag{16}$$

where $a = \lfloor n/2 \rfloor$. Then, the covariance matrix $\mathbf{C}$ is computed by

$$\mathbf{C} = \sum_{r=i-a}^{i+a} \sum_{c=j-a}^{j+a} [\mathbf{p}(r,c) - \mathbf{m}] [\mathbf{p}(r,c) - \mathbf{m}]^T \tag{17}$$

The surface normal is estimated as the eigenvector with the smallest eigenvalue of the matrix $\mathbf{C}$.

Although using a fixed sized window provides a simple way of finding neighboring points, it may also cause the estimation of normal vectors to become unreliable. This is the case when the surface within the fixed window contains noise, a crease edge, a jump edge, or simply missing data. Also, when the vertical and horizontal sampling resolutions of the range image are significantly different, the estimated normal vectors will be less robust with respect to the direction along which the sampling resolution is lower. Therefore, a region growing approach can be used for finding the neighboring points. That is, for each point of interest, a continuous region is defined such that the distance between the point of interest to each point in the region is less than a given threshold. Taking the points in the region as neighboring points reduces the difficulties mentioned above, but obviously requires more computations. The threshold for the region growing can be set, for example, as $2(v+h)$ where v and h are the vertical and horizontal sampling resolutions respectively.

Generating Triangular Mesh from Range Image

Generating triangular mesh from a range image is quite simple since a range image is maintained in a regular grid. Each sample point (entry) of a $m \times n$ range image is a potential vertex of a triangle. Four neighboring sample points are considered at a time, and two diagonal distances d_{14} and d_{23} as in Figure 6.7(a) are computed. If both distances are greater than a threshold, then no triangles are generated, and the next four points are considered. If one of the two distances is less than the threshold, say d_{14}, we have potentially two triangles connecting the points 1-3-4 and 1-2-4. A triangle is created when the distances of all three edges are below the threshold. Therefore, either zero, one, or two triangles are created with four neighboring points. When both diagonal distances are less than the threshold, the diagonal edge with the smaller distance is chosen. Figure 6.7(b) shows an example of the triangular mesh using this method.

The distance threshold is, in general, set to a small multiple of the sam-

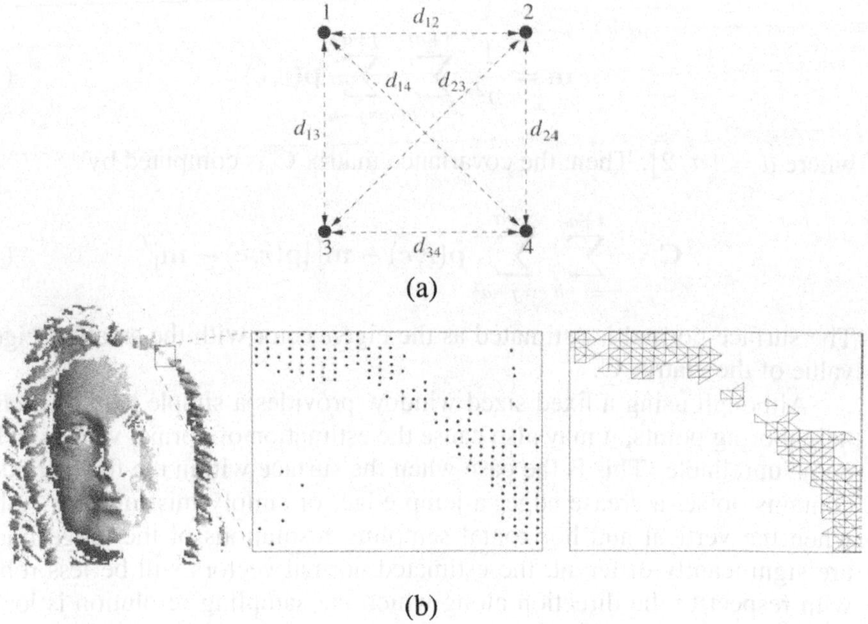

Fig. 6.7: Triangulation of range image

pling resolution. As illustrated in Figure 6.8, triangulation errors are likely to occur on object surfaces with high curvature, or on surfaces where the normal direction is close to the perpendicular to the viewing direction from the sensor. In practice, the threshold must be small enough to reject false edges even if it means that some of the edges that represent true surfaces can also be rejected. That is because we can always acquire another range image from a different viewing direction that can sample those missing surfaces more densely and accurately; however, it is not easy to remove false edges once they are created.

Experimental Result

To illustrate all the steps described above, we present the result images obtained in our lab. Figure 6.9 shows a photograph of our structured-light scanner. The camera is a Sony XC-7500 with pixel resolution of 659 by 494. The laser has $685nm$ wavelength with $50mW$ diode power. The rotary stage is Aerotech ART310, the linear stage is Aerotech ATS0260 with $1.25\mu m$ resolution and $1.0\mu m/25mm$ accuracy, and these stages are controlled by Aerotech Unidex 511.

Figure 6.10 shows the geometric data from a single linear scan acquired

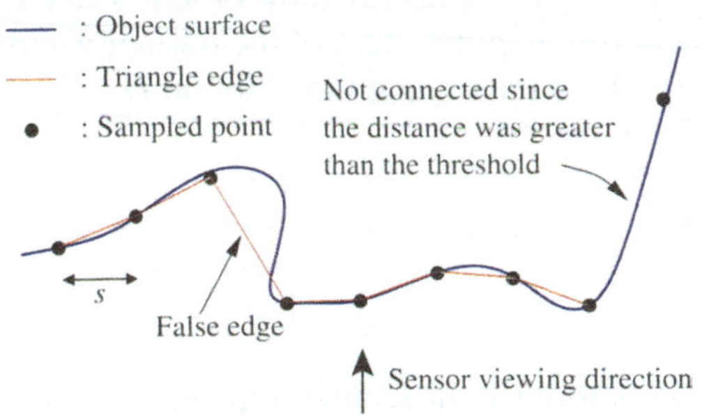

Fig. 6.8: Problems with triangulation

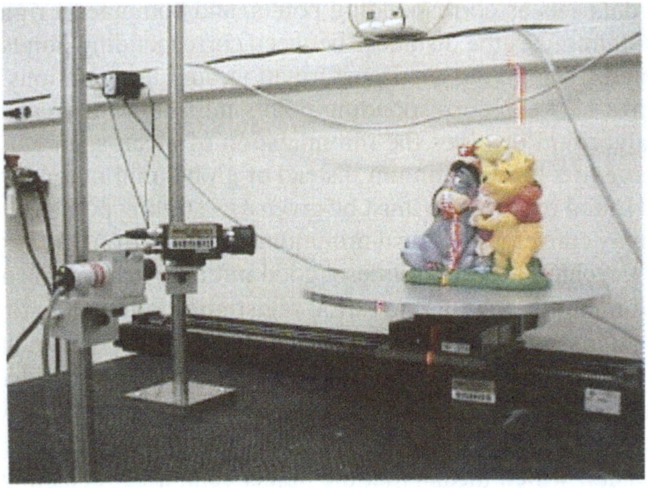

Fig. 6.9: Photograph of our structured-light scanner

by our structured-light scanner. Figure 6.10(a) shows the photograph of the object that was scanned, (b) shows the range image displayed as intensity values, (c) shows the computed 3D coordinates as point cloud, (d) shows the shaded triangular mesh, and finally (e) shows the normal vectors displayed as RGB colors where the X component of the normal vector corresponds to the R component, the Y to the G, and the Z to the B.

6.3 Registration

6.3.1 Overview

A single scan by a structured-light scanner typically provides a range image that covers only part of an object. Therefore, multiple scans from different viewpoints are necessary to capture the entire surface of the object. These multiple range images create a well-known problem called *registration* – aligning all the range images into a common coordinate system. Automatic registration is very difficult since we do not have any prior information about the overall object shape except what is given in each range image, and since finding the correspondence between two range images taken from arbitrary viewpoints is non-trivial.

The Iterative Closest Point (ICP) algorithm [8, 13, 62] made a significant contribution on solving the registration problem. It is an iterative algorithm for registering two data sets. In each iteration, it selects the closest points between two data sets as corresponding points, and computes a rigid transformation that minimizes the distances between corresponding points. The data set is updated by applying the transformation, and the iterations continued until the error between corresponding points falls below a preset threshold. Since the algorithm involves the minimization of mean-square distances, it may converge to a local minimum instead of global minimum. This implies that a good initial registration must be given as a starting point, otherwise the algorithm may converge to a local minimum that is far from the best solution. Therefore, a technique that provides a good initial registration is necessary.

One example for solving the initial registration problem is to attach the scanning system to a robotic arm and keep track of the position and the orientation of the scanning system. Then, the transformation matrices corresponding to the different viewpoints are directly provided. However, such a system requires additional expensive hardware. Also, it requires the object to be stationary, which means that the object cannot be repositioned for the purpose of acquiring data from new viewpoints. Another alternative for solving the initial registration is to design a graphical user interface that allows a human to interact with the data, and perform the registration manually.

Since the ICP algorithm registers two sets of data, another issue that should be considered is registering a set of multiple range data that mini-

(a) Photograph

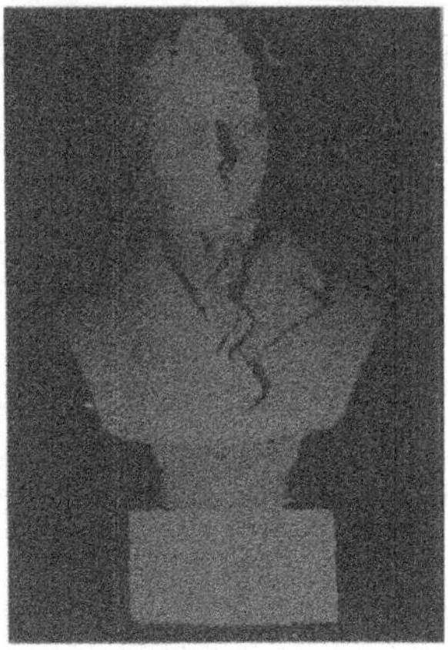

(b) Range image

(c) Point cloud

(d) Triangular mesh

(e) Normal vectors

Fig. 6.10: Geometric data acquired by our structured-light scanner
(a): The photograph of the figure that was scanned. (b): The range image displayed as intensity values. (c): The computed 3D coordinates as point cloud. (d): The shaded triangular mesh. (e): The normal vectors displayed as RGB colors where the X component of the normal vector corresponds to the R component, the Y to the G, and the Z to the B

mizes the registration error between all pairs. This problem is often referred to as *multi-view registration*, and we will discuss in more detail in Section 6.3.5.

6.3.2 Iterative Closest Point (ICP) Algorithm

The ICP algorithm was first introduced by Besl and McKay [8], and it has become the principle technique for registration of 3D data sets. The algorithm takes two 3D data sets as input. Let $\mathbf{P}$ and $\mathbf{Q}$ be two input data sets containing N_p and N_q points respectively. That is, $\mathbf{P} = \{\mathbf{p}_i\}$, $i = 1, ..., N_p$, and $\mathbf{Q} = \{\mathbf{q}_i\}$, $i = 1, ..., N_q$. The goal is to compute a rotation matrix $\mathbf{R}$ and a translation vector $\mathbf{t}$ such that the transformed set $\mathbf{P}' = \mathbf{R}\mathbf{P} + \mathbf{t}$ is best aligned with $\mathbf{Q}$. The following is a summary of the algorithm (See Figure 6.11 for a pictorial illustration of the ICP).

1. **Initialization**: $k = 0$ and $\mathbf{P}_k = \mathbf{P}$.

2. **Compute the closest point**: For each point in $\mathbf{P}_k$, compute its closest point in $\mathbf{Q}$. Consequently, it produces a set of closest points $\mathbf{C} = \{\mathbf{c}_i\}$, $i = 1, ..., N_p$ where $\mathbf{C} \subset \mathbf{Q}$, and $\mathbf{c}_i$ is the closest point to $\mathbf{p}_i$.

3. **Compute the registration**: Given the set of closest points $\mathbf{C}$, the mean square objective function to be minimized is:

$$f(\mathbf{R}, \mathbf{t}) = \frac{1}{N_p} \sum_{i=1}^{N_p} \|\mathbf{c}_i - \mathbf{R}\mathbf{p}_i - \mathbf{t}\|^2 \tag{18}$$

   Note that $\mathbf{p}_i$ is a point from the original set $\mathbf{P}$, not $\mathbf{P}_k$. Therefore, the computed registration applies to the original data set $\mathbf{P}$ whereas the closest points are computed using $\mathbf{P}_k$.

4. **Apply the registration**: $\mathbf{P}_{k+1} = \mathbf{R}\mathbf{P} + \mathbf{t}$.

5. **If the desired precision of the registration is met**: Terminate the iteration.
   Else: $k = k + 1$ and repeat steps 2-5.

Note that the 3D data sets $\mathbf{P}$ and $\mathbf{Q}$ do not necessarily need to be points. It can be a set of lines, triangles, or surfaces as long as closest entities can be computed and the transformation can be applied. It is also important to note that the algorithm assumes all the data in $\mathbf{P}$ lies inside the boundary of $\mathbf{Q}$. We will later discuss about relaxing this assumption.

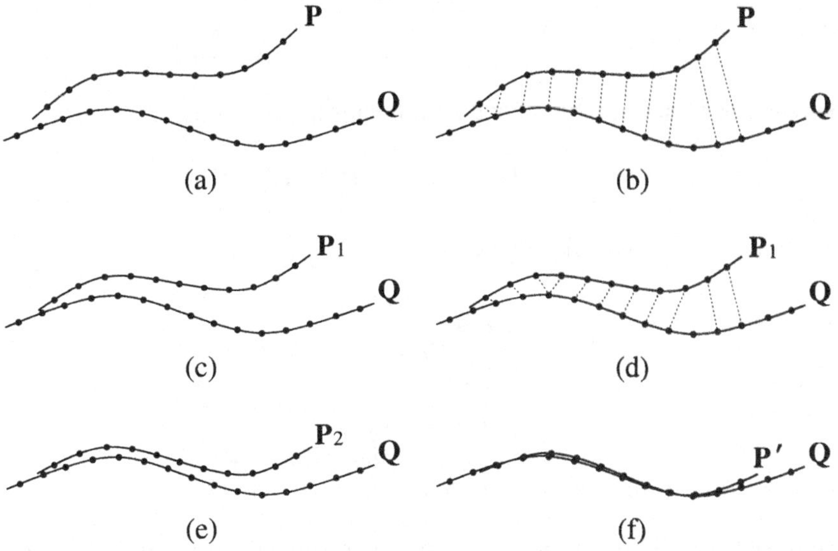

Fig. 6.11: Illustration of the ICP algorithm
(a): Initial $\mathbf{P}$ and $\mathbf{Q}$ to register. (b): For each point in $\mathbf{P}$, find a corresponding point, which is the closest point in $\mathbf{Q}$. (c): Apply $\mathbf{R}$ and $\mathbf{t}$ from Eq. (18) to $\mathbf{P}$. (d): Find a new corresponding point for each $\mathbf{P}_1$. (e): Apply new $\mathbf{R}$ and $\mathbf{t}$ that were computed using the new corresponding points. (f): Iterate the process until converges to a local minimum

Given the set of closest points **C**, the ICP computes the rotation matrix **R** and the translation vector **t** that minimizes the mean square objective function of Eq. (18). Among other techniques, Besl and McKay in their paper chose the solution of Horn [25] using unit quaternions. In that solution, the mean of the closet point set **C** and the mean of the set **P** are respectively given by

$$\mathbf{m}_c = \frac{1}{N_p}\sum_{i=1}^{N_p} \mathbf{c}_i \quad , \quad \mathbf{m}_p = \frac{1}{N_p}\sum_{i=1}^{N_p} \mathbf{p}_i.$$

The new coordinates, which have zero means are given by

$$\mathbf{c}_i' = \mathbf{c}_i - \mathbf{m}_c \quad , \quad \mathbf{p}_i' = \mathbf{p}_i - \mathbf{m}_p.$$

Let a 3×3 matrix **M** be given by

$$\begin{aligned} \mathbf{M} &= \sum_{i=1}^{N_p} \mathbf{p}_i' \mathbf{c}_i'^T \\ &= \begin{bmatrix} S_{xx} & S_{xy} & S_{xz} \\ S_{yx} & S_{yy} & S_{yz} \\ S_{zx} & S_{zy} & S_{zz} \end{bmatrix}, \end{aligned}$$

which contains all the information required to solve the least squares problem for rotation. Let us construct a 4×4 symmetric matrix **N** given by

$$\mathbf{N} = \begin{bmatrix} S_{xx}+S_{yy}+S_{zz} & S_{yz}-S_{zy} & S_{zx}-S_{xz} & S_{xy}-S_{yx} \\ S_{yz}-S_{zy} & S_{xx}-S_{yy}-S_{zz} & S_{xy}+S_{yx} & S_{zx}+S_{xz} \\ S_{zx}-S_{xz} & S_{xy}+S_{yx} & -S_{xx}+S_{yy}-S_{zz} & S_{yz}+S_{zy} \\ S_{xy}-S_{yx} & S_{zx}+S_{xz} & S_{yz}+S_{zy} & -S_{xx}-S_{yy}+S_{zz} \end{bmatrix}$$

Let the eigenvector corresponding to the largest eigenvalue of **N** be $\mathbf{e} = \begin{bmatrix} e_0 & e_1 & e_2 & e_3 \end{bmatrix}$ where $e_0 \geq 0$ and $e_0^2 + e_1^2 + e_2^2 + e_3^2 = 1$. Then, the rotation matrix **R** is given by

$$\mathbf{R} = \begin{bmatrix} e_0^2+e_1^2-e_2^2-e_3^2 & 2\left(e_1e_2-e_0e_3\right) & 2\left(e_1e_3-e_0e_2\right) \\ 2\left(e_1e_2+e_0e_3\right) & e_0^2-e_1^2+e_2^2-e_3^2 & 2\left(e_2e_3-e_0e_1\right) \\ 2\left(e_1e_3-e_0e_3\right) & 2\left(e_2e_3+e_0e_1\right) & e_0^2-e_1^2-e_2^2+e_3^2 \end{bmatrix}.$$

Once we compute the optimal rotation matrix **R**, the optimal translation vector **t** can be computed by

$$\mathbf{t} = \mathbf{m}_c - \mathbf{R}\mathbf{m}_p.$$

A complete derivation and proofs can be found in [25]. A similar method is also presented in [17].

The convergence of ICP algorithm can be accelerated by extrapolating the registration space. Let $\mathbf{r}_i$ be a vector that describes a registration (i.e., rotation and translation) at ith iteration. Then, its direction vector in the registration space is given by

$$\Delta\mathbf{r}_i = \mathbf{r}_i - \mathbf{r}_{i-1}, \tag{19}$$

and the angle between the last two directions is given by

$$\theta_i = \cos^{-1}\left(\frac{\Delta\mathbf{r}_i^T \Delta\mathbf{r}_{i-1}}{\|\Delta\mathbf{r}_i\| \, \|\Delta\mathbf{r}_{i-1}\|}\right). \tag{20}$$

If both θ_i and θ_{i-1} are small, then there is a good direction alignment for the last three registration vectors $\mathbf{r}_i$, $\mathbf{r}_{i-1}$, and $\mathbf{r}_{i-2}$. Extrapolating these three registration vectors using either linear or parabola update, the next registration vector $\mathbf{r}_{i+1}$ can be computed. They showed 50 iterations of normal ICP was accelerated to about 15 to 20 iterations using such a technique.

6.3.3 Variants of ICP

Since the introduction of the ICP algorithm, various modifications have been developed in order to improve its performance .

Chen and Medioni [12, 13] developed a similar algorithm around the same time. The main difference is its strategy for point selection and for finding the correspondence between the two data sets. The algorithm first selects initial points on a regular grid, and computes the local curvature of these points. The algorithm only selects the points on smooth areas, which they call "control points". Their point selection method is an effort to save computation time, and to have reliable normal directions on the control points. Given the control points on one data set, the algorithm finds the correspondence by computing the intersection between the line that passes through the control point in the direction of its normal and the surface of the other data set. Although the authors did not mention in their paper, the advantage of their method is that the correspondence is less sensitive to noise and to outliers. As illustrated in Fig. 6.12, the original ICP's correspondence method may select outliers in the data set $\mathbf{Q}$ as corresponding points since the distance is the only constraint. However, Chen and Medioni's method is less sensitive to noise since the normal directions of the control points in $\mathbf{P}$ are reliable, and the noise in $\mathbf{Q}$ have no effect in finding the correspondence. They also briefly discussed the issues in registering multiple range data (i.e., multi-view registration). When registering multiple range data, instead of registering with a single neighboring range data each time, they suggested to register with the previously registered data as a whole. In this way, the information from all the previously registered data can be used. We will elaborate

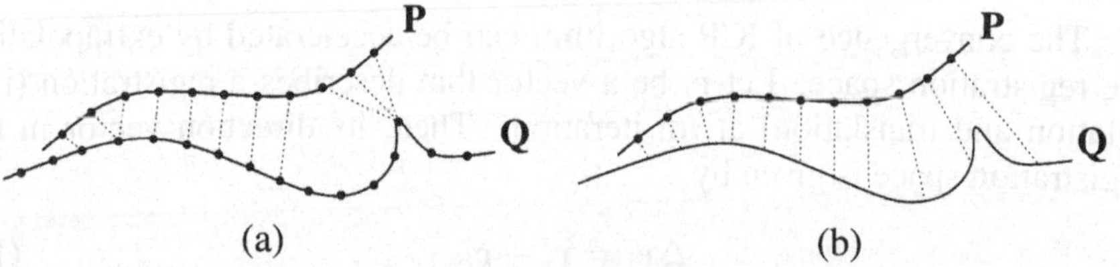

Fig. 6.12: Advantage of Chen and Medioni's algorithm.

(a): Result of the original ICP's correspondence method in the presence of noise and outliers.
(b): Since Chen and Medioni's algorithm uses control points on smooth area and its normal direction, it is less sensitive to noise and outliers

the discussion in multi-view registration in a separate section later.

Zhang [62] introduced a dynamic thresholding based variant of ICP, which rejects some corresponding points if the distance between the pair is greater than a threshold D_{max}. The threshold is computed dynamically in each iteration by using statistics of distances between the corresponding points as follows:

```
if μ < D                  /* registration is very good */
    D_max = μ + 3σ
else if μ < 3D            /* registration is good */
    D_max = μ + 2σ
else if μ < 6D            /* registration is not good */
    D_max = μ + σ
else                      /* registration is bad */
    D_max = ξ
```

where μ and σ are the mean and the standard deviation of distances between the corresponding points. D is a constant that indicates the expected mean distance of the corresponding points when the registration is good. Finally, ξ is a maximum tolerance distance value when the registration is bad. This modification relaxed the constraint of the original ICP, which required one data set to be a complete subset of the other data set. As illustrated in Figure 6.13, rejecting some corresponding pairs that are too far apart can lead to a better registration, and more importantly, the algorithm can be applied to partially overlapping data sets. The author also suggested that the points be stored in a k-D tree for efficient closest-point search.

Turk and Levoy [57] added a weight term (i.e., confidence measure) for each 3D point by taking a dot product of the point's normal vector and the

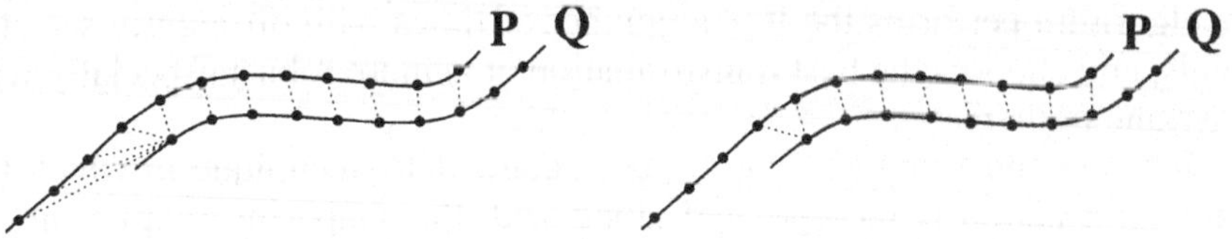

Fig. 6.13: Advantage of Zhang's algorithm.
(a): Since the original ICP assumes $\mathbf{P}$ is a subset of $\mathbf{Q}$, it finds corresponding points for all $\mathbf{P}$. (b): Zhang's dynamic thresholding allows $\mathbf{P}$ and $\mathbf{Q}$ to be partially overlapping

vector pointing to the light source of the scanner. This was motivated by the fact that structured-light scanning acquires more reliable data when the object surface is perpendicular to the laser plane. Assigning lower weights to unreliable 3D points (i.e., points on the object surface nearly parallel with the laser plane) helps to achieve a more accurate registration. The weight of a corresponding pair is computed by multiplying the weights of the two corresponding points. Let the weights of corresponding pairs be $\mathbf{w} = \{w_i\}$, then the objective function in Eq. (18) is now a weighted function:

$$f(\mathbf{R}, \mathbf{t}) = \frac{1}{N_p} \sum_{i=1}^{N_p} w_i \left\| \mathbf{c}_i - \mathbf{R}\mathbf{p}_i - \mathbf{t} \right\|^2 \tag{21}$$

For faster and efficient registration, they proposed to use increasingly more detailed data from a hierarchy during the registration process where less detailed data are constructed by sub-sampling range data. Their modified ICP starts with the lowest-level data, and uses the resulting transformation as the initial position for the next data in the hierarchy. The distance threshold is set as twice of sampling resolution of current data. They also discarded corresponding pairs in which either points is on a boundary in order to make reliable correspondences.

Masuda *et al.* [38, 37] proposed an interesting technique in an effort to add robustness to the original ICP. The motivation of their technique came from the fact that a local minimum obtained by the ICP algorithm is predicated by several factors such as initial registration, selected points and corresponding pairs in the ICP iterations, and that the outcome would be more unpredictable when noise and outliers exist in the data. Their algorithm consists of two main stages. In the first stage, the algorithm performs the ICP a number of times, but in each trial the points used for ICP calculations are selected differently based on random sampling. In the second stage, the algorithm selects the transformation that produced the minimum median distance between the corresponding pairs as the final resulting transformation. Since the algorithm performs the ICP a number of times with differently selected points, and chooses the best transformation, it is more robust especially with noise and outliers.

Johnson and Kang [29] introduced "color ICP" technique in which the color information is incorporated along with the shape information in the closest-point (i.e., correspondence) computation. The distance metric d between two points $\mathbf{p}$ and $\mathbf{q}$ with the 3D location and the color are denoted as (x, y, z) and (r, g, b) respectively can be computed as

$$d^2(\mathbf{p}, \mathbf{q}) = d_e^2(\mathbf{p}, \mathbf{q}) + d_c^2(\mathbf{p}, \mathbf{q}) \tag{22}$$

where

$$d_e(\mathbf{p}, \mathbf{q}) = \sqrt{(x_p - x_q)^2 + (y_p - y_q)^2 + (z_p - z_q)^2}, \tag{23}$$

$$d_c(\mathbf{p}, \mathbf{q}) = \sqrt{\lambda_1(r_p - r_q)^2 + \lambda_2(g_p - g_q)^2 + \lambda_3(b_p - b_q)^2} \tag{24}$$

and $\lambda_1, \lambda_2, \lambda_3$ are constants that control the relative importance of the different color components and the importance of color overall vis-a-vis shape. The authors have not discussed how to assign values to the constants, nor the effect of the constants on the registration. A similar method was also presented in [21].

Other techniques employ using other attributes of a point such as normal direction [53], curvature sign classes [19], or combination of multiple attributes [50], and these attributes are combined with the Euclidean distance in searching for the closest point. Following these works, Godin *et al.* [20] recently proposed a method for the registration of attributed range data based on a random sampling scheme. Their random sampling scheme differs from that of [38, 37] in that it uses the distribution of attributes as a guide for point selection as opposed to uniform sampling used in [38, 37]. Also, they use attribute values to construct a compatibility measure for the closest point search. That is, the attributes serve as a boolean operator to either accept or reject a correspondence between two data points. This way, the difficulty of choosing constants in distance metric computation, for example $\lambda_1, \lambda_2, \lambda_3$ in Eq. (24), can be avoided. However, a threshold for accepting and rejecting correspondences is still required.

6.3.4 Initial Registration

Given two data sets to register, the ICP algorithm converges to different local minima depending on the initial positions of the data sets. Therefore, it is not guaranteed that the ICP algorithm will converge to the desired global minimum, and the only way to confirm the global minimum is to find the minimum of all the local minima. This is a fundamental limitation of the ICP that it requires a good initial registration as a starting point to maximize the probability of converging to a correct registration. Besl and McKay in their ICP paper [8] suggested to use a set of initial registrations chosen by sampling of quaternion states and translation vector. If some geometric properties such as principle components of the data sets provide distinctness, such information may be used to help reduce the search space.

As mentioned before, one can provide initial registrations by a tracking system that provides relative positions of each scanning viewpoint. One can also provide initial registrations manually through human interaction. Some researchers have proposed other techniques for providing initial registrations [11, 17, 22, 28], but it is reported in [46] that these methods do not work reliably for arbitrary data.

Recently, Huber [26] proposed an automatic registration method in which no knowledge of data sets is required. The method constructs a globally consistent model from a set of pairwise registration results. Although the experiments showed good results considering the fact that the method does not require any initial information, there was still some cases where incorrect registration was occurred.

6.3.5 Multi-view Registration

Although the techniques we have reviewed so far only deal with pairwise registration – registering two data sets, they can easily be extended to multi-view registration – registering multiple range images while minimizing the registration error between all possible pairs. One simple and obvious way is to perform a pairwise registration for each of two neighboring range images sequentially. This approach, however, accumulates the errors from each registration, and may likely have a large error between the first and the last range image.

Chen and Medioni [13] were the first to address the issues in multi-view registration. Their multi-view registration goes as follows: First, a pairwise registration between two neighboring range images is carried out. The resulting registered data is called a meta-view. Then, another registration between a new unregistered range image and the meta-view is performed, and the new data is added to the meta-view after the registration. This process is continued until all range images are registered. The main drawback of the meta-view approach is that the newly added images to the meta-view may contain information that could have improved the registrations performed previously.

Bergevin *et al.* [5, 18] noticed this problem, and proposed a new method that considers the network of views as a whole and minimizes the registration errors for all views simultaneously. Given N range images from the viewpoints $V_1, V_2, ..., V_N$, they construct a network such that $N - 1$ viewpoints are linked to one central viewpoint in which the reference coordinate system

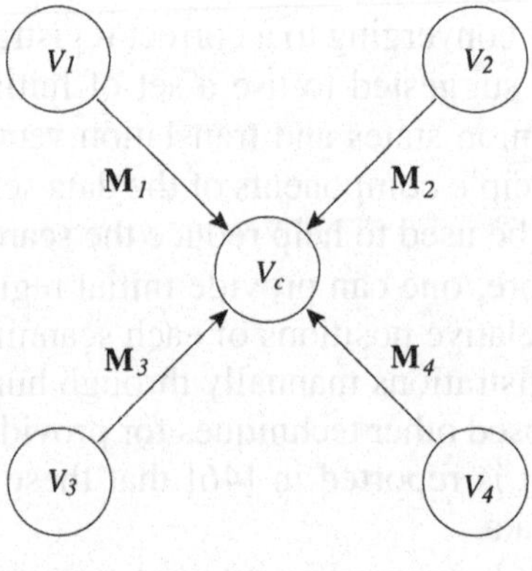

Fig. 6.14: Network of multiple range data was considered in the multi-view registration method by Bergevin et al. [5, 18]

is defined. For each link, an initial transformation matrix $\mathbf{M}_{i,0}$ that brings the coordinate system of V_i to the reference coordinate system is given. For example, consider the case of 5 range images shown in Fig. 6.14 where viewpoints V_1 through V_4 are linked to a central viewpoint V_c. During the algorithm, 4 incremental transformation matrices $\mathbf{M}_{1,k}, ..., \mathbf{M}_{4,k}$ are computed in each iteration k. In computing $\mathbf{M}_{1,k}$, range images from V_2, V_3 and V_4 are transformed to the coordinate system of V_1 by first applying its associated matrix $\mathbf{M}_{i,k-1}$, $i = 2, 3, 4$ followed by $\mathbf{M}_{1,k-1}^{-1}$. Then, it computes the corresponding points between the range image from V_1 and the three transformed range images. $\mathbf{M}_{1,k}$ is the transformation matrix that minimizes the distances of all the corresponding points for all the range images in the reference coordinate system. Similarly, $\mathbf{M}_{2,k}$, $\mathbf{M}_{3,k}$ and $\mathbf{M}_{4,k}$ are computed, and all these matrices are applied to the associated range images simultaneously at the end of iteration k. The iteration continues until all the incremental matrices $\mathbf{M}_{i,k}$ become close to identity matrices.

Benjemaa and Schmitt [4] accelerated the above method by applying each incremental transformation matrix $\mathbf{M}_{i,k}$ immediately after it is computed instead of applying all simultaneously at the end of the iteration. In order to not favor any individual range image, they randomized the order of registration in each iteration.

Pulli [45, 46] argued that these methods cannot easily be applied to large data sets since they require large memory to store all the data, and since the methods are computationally expensive as $N - 1$ ICP registrations are performed. To get around these limitations, his method first performs pairwise registrations between all neighboring views that result in overlapping range images. The corresponding points discovered in this manner are used in the next step that does multi-view registration. The multi-view registration process is similar to that of Chen and Medioni except for the fact that the corresponding points from the previous pairwise registration step are used as permanent corresponding points throughout the process. Thus, searching for corresponding points, which is computationally most demanding, is avoided, and the process does not require large memory to store all the data. The author claimed that his method, while being faster and less demanding on memory, results in similar or better registration accuracy compared to the previous methods.

6.3.6 Experimental Result

We have implemented a modified ICP algorithm for registration of our range images. Our algorithm uses Zhang's dynamic thresholding for rejecting correspondences. In each iteration, a threshold D_{max} is computed as

$$D_{max} = m + 3\sigma$$

where m and σ are the mean and the standard deviation of the distances of the corresponding points. If the Euclidean distance between two corresponding points exceeds this threshold, the correspondence is rejected. Our algorithm also uses the bucketing algorithm (i.e., Elias algorithm) for fast corresponding point search. Figure 6.15 shows an example of a pairwise registration. Even though the initial positions were relatively far from the correct registration, it successfully converged in 53 iterations. Notice in the final result (Figure 6.15(d)) that the overlapping surfaces are displayed with many small patches, which indicates that the two data sets are well registered.

We acquired 40 individual range images from different viewpoints to capture the entire surface of the bunny figure. Twenty range images covered about 90% of the entire surface. The remaining 10% of surface was harder to view on account of either self-occlusions or because the object would need to be propped so that those surfaces would become visible to the sensor. Additional 20 range images were gathered to get data on such surfaces.

Our registration process consists of two stages. In the first stage, it performs a pairwise registration between a new range image and all the previous range images that are already registered. When the new range image's initial registration is not available, for example when the object is repositioned, it first goes through a human assisted registration process that allows a user to visualize the new range image in relation to the previously registered range images. The human is able to rotate one range image vis-a-vis the other and provide corresponding points. See Figure 6.16 for an illustration of the human assisted registration process. The corresponding points given by the human are used to compute an initial registration for the new range image. Subsequently, registration proceeds as before.

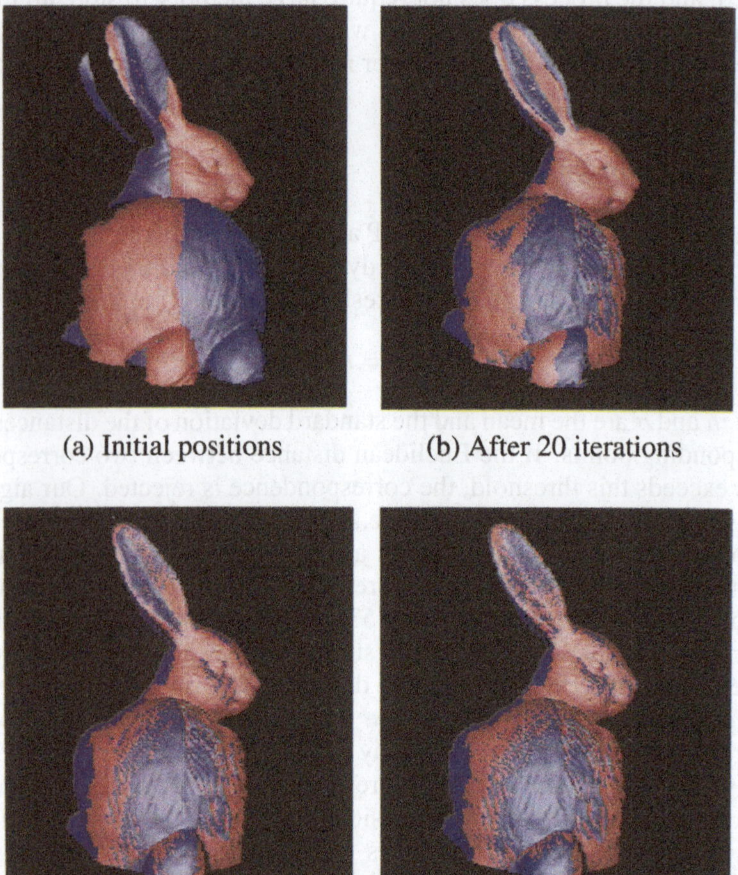

(a) Initial positions (b) After 20 iterations

(c) After 40 iterations (d) Final after 53 iterations

Fig. 6.15: Example of a pairwise registration using the ICP algorithm

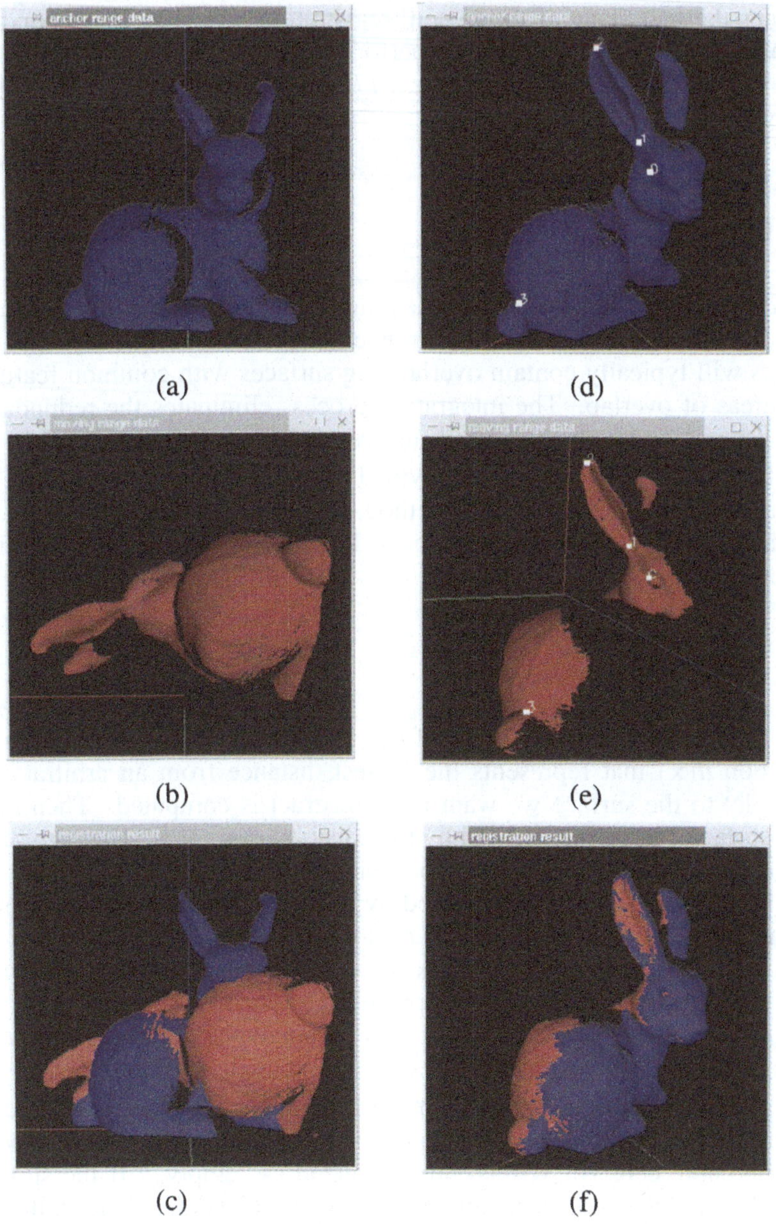

Fig. 6.16: Human assisted registration process
(a),(b),(c): Initial Positions of two data sets to register. (d),(e): User can move around the data and click corresponding points. (f): The given corresponding points are used to compute an initial registration

Registration of all the range images in the manner described above constitutes the first stage of the overall registration process. The second stage then fine-tunes the registration by performing a multi-view registration using the method presented in [4]. Figure 6.17 shows the 40 range images after the second stage.

6.4 Integration

Successful registration aligns all the range images into a common coordinate system. However, the registered range images taken from adjacent viewpoints will typically contain overlapping surfaces with common features in the areas of overlap. The integration process eliminates the redundancies, and generates a single connected surface model.

Integration methods can be divided into five different categories: volumetric method, mesh stitching method, region-growing method, projection method, and sculpting-based method. In the next sub-sections we will explain each of these categories.

6.4.1 Volumetric Methods

The volumetric method consists of two stages. In the first stage, an implicit function $d(\mathbf{x})$ that represents the closest distance from an arbitrary point $\mathbf{x} \in \Re^3$ to the surface we want to reconstruct is computed. Then the object surface can be represented by the equation $d(\mathbf{x}) = 0$. The sign of $d(\mathbf{x})$ indicates whether $\mathbf{x}$ lies outside or inside the surface. In the second stage, the isosurface – the surface defined by $d(\mathbf{x}) = 0$ – is extracted by triangulating the zero-crossing points of $d(\mathbf{x})$ using the marching cubes algorithm [36, 39]. The most important task here is to reliably compute the function $d(\mathbf{x})$ such that this function best approximates the true surface of the object. Once $d(\mathbf{x})$ is approximated, other than the marching cubes algorithm, such as marching triangles algorithm, can be used to extract the isosurface.

The basic concept of the volumetric method is illustrated in Figure 6.18. First, a 3D volumetric grid that contains the entire surface is generated, and all the cubic cells (or voxels) are initialized as "empty". If the surface is found "near" the voxel (the notion of "near" will be defined later), the voxel is set to "non-empty" and $d(\mathbf{x})$ for each of the 8 vertices of the voxel is computed by the signed distance between the vertex to the closest surface point. The sign of $d(\mathbf{x})$ is positive if the vertex is outside the surface, and negative otherwise. After all the voxels in the grid are tested, the triangulation is performed as follows. For each non-empty voxel, zero crossing points of $d(\mathbf{x})$, if any, are computed. The computed zero crossing points are then triangu-

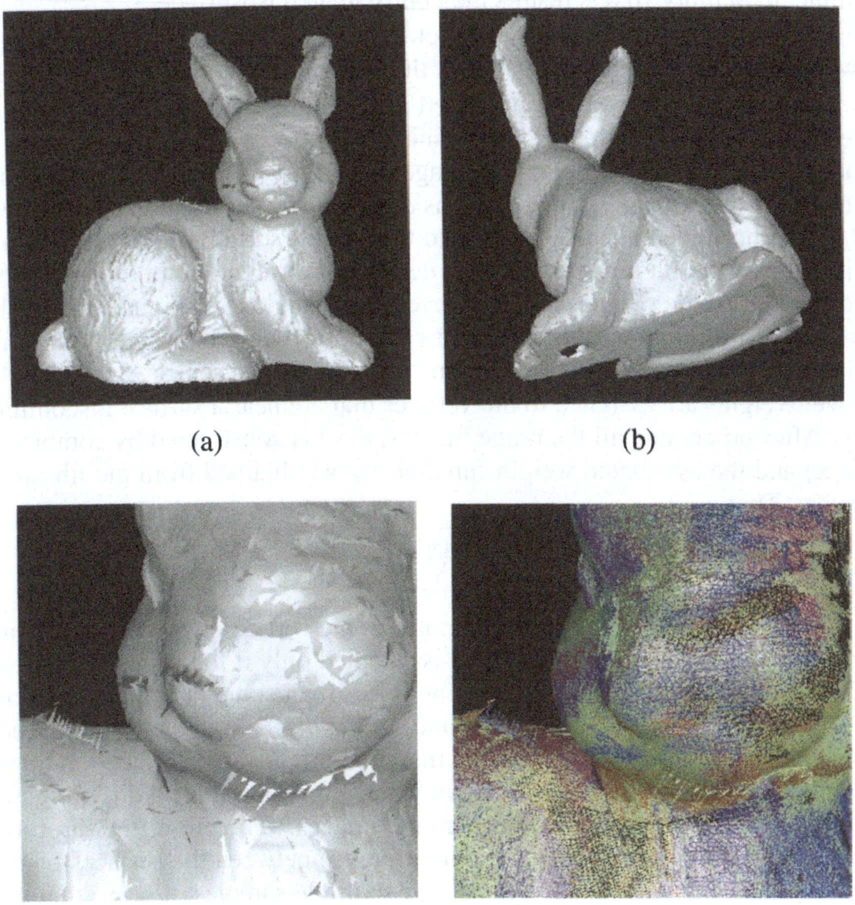

(a) (b)

(c) (d)

Fig. 6.17: 40 range images after the second stage of the registration process (a),(b): Two different views of the registered range images. All the range images are displayed as shaded triangular mesh. (c): Close-up view of the registered range images. (d): The same view as (c) displayed with triangular edges. Each color represents an individual range image lated by applying one of the 15 cases in the marching cubes look-up-table.[1] For example, the upper-left voxel in Figure 6.18(d) corresponds to the case number 2 of the look-up-table, and the upper-right and the lower-left voxels both correspond to the case number 8. Triangulating zero crossing points of all the non-empty voxels results the approximated isosurface.

We will now review three volumetric methods. The main difference between these three methods lies in how the implicit function $d(\mathbf{x})$ is computed.

Curless and Levoy [14] proposed a technique tuned for range images generated by a structured-light scanner. Suppose we want to integrate n range images where all the range images are in the form of triangular mesh. For each range image i, two functions $d_i(\mathbf{x})$ and $w_i(\mathbf{x})$ are computed where $d_i(\mathbf{x})$ is the signed distance from $\mathbf{x}$ to the nearest surface along the viewing direction of the ith range image and $w_i(\mathbf{x})$ is the weight computed by interpolating the three vertices of the intersecting triangle (See Figure 6.19). The weight of each vertex is computed as the dot product between the normal direction of the vertex and the viewing direction of the sensor. Additionally, lower weights are assigned to the vertices that are near a surface discontinuity. After processing all the range images, $d(\mathbf{x})$ is constructed by combining $d_i(\mathbf{x})$ and the associated weight function $w_i(\mathbf{x})$ obtained from the ith range image. That is,

$$d(\mathbf{x}) = \frac{\sum_{i=1}^{n} w_i(\mathbf{x}) d_i(\mathbf{x})}{\sum_{i=1}^{n} w_i(\mathbf{x})}$$

We said earlier that $d(\mathbf{x})$ is computed at the vertices of a voxel if the surface is "near". In other words, $d(\mathbf{x})$ is sampled only if the distance between the vertex to the nearest surface point is less than some threshold. Without imposing this threshold, computing and storing $d(\mathbf{x})$ for all the voxels in each range image will be impractical, but more importantly, the surfaces on opposite sides will interfere with each other since the final $d(\mathbf{x})$ is the weighted average of $d_i(\mathbf{x})$ obtained from n range images. Therefore, the threshold must be small enough to avoid the interference between the surfaces on opposite sides, but large enough to acquire multiple samples of $d_i(\mathbf{x})$ that will contribute to a reliable computation of $d(\mathbf{x})$ and subsequent zero crossing points. Considering this tradeoff, a practical suggestion would be to set the threshold as half the maximum uncertainty of the range measurement.

Hoppe *et al.* [24] were the first to propose the volumetric method. Their algorithm is significant in that it assumes the input data is unorganized. That is, neither the connectivity nor the normal direction of points is known in advance. Therefore, the method first estimates the oriented tangent plane for each data point. The tangent plane is computed by fitting the best plane in

[1] Since there are 8 vertices in a voxel, there are 256 ways in which the surface can intersect the voxel. These 256 cases can be reduced to 15 general cases by applying the reversal symmetry and the rotational symmetry.

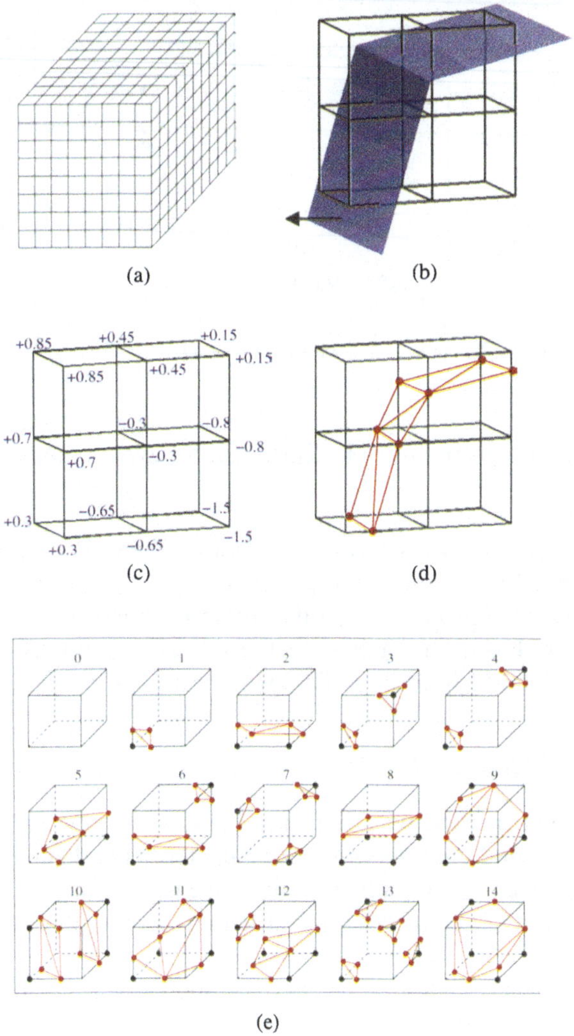

Fig. 6.18: Volumetric Method

(a): 3D volumetric grid. (b): Four neighboring cubes near the surface. The arrow points to the outside of the surface. (c): Signed distance function $d(\mathbf{x})$ is sampled at each vertex. (d): Zero-crossing points of $d(\mathbf{x})$ (red circles) are triangulated by the marching cubes algorithm. (e): 15 general cases of the marching cubes algorithm

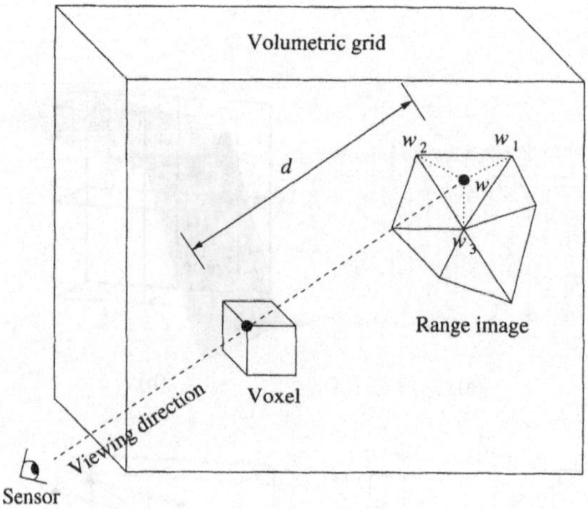

Fig. 6.19: Computing $d(\mathbf{x})$ in Curless and Levoy's method [14]

the least squares sense on k nearest neighbors. Then, $d(\mathbf{x})$ is the distance between $\mathbf{x}$ and its closest point's tangent plane.

Wheeler *et al.* [59] proposed a similar method called "consensus-surface algorithm". Their algorithm emphasizes the selection of points used to compute the signed-distance in order to deal with noise and outliers.

6.4.2 Mesh Stitching Methods

The Mesh stitching method was first introduced by Soucy and Laurendeau [51, 52]. Their method consists of three main steps: (1) determining redundant surface regions, (2) reparameterizing those regions into non-redundant surface regions, and (3) connecting (or stitching) all the non-redundant surface regions.

Redundant surface regions represent common surface regions sampled by two or more range images. The content of each of the redundant surface regions can be determined by finding all possible pairs of range images and their redundant surface regions. For example, consider Figure 6.20 where 3 range images V_1, V_2 and V_3 have 4 different redundant surface regions. If we find the pairwise redundant surface regions of V_1V_2, V_1V_3, and V_2V_3, it is possible to determine for each point, which range images have sampled that point. Therefore, the contents of 4 redundant surface regions are implicitly available.

Now, we will describe how the redundant surface region between a pair of range images, say V_1 and V_2, can be found. Two conditions are imposed to determine if a point in V_1 is redundant with V_2: First, the point must be near

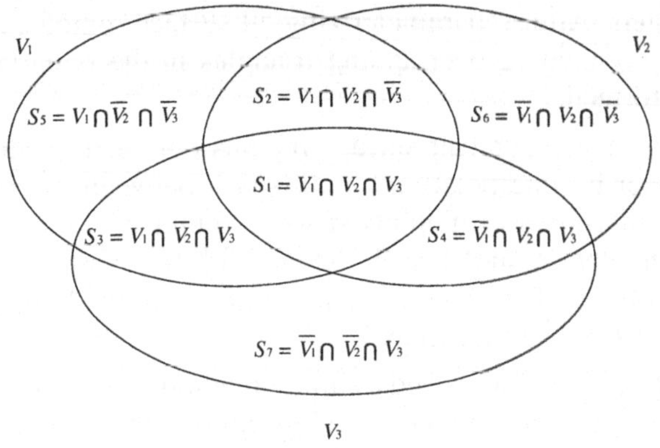

Fig. 6.20: Redundant surfaces of three different range images

the surface of V_2, and second, the point must be visible from the viewing direction of V_2. The Spatial Neighborhood Test (SNT), which tests the first condition, checks whether the distance between the point and the surface of V_2 is within the uncertainty of the range sensor. The Surface Visibility Test (SVT), which tests the second condition, checks if the dot product between the normal direction of the point and the viewing direction (i.e., optical axis) of the V_2 is positive. All the points in V_1 that satisfy the two tests are assumed to be in the redundant surface with V_2. Unfortunately, the SNT and the SVT yield unreliable results in the regions where surface discontinuities occur or when noise is present. Therefore, a heuristic region-growing technique that fine-tunes the estimated redundant surfaces is used. By observing that the boundaries of the redundant surface correspond to the surface discontinuity at least in one of the range images, each of the estimated redundant regions is expanded until it reaches the surface discontinuity of one of the range images. In order to prevent small isolated regions to grow freely, an additional constraint that the expanded region must contain at least 50 percent of the original seed region is imposed.

After the redundant surface regions are determined, those regions are reparameterized into non-redundant surfaces. For each redundant surface region, a plane grid is defined; the plane grid has the same sampling resolution as that of a range image, and passes through the center of mass of the redundant surface region with the normal direction given by the average normal of all the points in the region. All the points in the region are then projected onto this plane grid. Associated with each vertex in the grid is the average of the perpendicular coordinate values for all the points that projected onto the cell represented by that vertex. The grid coordinates together with the computed perpendicular coordinates define new non-redundant surface points that are then triangulated. After reparameterizing all surface regions, a process that eliminates any remaining overlapping triangles in the boundary of surface regions is performed.

Finally, the non-redundant surface regions obtained in this manner are stitched together by interpolating empty space between the non-redundant surfaces. The interpolation of empty space is obtained by the constrained 2D Delaunay triangulation on the range image grid that sampled that particular empty space continuously. The result after interpolating all the empty spaces is the final connected surface model.

Turk and Levoy [57] proposed a similar method called "mesh zippering". The main difference between the two algorithms is the order of determining the connectivity and the geometry. The previous algorithm first determines the geometry by reparameterizing the projected points on the grid, then determines the connectivity by interpolating into the empty spaces between the re-parameterized regions. By contrast, Turk and Levoy's algorithm first determines the connectivity by removing the overlapping surfaces and stitching (or zippering) the borders. Then, it determines the geometry by adjusting surface points as weighted averages of all the overlapping surface points. The mesh zippering algorithm is claimed to be less sensitive to the artifacts of the stitching process since the algorithm first determines the connectivity followed by the geometry.

Let us describe the mesh zippering method in more detail with the illustrations in Figure 6.21. In (a), two partially overlapping surfaces are shown as red and blue triangles. From (b) to (d), the redundant triangles shown as green triangles are removed one by one from each surface until both surfaces remain unchanged. A triangle is redundant if all three distances between its vertices to the other surface are less than a predefined threshold where the threshold is typically set to a small multiple of the range image resolution. After removing the redundant triangles, it finds the boundary edges of one of the two surfaces; the boundary edges of the blue triangles are shown as green lines in (e). Then, the intersections between these boundary edges and the other surface are determined; the intersecting points are depicted as black circles in (f). Since it is unlikely that the boundary edges will exactly intersect the surface, a "thickened wall" is created for each boundary edge; a thickened wall is made of four triangles, and it is locally perpendicular to the boundary edge points of one of the surfaces. The problem now becomes finding intersecting points between the boundary edge wall and the surface. From this point, all the red triangle edges that are beyond the boundary edges are discarded as shown in (g). In (h), the intersecting points are added as new vertices, and triangulated through a constrained triangulation routine [6]. After zippering all the surfaces together, the final step fine-tunes the geometry by considering all the information of the surfaces including those that were discarded in the zippering process. The final position of each surface point is computed as the weighted average of all the overlapping surfaces along the normal direction of the point. The weight of each point is computed as a dot product between the normal direction of the point and its corresponding range image's viewing direction.

6.4.3 Region-Growing Methods

We introduce two region-growing based integration methods. The first method [23], called "marching triangles" consists of two stages. In the first stage, similar to the volumetric method, it defines an implicit surface representation as the zero crossings of a function $d(\mathbf{x})$, which defines the signed distance to the nearest point on the surface for any point $\mathbf{x}$ in 3D space. In the second stage, instead of using the marching cubes algorithm, the marching triangles algorithm is used to triangulate the zero crossings of $d(\mathbf{x})$. The marching triangles algorithm starts with a seed triangle, adds a neighbor triangle based on the 3D Delaunay surface constraint, and continues the process until all the points have been considered.

The second method [7], which is more recently developed, is called "ball-pivoting algorithm (BPA)". The basic principle of the algorithm is that three points form a triangle if a ball of a certain radius ρ touches all of them without containing any other points. Starting with a seed triangle, the ball pivots around an edge until it touches another point, then forms a new triangle. The process continues until all points have been considered. The BPA is related to the α-shape[2] [16], thus provides a theoretical guarantee to reconstruct a surface homeomorphic to the original surface within a bounded distance if sufficiently dense and uniform sampling points are given. It is also shown that the BPA can be applied to a large set of data proving that it is efficient in computation and memory usage. The main disadvantage of this method is that the size of radius ρ must be given manually, and a combination of multiple processes with different ρ values may be necessary to generate a correct integrated model.

[2] The α-shape of a finite point set S is a polytope uniquely determined by S and the parameter α that controls the level-of-detail. A subset $T \subseteq S$ of size $|T| = k + 1$ with $0 \leq k \leq 2$ belongs to a set $F_{k,\alpha}$ if a sphere of radius α contains T without containing any other points in S. The α-shape is described by the polytope whose boundary consists of the triangles connecting the points in $F_{2,\alpha}$, the edges in $F_{1,\alpha}$, and vertices in $F_{0,\alpha}$. If $\alpha = \infty$, the α-shape is identical to the convex hull of S, and if $\alpha = 0$, the α-shape is the point set S itself.

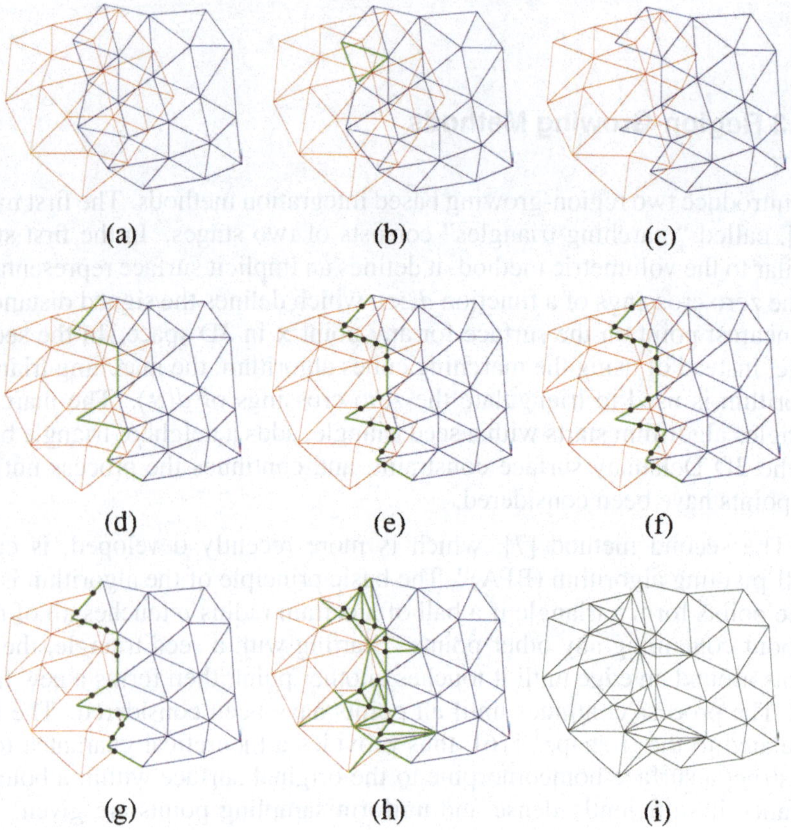

Fig. 6.21: Mesh Zippering algorithm
(a): Two overlapping surfaces. (b): A redundant triangle from the blue surface is removed. (c): A redundant triangle from the red surface is removed. (d): Steps (b) and (c) are continued until both surfaces remain unchanged. (e): After removing all the redundant triangles, the boundary edges blue surface is found. (f): The intersections between the boundary edge and the edges from the red surface are determined. (g): All the edges from the red surface that are beyond the boundary edges are discarded. (h): The intersecting points are added as new vertices and are triangulated. (i): The final position of each point is adjusted by considering all the surfaces including those that were discarded during the zippering process.

6.4.4 Projection Methods

The Projection method [13, 44, 58], one of the earlier integration methods, simply projects the data onto a cylindrical or a spherical grid. Multiple data projections onto a same grid are averaged, and the resulting data is reparameterized. Although this method provides a simple way of integration, it suffers from the fundamental limitation that it can only handle convex objects.

6.4.5 Sculpting Based Methods

The Sculpting based method [1, 2, 10, 16] typically computes tetrahedra volumes from the data points by the 3D Delaunay triangulation. Then, it progressively eliminates tetrahedra until the original shape is extracted. Since the method is based on the Delaunay triangulation, it guarantees that the resulting surface is topologically correct as long as the data points are dense and uniformly distributed. Also, it can be applied to a set of unorganized points. However, the method has difficulty with constructing sharp edges, and it suffers from the expensive computations needed for calculating 3D Delaunay triangulations.

6.4.6 Experimental Result

In order to illustrate the results from integration, let's take as example the Curless and Levoy's volumetric integration method[14]. For that, we used 40 range images of a bunny figurine that were acquired by our structured-light scanning system. All the range images were registered as described in the previous chapter. The integration was performed at 0.5mm resolution (i.e., the size of a voxel of the grid is 0.5mm), which is an approximate sampling resolution of each range image.

Figure 6.22 shows four different views of the resulting model. The total number of points and triangles in the 40 range images were 1,601,563 and 3,053,907, respectively, and these were reduced to 148,311 and 296,211 in the integrated model. Figure 6.23(a) shows a close-up view around the bunny's nose area before the integration where different colored triangles represent different range images. Figure 6.23(b) shows the same view after the integration.

6.5 Acquisition of Reflectance Data

Successful integration of all range images results in a complete geometric model of an object. This model itself can be the final output if only the shape of the object is desired. But since a photometrically correct visualization is

(a) (b) (c) (d)

Fig. 6.22: Integrated model visualized from four different viewpoints

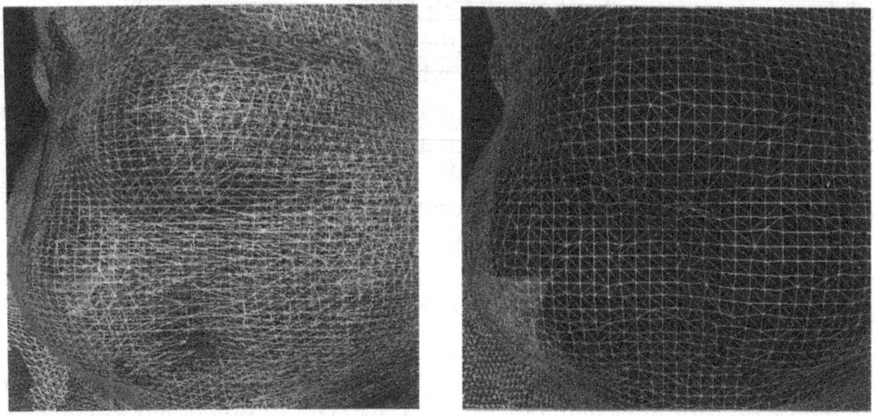

(a) (b)

Fig. 6.23: Close-up view of the model before and after integration

frequently required, we must also obtain the reflectance data of the object surface.

In general, there are two approaches for acquiring the reflectance data. The first approach employs one of the many parametric reflectance models and estimates the reflectance parameters for each data point by using multiple images taken from different viewpoints and under different lighting conditions [27, 30, 33, 48, 49]. Once the reflectance parameters are estimated, it is possible to visualize the object under any novel lighting condition from any novel viewpoint. We will describe this approach in greater depth in Section 6.5.2.

The second approach, instead of using a parametric reflectance model, utilizes only a set of color images of the object. Some methods [15, 47] exploit the use of view dependent texture maps. For each viewing direction of the 3D model, a synthetic image for texture mapping is generated by interpolating the input images that were taken from the directions close to the current viewing direction. The synthetic image simulates what would have been the image taken from the current viewing direction, thus it provides a correct texture to the 3D model. Other methods [42, 60] store a series of N textures for each triangle where the textures are obtained from the color images taken from different viewpoints under known light source directions. The N textures are compressed by applying the Principal Components Analysis, and a smaller number of textures that approximate basis functions of the viewing space are computed. These basis functions are then interpolated to represent the texture of each triangle from a novel viewpoint.

Although the second approach provides realistic visualization from an arbitrary viewpoint without estimating reflectance parameters for each data point, one of the major drawbacks is the fact that it can only render the object under the same lighting condition in which the input images were taken. On the other hand, the first approach provides the underlying reflectance properties of the object surface, and thus makes it possible to visualize the object under a novel lighting condition. We will first describe some of the well known reflectance models that are commonly used followed by the methods for estimating reflectance parameters.

6.5.1 Reflectance Models

The true reflectance property of an object is based on many complex physical interactions of light with object materials. The Bidirectional Reflectance Distribution Function (BRDF) developed by Nicodemus *et al.* [41] provides a general mathematical function for describing the reflection property of a surface as a function of illumination direction, viewing direction, surface normal, and spectral composition of the illumination used. For our application, we can use the following definition for each of the primary color components:

$$f_r(\theta_i, \phi_i; \theta_r, \phi_r) = \frac{dL_r(\theta_r, \phi_r)}{dE_i(\theta_i, \phi_i)} \qquad (25)$$

where L_r is reflected radiance, E_i is incident irradiance, θ_i and ϕ_i specify the incident light direction, and θ_r and ϕ_r specify the reflected direction.

Many researchers have proposed various parametric models to represent the BRDF, each having different strengths and weaknesses. Two of the well known models are those developed by Beckmann and Spizzichino [3], and Torrance and Sparrow [54]. The Beckmann-Spizzichino model was derived using basic concepts of electromagnetic wave theory, and is more general than the Torrance-Sparrow model in the sense that it describes the reflection from smooth to rough surfaces. The Torrance-Sparrow model was developed to approximate reflectance on rough surfaces by geometrically analyzing a path of light ray on rough surfaces. The Torrance-Sparrow model, in general, is more widely used than the Beckman-Spizzichino model because of its simpler mathematical formula.

Torrance-Sparrow Model

The Torrance-Sparrow model assumes that a surface is a collection of planar micro-facets as shown in Figure 6.24. An infinitesimal surface patch dA consists of a large set of micro-facets where each facet is assumed to be one side of a symmetric V-cavity. The set of micro-facets has a mean normal vector of $\mathbf{n}$, and a random variable α is used to represent the angle between each micro-facet's normal vector and the mean normal vector. Assuming

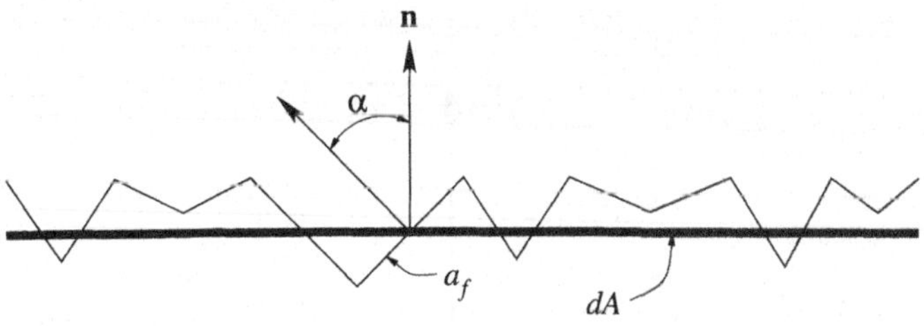

Fig. 6.24: Surface model

the surface patch is isotropic (i.e., rotationally symmetric about surface normal), the distribution of α can be expressed as a one-dimensional Gaussian distribution with mean value of zero and standard deviation of σ_α:

$$P(\alpha) = ce^{-\frac{\alpha^2}{2\sigma_\alpha^2}} \tag{26}$$

where c is a constant. The standard deviation σ_α represents the roughness of surface – the larger σ_α, the rougher surface, and vice versa.

Figure 6.25 shows the coordinate system used in the Torrance-Sparrow model. A surface patch dA is located at the origin of the coordinate system with its normal vector coinciding with the Z axis. The surface is illuminated by the incident beam that lies on the YZ plane with a polar angle of θ_i, and a particular reflected beam which we are interested in travels along the direction (θ_r, ϕ_r). Unit solid angles $d\omega_i$ and $d\omega_r$ are used to denote the directions of the incident beam and the reflected beam respectively. The bisector between the incident direction and the reflected direction is described by a unit solid angle $d\omega'$ which has a polar angle of α.

Only the micro-facets in dA with normal vectors within $d\omega'$ can reflect the incident light specularly to the direction (θ_r, ϕ_r). Let $P(\alpha)d\omega'$ be the number of facets per unit surface area whose normal vectors are contained within $d\omega'$ where $P(\alpha)$ was defined in Eq. (26). Then, the number of facets in dA with normal vectors lying within $d\omega'$ is

$$P(\alpha)d\omega' dA.$$

Let a_f be the area of each micro-facet. Then, the total reflecting area of the facets is

$$a_f P(\alpha)d\omega' dA,$$

and the projected area in the incident direction is

$$a_f P(\alpha)d\omega' dA \cos\theta_i'.$$

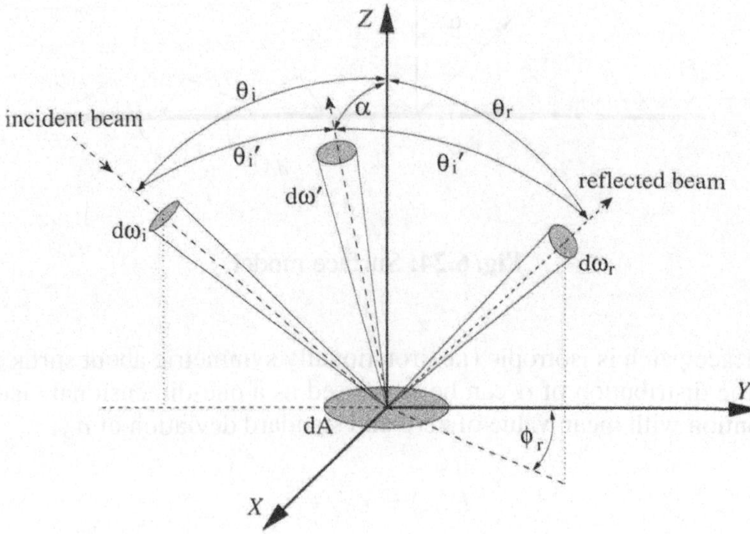

Fig. 6.25: Coordinate system for the Torrance-Sparrow model

Thus, the incident radiance of the specularly reflecting facets in dA is

$$L_i = \frac{d^2\Phi_i}{d\omega_i(a_f P(\alpha) d\omega' dA)\cos\theta_i'} \tag{27}$$

where Φ_i is the incident flux. Since a surface is not a perfect reflector, only a fraction of the incident flux is reflected. Therefore, Torrance and Sparrow considered two phenomena for relating the incident flux and the reflected flux. First, they considered *Fresnel reflection coefficient* $F'(\theta_i', \eta')$ [3], which determines the fraction of incident light that is reflected by a surface. θ_i' represents the incident angle and η' represents the complex index of refraction of the surface. The Fresnel reflection coefficient is sufficient for relating the incident and reflected flux when facet shadowing and masking (See Figure 6.26) are neglected. For the second phenomenon, Torrance and Sparrow considered the effects of facet shadowing and masking, and introduced the *geometrical attenuation factor* $G(\theta_i, \theta_r, \phi_r)$[3]. On the basis of these two phenomena, the incident flux Φ_i and the reflected flux Φ_r can be related as

$$d^2\Phi_r = F'(\theta_i', \eta')G(\theta_i, \theta_r, \phi_r)d^2\Phi_i. \tag{28}$$

[3]Readers are referred to Torrance and Sparrow's paper [54] for the detailed description of geometrical attenuation factor.

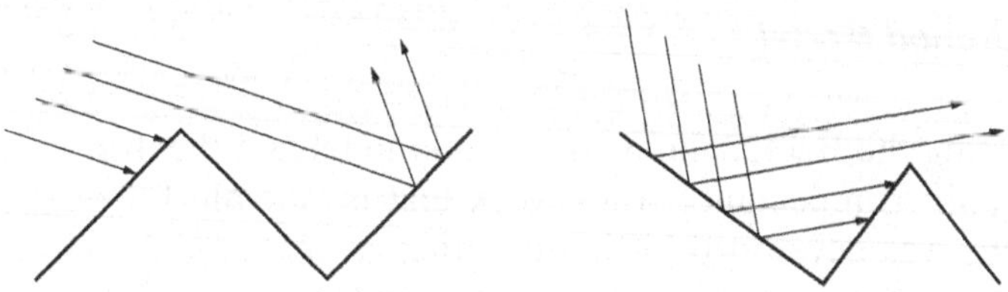

Fig. 6.26: Facet shadowing and masking

Since the radiance reflected in the direction (θ_r, ϕ_r) is given by

$$L_r = \frac{d^2\Phi_r}{d\omega_r dA \cos\theta_r},$$

using Eq. (27) and Eq. (28), the above equation can be rewritten as

$$L_r = \frac{F'(\theta_i', \eta')G(\theta_i, \theta_r, \phi_r)L_i d\omega_i(a_f P(\alpha) d\omega' dA)\cos\theta_i'}{d\omega_r dA \cos\theta_r}. \tag{29}$$

The solid angles $d\omega_r$ and $d\omega'$ are related as

$$d\omega' = \frac{d\omega_r}{4\cos\theta_i'},$$

thus by rewriting Eq. (29), we have

$$L_r = K_{spec}\frac{L_i d\omega_i}{\cos\theta_r} e^{-\frac{\alpha^2}{2\sigma_\alpha^2}} \tag{30}$$

where

$$K_{spec} = \frac{c a_f F'(\theta_i', \eta')G(\theta_i, \theta_r, \phi_r)}{4}$$

In order to account for the diffusely reflecting light, Torrance and Sparrow added the Lambertian model to Eq. (30):

$$L_r = K_{diff} L_i d\omega_i \cos\theta_i + K_{spec}\frac{L_i d\omega_i}{\cos\theta_r} e^{-\frac{\alpha^2}{2\sigma_\alpha^2}}, \tag{31}$$

This equation describes the general Torrance-Sparrow reflection model.

Nayar's Unified Model

By comparing the Beckmann-Spizzichino model and the Torrance-Sparrow model, a unified reflectance framework that is suitable for machine vision applications was developed [40]. In particular, this model consists of three components: diffuse lobe; specular lobe; and specular spike. The diffuse lobe represents the internal scattering mechanism, and is distributed around the surface normal. The specular lobe represents the reflection of incident light, and is distributed around the specular direction. Finally, the specular spike represents mirror-like reflection on smooth surfaces, and is concentrated along the specular direction. In machine vision, we are interested in image irradiance (intensity) values. Assuming that the object distance is much larger than both the focal length and the diameter of lens of imaging sensor (e.g., CCD camera), then it is shown that image irradiance is proportional to surface radiance . Therefore, the image intensity is given as a linear combination of the three reflection components:

$$I = I_{dl} + I_{sl} + I_{ss} \tag{32}$$

Two specific reflectance models were developed – one for the case of fixed light source with moving sensor and the other for the case of moving light source with fixed sensor. Figure 6.27 illustrates the reflectance model for the case of fixed light source and moving sensor. In this case, the image intensity observed by the sensor is given by

$$I = C_{dl} + C_{sl} \frac{1}{\cos\theta_r} e^{-\frac{\alpha^2}{2\sigma_\alpha^2}} + C_{ss}\delta(\theta_i - \theta_r)\delta(\phi_r) \tag{33}$$

where the constants C_{dl}, C_{sl} and C_{ss} represent the strengths of the diffuse lobe, specular lobe and specular spike respectively, and δ is a delta function. Pictorially, the strength of each reflection component is the magnitude of intersection point between the component contour and the viewing ray from the sensor. Notice that the strength of diffuse lobe is the same for all directions. Notice also that the peak of specular lobe is located at the angle slightly greater than the specular direction. This phenomenon is called *off-specular peak*, and it is caused by $\frac{1}{\cos\theta_r}$ in the specular lobe component term in Eq. (33). The off angle between the specular direction and the peak direction of specular lobe becomes larger for rougher surfaces.

Figure 6.28 illustrates the reflectance model for the case of moving source light and fixed sensor, and the image intensity observed by the sensor in this case is given by

$$I = K_{dl}\cos\theta_i + K_{sl} e^{-\frac{\alpha^2}{2\sigma_\alpha^2}} + K_{ss}\delta(\theta_i - \theta_r)\delta(\phi_r) \tag{34}$$

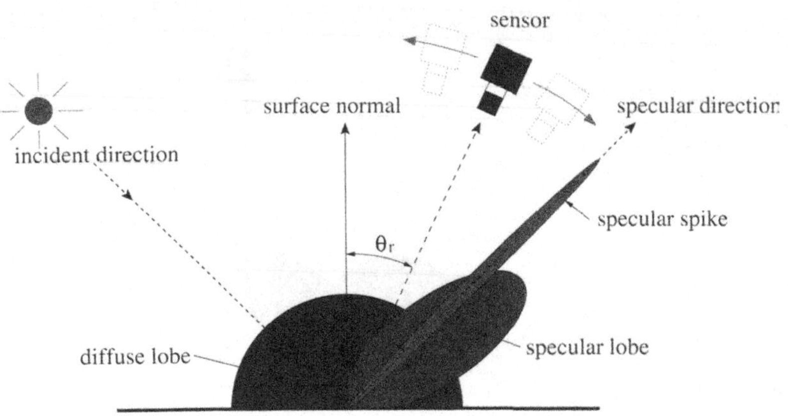

Fig. 6.27: Reflectance model for the case of fixed light source and moving sensor

It is important to note that the pictorial illustration of the strength of diffuse lobe component is different from the previous case whereas the strengths of specular lobe and specular spike are the same. Specifically, the strength of the diffuse lobe component is the magnitude of the intersection point between the diffuse lobe contour and the incident light ray, not the viewing ray as in the previous case. Notice that θ_r is constant since the sensor is fixed. Therefore, $\frac{1}{\cos\theta_r}$ can be added to the constant term of the specular lobe component (i.e., K_{sl}). Consequently, the off-specular peak is no longer observed in this case. Eq. (34) is useful for acquiring reflectance property of an object using the photometric stereo method.

The specular lobe constants C_{sl} in Eq. (33) and K_{sl} in Eq. (34) represent K_{spec} in Eq. (31). Clearly, K_{spec} is not a constant since it is a function of the Fresnel reflection coefficient $F'(\theta_i', \eta')$ and the geometrical attenuation factor $G(\theta_i, \theta_r, \phi_r)$. However, the Fresnel reflection coefficient is nearly constant until θ_i' becomes 90°, and the geometrical attenuation factor is 1 as long as both θ_i and θ_r are within 45°. Thus, assuming that θ_i' is less than 90° and θ_i and θ_r are less than 45°, C_{sl} and K_{sl} can be considered to be constants.

Ambient-Diffuse-Specular Model

The ambient-diffuse-specular model, despite its inaccuracy in representing reflectance properties of real object surfaces, is currently the most commonly used reflectance model in the computer graphics community. The main attraction of this model is its simplicity. It describes the reflected light on

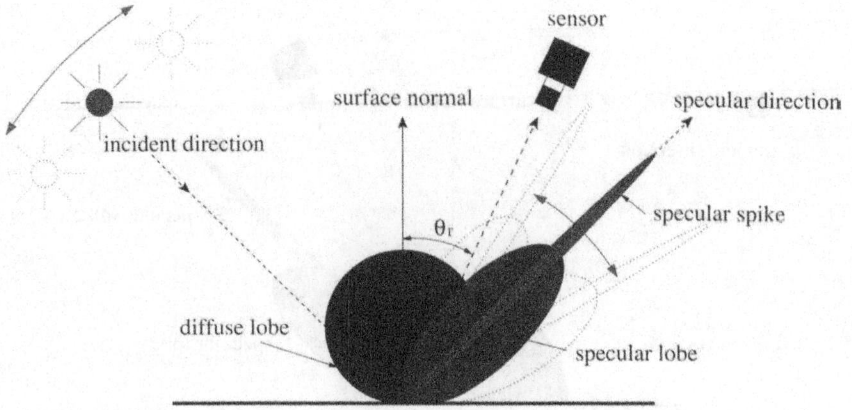

Fig. 6.28: Reflectance model for the case of moving light source and fixed sensor

the object point as a mixture of ambient, diffuse (or body), and specular (or surface) reflection. Roughly speaking, the ambient reflection represents the global reflection property that is constant for entire scene, the diffuse reflection represents the property that plays the most important role in determining what is perceived as the "true" color, and the specular reflection represents bright spots, or highlights caused by the light source. Most commonly used computer graphics applications (e.g., OpenGL) formulate the ambient-diffuse-specular model as

$$I = I_a K_a + I_l K_d \cos\theta + I_l K_s \cos^n \alpha \tag{35}$$

where I_a and I_l are the intensities of ambient light and light source respectively, and K_a, K_d and K_s are constants that represent the strengths of ambient, diffuse and specular components respectively . θ is the angle between the light source direction and the surface normal direction of the object point, α is the angle between the surface normal and the bisector of the light source and the viewing direction, and n is a constant that represents the "shininess" of the surface.

Let **L** be the light source direction, **N** the surface normal, **E** the viewing direction, and **H** the bisector of **L** and **E** (see Figure 6.29), then assuming all vectors are unit vectors, we can rewrite Eq. (35) as

$$I = I_a K_a + I_l K_d(\mathbf{L} \cdot \mathbf{N}) + I_l K_s(\mathbf{H} \cdot \mathbf{N})^n \tag{36}$$

where $\cdot$ is a dot product.

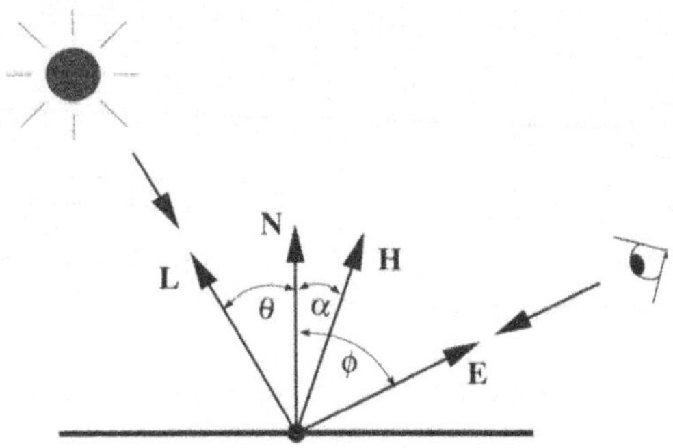

Fig. 6.29: Basic light reflection model

6.5.2 Reflection Model Parameter Estimation

Ikeuchi and Sato [27] presented a system for determining reflectance properties of an object using a single pair of range and intensity images. The range and intensity images are acquired by the same sensor, thus the correspondence between the two images are directly provided. That is, 3D position, normal direction, and intensity value for each data point are available. The reflectance model they used is similar to that of Nayar's unified model [40], but only considered the diffuse lobe and the specular lobe:

$$I = K_d \cos\theta_i + K_s \frac{1}{\cos\theta_r} e^{-\frac{\alpha^2}{2\sigma_\alpha^2}} \tag{37}$$

Assuming the object's reflectance property is uniform over the surface, their system estimates four variables: light source direction $\mathbf{L} = [L_x\ L_y\ L_z]^T$, diffuse component constant K_d, specular component constant K_s and surface roughness σ_α. Let $I(i,j)$ be the intensity value at ith row and jth column of the intensity image. The corresponding data point's normal is denoted as $\mathbf{N}(i,j) = [N_x(i,j)\ N_y(i,j)\ N_z(i,j)]^T$. Assuming the intensity image has no specular components, we have

$$\begin{aligned} I(i,j) &= K_d(\mathbf{L} \cdot \mathbf{N}(i,j)) \\ &= aN_x(i,j) + bN_y(i,j) + cN_z(i,j) \\ &= \mathbf{A} \cdot \mathbf{N}(i,j) \end{aligned}$$

where $a = K_d L_x$, $b = K_d L_y$, $c = K_d L_z$ and $\mathbf{A} = [a\ b\ c]^T$. Then, $\mathbf{A}$ is initially estimated using a least square fitting by minimizing the following equation:

$$e_1 = \sum_{i,j} [I(i,j) - aN_x(i,j) - bN_y(i,j) - cN_z(i,j)]^2 .$$

The estimated vector $\mathbf{A}^* = [a^* \ b^* \ c^*]^T$ is used to determine the ideal diffuse brightness I' for each data point:

$$I'(i,j) = a^* N_x(i,j) + b^* N_y(i,j) + c^* N_z(i,j).$$

Based on the computed ideal diffuse brightness values, the pixels are categorized into three groups using a threshold: if the observed intensity is much greater than the ideal diffuse intensity, it is considered to be a *highlight pixel*; if the observed intensity is much less than the ideal diffuse intensity, it is considered to be a *shadow pixel*; and all other pixels are categorized as *diffuse pixels*. Using only the diffuse pixels, the vector $\mathbf{A}$ and the ideal diffuse intensity values $I'(i,j)$ are recomputed, and the process is repeated until $\mathbf{A}^*$ converges. At the end, the diffuse component constant K_d and the direction of light source $\mathbf{L}$ are given by

$$\begin{aligned} K_d &= \sqrt{a^{*2} + b^{*2} + c^{*2}} \\ \mathbf{L} &= \left[\frac{a^*}{K_d} \ \frac{b^*}{K_d} \ \frac{c^*}{K_d} \right]^T \end{aligned}$$

The next step consists of estimating the specular parameters K_s and σ_α. In order to estimate the specular parameters, the highlight pixels determined in the previous process are additionally divided into two subgroups, *specular* and *interreflection* pixels, based on the angle α (Recall that α is the angle between surface normal and the bisector of source light and viewing direction). If $\alpha(i,j)$ is less than a threshold, the pixel is categorized as a specular pixel, and otherwise, it is considered to be a interreflection pixel. Intuitively, this criterion is due to the fact that mirror-like or close to mirror-like reflecting data points must have small α values. If a point contains high intensity value with relatively large α, we may assume that the main cause of the high intensity is not from the source light, but from interreflected lights. Let $d(i,j)$ be a portion of intensity $I(i,j)$ contributed by the specular component, and it is given by

$$\begin{aligned} d(i,j) &= K_s \frac{1}{\cos\theta_r(i,j)} e^{-\frac{\alpha^2(i,j)}{2\sigma_\alpha^2}} \\ &= I(i,j) - \mathbf{A} \cdot \mathbf{N}(i,j) \end{aligned}$$

The specular parameters K_s and σ_α are estimated by employing a two-step fitting method. The first step assumes that K_s is known, and estimates σ_α by minimizing

$$e_2 = \sum_{i,j} \left[\ln d'(i,j) - \ln K_s + \ln(\cos\theta_r(i,j)) + \frac{\alpha^2(i,j)}{2\sigma_\alpha^2} \right]^2$$

where $d'(i,j) = I(i,j) - \mathbf{A}^* \cdot \mathbf{N}(i,j)$. Given σ_α, the second step estimates K_s by minimizing

$$e_3 = \sum_{i,j} \left[d'(i,j) - K_s \frac{1}{\cos\theta_r(i,j)} e^{-\frac{\alpha^2(i,j)}{2\sigma_\alpha^2}} \right]^2 .$$

By repeating the two steps, the specular parameters K_s and σ_α are estimated.

Sato and Ikeuchi [48] extended the above system for multiple color images of a geometric model generated from multiple range images. They first acquire a small number of range images by rotating the object on a rotary stage, and generate a 3D model. Then, they acquire color images of the object, but this time they acquire more color images than the range images by rotating the object with a smaller interval between images.[4] The correspondence between the 3D model and the color images are known since the same sensor is used for both range and color image acquisition, and since each image was taken at a known rotation angle without moving the object. Specifically, the 3D model can be projected onto a color image using the 4 by 3 camera projection matrix rotated by the angle in which the image was taken. The light source is located near the sensor, thus they assume that the light source direction is the same as the viewing direction. Consequently, the angles θ_r, θ_i and α are all the same, and the reflectance model is given by

$$I = K_d \cos\theta + K_s \frac{1}{\cos\theta} e^{-\frac{\theta^2}{2\sigma_\alpha^2}} \tag{38}$$

In order to estimate the reflectance parameters, they first separate the diffuse components from the specular components. Let $\mathbf{M}$ be a series of intensity values of a data point observed from n different color images:

$$\mathbf{M} = \begin{bmatrix} I_1 \\ I_2 \\ \vdots \\ I_n \end{bmatrix} = \begin{bmatrix} I_{1,R} & I_{1,G} & I_{1,B} \\ I_{2,R} & I_{2,G} & I_{2,B} \\ \vdots & \vdots & \vdots \\ I_{n,R} & I_{n,G} & I_{n,B} \end{bmatrix}$$

[4] 8 range images (45° interval) and 120 color images (3° interval) were acquired in the example presented in their paper

where the subscripts R, G and B represent three primary colors. Using Eq. (38), $\mathbf{M}$ can be expressed as

$$\begin{aligned}\mathbf{M} &= \begin{bmatrix} \cos\theta_1 & E(\theta_1) \\ \cos\theta_2 & E(\theta_2) \\ \vdots & \vdots \\ \cos\theta_n & E(\theta_n) \end{bmatrix} \begin{bmatrix} K_{d,R} & K_{d,G} & K_{d,B} \\ K_{s,R} & K_{s,G} & K_{s,B} \end{bmatrix} \\ &= \begin{bmatrix} \mathbf{G}_d & \mathbf{G}_s \end{bmatrix} \begin{bmatrix} \mathbf{K}_d^T \\ \mathbf{K}_s^T \end{bmatrix} \\ &= \mathbf{GK}\end{aligned}$$

where $E(\theta_i) = \frac{1}{\cos\theta_i} e^{-\frac{\theta_i^2}{2\sigma_\alpha^2}}$. By assuming $\mathbf{K}_s$ is pure white (i.e. $\mathbf{K}_s = [1\ 1\ 1]^T$) and $\mathbf{K}_d$ is the color value with the largest θ (i.e., $\mathbf{K}_d = [I_{i,R}\ I_{i,G}\ I_{i,B}]^T$ where $\theta_i = \max(\theta_1, \theta_2, ..., \theta_n)$), $\mathbf{G}$ can be computed by

$$\mathbf{G} = \mathbf{MK}^+$$

where $\mathbf{K}^+$ is a 3×2 pseudo-inverse of $\mathbf{K}$. With the computed $\mathbf{G}$, we can separate the diffuse components $\mathbf{M}_d$ and the specular components $\mathbf{M}_s$ by

$$\begin{aligned}\mathbf{M}_d &= \mathbf{G}_d\mathbf{K}_d^T \\ \mathbf{M}_s &= \mathbf{G}_s\mathbf{K}_s^T .\end{aligned}$$

Then, the diffuse reflectance parameter $\mathbf{K}_d$ and the specular reflectance parameters $\mathbf{K}_s$ and σ_α can be estimated by applying two separate fitting processes on $\mathbf{M}_d$ and $\mathbf{M}_s$. However, the authors pointed out that the diffuse reflectance parameter was reliably estimated for each data point while the estimation of specular reflectance parameters, on the other hand, was unreliable because the specular component is usually observed from a limited range of viewing directions, and even if the specular component is observed, the parameter estimation can become unreliable if it is not observed strongly. Therefore, the specular reflectance parameters are estimated for each segmented region based on the hue value assuming that all the data points in each region are characterized by common specular reflectance parameters.[5]

In Sato *et al.* [49], instead of estimating common specular reflectance parameters for each segmented region, the authors simply select data points where specular component is observed sufficiently, and estimate parameters only on those points. The estimated parameters are then linearly interpolated over the entire object surface.

[5]The specular reflectance parameters were estimated in 4 different regions in the example in the paper.

Kay and Caelli [30] follow the idea of photometric stereo [61] and take multiple intensity images of a simple object from a single viewpoint but each time with a different light source position. The acquired intensity images along with a range image acquired from the same viewpoint are used to estimate reflectance parameters for each data point. Since all the intensity images and the range image are acquired from the same viewpoint, the problem of registration is avoided. Like [27], they also categorize each data point by the amount of information needed for the parameter estimation. If a data point contains sufficient information, the reflectance parameters are computed by fitting the data to the reflectance model similar to Eq (37), otherwise the parameters are interpolated.

Lensch *et al.* [33] first generate a 3D model of an object, and acquire several color images of the object from different viewpoints and light source positions. In order to register the 3D model and the color images, a silhouette based registration method described in their earlier paper [32] is used. Given the 3D model, and multiple radiance samples of each data point obtained from color images, reflectance parameters are estimated by fitting the data into the reflectance model proposed by Lafortune *et al.* [31]. For reliable estimation, the reflectance parameters are computed for each cluster of similar material. The clustering process initially begins by computing a set of reflectance parameters, $\mathbf{a}$, that best fits the entire data. The covariance matrix of the parameters obtained from the fitting, which provides the distribution of fitting error, is used to generate two new clusters. Specifically, two sets of reflectance parameters, $\mathbf{a}_1$ and $\mathbf{a}_2$, are computed by shifting $\mathbf{a}$ in the parameter space along the eigenvector corresponding to the largest eigenvalue of the covariance matrix. That is,

$$\mathbf{a}_1 = \mathbf{a} + \tau\mathbf{e} \qquad \mathbf{a}_2 = \mathbf{a} - \tau\mathbf{e}$$

where $\mathbf{e}$ is the largest eigenvector and τ is a constant. The data are then redistributed into two clusters based on the magnitude of fitting residuals to $\mathbf{a}_1$ and $\mathbf{a}_2$. However, due to the data noise and improper scaling of τ, the split will not be optimal, and the two new clusters may not be clearly separated. Thus, the splitting process includes an iteration of redistributing the data based on $\mathbf{a}_1$ and $\mathbf{a}_2$, and recomputing $\mathbf{a}_1$ and $\mathbf{a}_2$ by fitting the data of the corresponding cluster. The iteration terminates when the members of both clusters do not change any more. The splitting process is repeatedly performed on a new cluster until the number of clusters reaches a prespecified number. The clustering process results reflectance parameters for each cluster of similar material. Although applying a single set of reflectance parameters for each cluster would yield a plausible result, the authors provided a method for generating point by point variations within a cluster. The idea is to represent each point by a linear combination of the elements of the basis set of reflectance parameters. The basis set include the original reflectance parameters computed for the cluster, the reflectance parameters of neighboring clusters, the reflectance parameters of similar clusters, and reflectance parameters generated by slightly increasing or decreasing the original values. The authors pointed out that the use of a linear basis set in most cases does not improve upon the results achieved with the original reflectance parameter set.

Levoy *et al.* [35], in their Digital Michelangelo Project, also employ two different passes for the acquisition of geometric data and color images. The registration problem between the color images and the geometric model is solved by maintaining the position and the orientation of the camera with respect to the range sensor at all times. Since the acquisition process had to be performed inside the museum, the lighting condition could not be controlled. This implies that the ambient light had be considered as well. To get around this problem, they took two images from identical camera position, but one image only under the ambient light, and the other under the ambient light together with the calibrated light source. Then, subtracting the first image from the second results an image that represents what the camera would have seen only with the calibrated light source. After acquiring color images covering the entire surface, the systematic camera distortions of the images such as geometric distortion and chromatic aberration are corrected. Next, pixels that were occluded with respect to the camera or the light source are discarded. Finally, the remaining pixels are projected onto the merged geometric data for estimating the reflection parameters. They followed an approach similar to that described in [49], except that they only extracted diffuse reflection parameters. To eliminate specular contributions, they additionally discarded pixels that were observed with small α (i.e., close to mirror reflection direction).

6.6 Conclusion

In this report, we have presented the state-of-the-art methods for constructing geometrically and photometrically correct 3D models of real-world objects using range and intensity images. We have described four general steps involved in 3D modeling where each respective step continues to be an active research area on its own in the computer vision and computer graphics communities.

Although recent research efforts established the feasibility of constructing photo-realistic 3D models of physical objects, the current techniques are capable of modeling only a limited range of objects. One source of this limitation is severe self-occlusions, which make certain areas of object very difficult to be reached by the sensors. Another source of difficulty is the fact that many real-world objects have complex surface materials that cause problems particularly in range data acquisition and in reflectance property estimation. Various surface properties that cause difficulties in range data acquisition include specular surfaces, highly absorptive surfaces, translucent surfaces and transparent surfaces. In order to ensure that the object surface is ideal for range imaging, some researchers have simply painted the object or coated the object with removable powder. Obviously, such approaches may not be desirable or even possible outside laboratories. Park and Kak [43] recently developed a new range imaging method that accounts for the effects of mutual reflections, thus providing a way to construct accurate 3D models even of specular objects.

Complex surface materials also cause problems in reflectance property estimation. As we mentioned earlier, a large number of samples is needed in order to make a reliable estimation of reflectance property, and acquiring sufficient samples for each point of the object is very difficult. Therefore, some methods assume the object to have uniform reflectance property while other methods estimate reflectance properties only for the points with sufficient samples and linearly interpolate the parameters throughout the entire surface. Or yet, some methods segment the object surface into groups of similar materials and estimate reflectance property for each group. As one can expect, complex surface materials with high spatial variations can cause unreliable estimation of reflectance property.

The demand for constructing 3D models of various objects has been steadily growing and we can naturally predict that it will continue to grow in the future. Considering all innovations in 3D modeling we have seen in recent years, we believe the time when machines take a random object and automatically generate its replica is not too far away.

References

[1] N. Amenta, M. Bern, and M. Kamvysselis. A new voronoi-based surface reconstruction algorithm. In *SIGGRAPH'98*, pages 415–412, 1998.

[2] C. Bajaj, F. Bernardini, and G. Xu. Automatic reconstruction of surfaces and scalar fields from 3D scans. In *SIGGRAPH'95*, pages 109–118, 1995.

[3] P. Beckmann and A. Spizzichino. *The Scattering of Electromagnetic Waves from Rough Surfaces*. Pergamon Press, 1963.

[4] R. Benjemaa and F. Schmitt. Fast global registration of 3D sampled surfaces using a multi-Z-buffer technique. In *Conference on Recent Advances in 3-D Digitial Imaging and Modeling*, pages 113–120, 1997.

[5] R. Bergevin, M. Soucy, H Gagnon, and D. Laurendeau. Towards a general multiview registration technique. *IEEE Transactions on Pattern Analysis and Machine Intelligence*, 18(5):540–547, 1996.

[6] M. Bern and D. Eppstein. Mesh generation and optimal triangulation. Technical Report P92-00047, Xerox Palo Alto Research Center, 1992.

[7] F. Bernardini, J. Mittleman, H. Rushmeier, C. Silva, and G. Taubin. The ball-pivoting algorithm for surface reconstruction. *IEEE Transactions on Visualization and Computer Graphics*, 5(4):349–359, 1999.

[8] P. J. Besl and N. D. McKay. A method for registration of 3-D shapes. *IEEE Transactions on Pattern Analysis and Machine Intelligence*, 14(2):239–256, 1992.

[9] F. Blais and M. Rioux. Real-time numerical peak detector. *Signal Processing*, 11:145–155, 1986.

[10] J-D Boissonnat. Geometric structure for three-dimensional shape representation. *ACM Transactions on Graphics*, 3(4):266–286, 1984.

[11] C. Chen, Y. Hung, and J. Chung. A fast automatic method for registration of partially-overlapping range images. In *IEEE International Conference on Computer Vision*, pages 242–248, 1998.

[12] Y. Chen and G. Medioni. Object modeling by registration of multiple range images. In *IEEE International Conference on Robotics and Automation*, pages 2724–2729, 1991.

[13] Y. Chen and G. Medioni. Object modeling by registration of multiple range images. *Image and Vision Computing*, 14(2):145–155, 1992.

[14] B. Curless and M. Levoy. A volumetric method for building complex models from range images. In *SIGGRAPH'96*, pages 303–312, 1996.

[15] P. E. Debevec, C. J. Taylor, and J. Malik. Modeling and rendering architecture from photographs: A hybrid geometry- and image-based approach. In *SIGGRAPH'96*, pages 11–20, 1996.

[16] H. Edelsbrunner and E. P. Mucke. Three-dimensional alpha shapes. *ACM Transactions on Graphics*, 13(1):43–72, 1994.

[17] O. Faugeras and M. Hebert. The representation, recognition, and locating of 3D shapes from range data. *The International Journal of Robotics Research*, 5(3):27–52, 1986.

[18] H. Gagnon, M. Soucy, R. Bergevin, and D. Laurendeau. Registration of multiple range views for automatic 3-D modeling building. In *IEEE Computer Vision and Pattern Recognition*, pages 581–586, 1994.

[19] G. Godin and P. Boulanger. Range image registration through invariant computation of curvature. In *ISPRS Workshop: From Pixels to Sequences*, pages 170–175, 1995.

[20] G. Godin, D. Laurendeau, and R. Bergevin. A method for the registration of attributed range images. In *Third International Conference on 3-D Digitial Imaging and Modeling*, pages 179–186, 2001.

[21] G. Godin, M. Rioux, and R. Baribeau. 3-D registration using range and intensity information. In *SPIE Videometrics III*, pages 279–290, 1994.

[22] M. Hebert, K. Ikeuchi, and H. Delingette. A spherical representation for recognition of free-form surfaces. *IEEE Transactions on Pattern Analysis and Machine Intelligence*, 17(7):681–690, 1995.

[23] A. Hilton, A. Stoddart, J. Illingworth, and T. Windeatt. Marching triangles: Range image fusion for complex object modeling. In *IEEE International Conference on Image Processing*, pages 381–384, 1996.

[24] H. Hoppe, T. DeRose, T. Duchamp, J McDonald, and W. Stuelzle. Surface reconstruction from unorganized points. In *SIGGRAPH'92*, pages 71–78, 1992.

[25] B. K. P. Horn. Closed-form solution of absolute orientation using unit quaternions. *Optical Society of America A*, 4(4):629–642, April 1987.

[26] D. Huber and M. Hebert. Fully automatic registration of multiple 3d data sets. *Image and Vision Computing*, 21(7):637–650, July 2003.

[27] K. Ikeuchi and K. Sato. Determining reflectance properties of an object using range and brightness images. *IEEE Transactions on Pattern Analysis and Machine Intelligence*, 13(11):1139–1153, November 1991.

[28] A. Johnson and M. Hebert. Surface registration by matching oriented points. In *Conference on Recent Advances in 3-D Digitial Imaging and Modeling*, pages 121–128, 1997.

[29] A. Johnson and S. Kang. Registration and integration of textured 3-D data. In *Conference on Recent Advances in 3-D Digitial Imaging and Modeling*, pages 234–241, 1997.

[30] G. Kay and T. Caelli. Inverting an illumination model from range and intensity maps. *CVGIP: Image Understanding*, 59(2):183–201, March 1994.

[31] E. P. E. Lafortune, S.-C. Foo, K. E. Torrance, and D. P. Greenberg. Non-linear approximation of reflectance functions. In *SIGGRAPH'97*, pages 117–126, 1997.

[32] H. P. A. Lensch, W. Heidrich, and H.-P. Seidel. Automated texture registration and stitching for real world models. In *The 8th Pacific Conference on Computer Graphics and Applications*, pages 317–326, 2000.

[33] H. P. A. Lensch, J. Kautz, M. Goesele, W. Heidrich, and H.-P. Seidel. Image-based reconstruction of spatially varying materials. In *The 12th Eurographics Rendering Workshop*, 2001.

[34] R. Lenz and R. Tsai. Techniques for calibration of the scale factor and image center for high frequency 3-D machine vision metrology. *IEEE Transactions on Pattern Analysis and Machine Intelligence*, 10(5):713–720, 1988.

[35] M. Levoy, K. Pulli, B. Curless, Z. Rusinkiewicz, D. Koller, L. Pereira, M. Ginzton, S. Anderson, J. Davis, J. Ginsberg, J. Shade, and D. Fulk. The digital michelangelo project: 3D scanning of large statues. In *SIGGRAPH'00*, pages 131–144, 2000.

[36] W. E. Lorensen and H. E. Cline. Marching cubes: A high resolution 3D surface construction algorithm. In *SIGGRAPH'87*, pages 163–169, 1987.

[37] T. Masuda, K. Sakaue, and N. Yokoya. Registration and integration of multiple range images for 3-D model construction. In *IEEE International Conference on Pattern Recognition*, pages 879–883, 1996.

[38] T. Masuda and N Yokoya. A robust method for registration and segmentation of multiple range images. In *IEEE CAD-Based Vision Workshop*, pages 106–113, 1994.

[39] C. Montani, R. Scateni, and R. Scopigno. A modified look-up table for implicit disambiguation of marching cubes. *Visual Computer*, 10(6):353–355, 1994.

[40] S. K. Nayar, K. Ikeuchi, and T. Kanade. Surface reflection: Physical and geometrical perspectives. *IEEE Transactions on Pattern Analysis and Machine Intelligence*, 13(7):611–634, July 1991.

[41] F. E. Nicodemus, J. C. Richmond, J. J. Hsia, I. W. Ginsberg, and T. Limperis. Geometrical considerations and nomenclature for reflectance. Technical Report BMS Monograph 160, National Bureau of Standards, October 1977.

[42] K. Nishino, Y. Sato, and K. Ikeuchi. Eigen-texture method: Appearance compression based on 3D model. In *IEEE Conference on Computer Vision and Pattern Recognition*, volume 1, pages 618–624, 1999.

[43] J. Park and A. C. Kak. Multi-peak range imaging for accurate 3D reconstruction of specular objects. In *6th Asian Conference on Computer Vision*, 2004.

[44] M. Potmesil. Generating models of solid objects by matching 3D surface segments. In *The 8th International Joint Conference on Artificial Intelligence (IJCAI)*, pages 1089–1093, 1983.

[45] K. Pulli. *Surface Reconstruction and Display from Range and Color Data*. PhD thesis, University of Washington, 1997.

[46] K. Pulli. Multiview registration for large data sets. In *Second International Conference on 3-D Digitial Imaging and Modeling*, pages 160–168, 1999.

[47] K. Pulli, M. Cohen, T. Duchamp, H. Hoppe, L. Shapiro, and W. Stuetzle. View-based rendering: Visualizing real objects from scanned range and color data. In *8th Eurographics Workshop on Rendering*, pages 23–34, 1997.

[48] Y. Sato and K. Ikeuchi. Reflectance analysis for 3D computer graphics model generation. *Graphical Models and Image Processing*, 58(5):437–451, September 1996.

[49] Y. Sato, M. Wheeler, and K. Ikeuchi. Object shape and reflectance modeling from observation. In *SIGGRAPH'97*, pages 379–387, 1997.

[50] T. Schuts, T. Jost, and H. Hugli. Multi-featured matching algorithm for free-form 3D surface registration. In *IEEE International Conference on Pattern Recognition*, pages 982–984, 1998.

[51] M. Soucy and D. Laurendeau. Multi-resolution surface modeling from multiple range views. In *IEEE Computer Vision and Pattern Recognition*, pages 348–353, 1992.

[52] M. Soucy and D. Laurendeau. A general surface approach to the integration of a set of range views. *IEEE Transactions on Pattern Analysis and Machine Intelligence*, 17(4):344–358, 1995.

[53] A. Stoddart, S. Lemke, A. Hilton, and T. Penn. Estimating pose uncertainty for surface registration. In *British Machine Vision Conference*, pages 23–32, 1996.

[54] K. E. Torrance and E. M. Sparrow. Theory for off-specular reflection from roughened surfaces. *Journal of the Optical Society of America A*, 57(9):1105–1114, September 1967.

[55] E. Trucco, R. B. Fisher, A. W. Fitzgibbon, and D. K. Naidu. Calibration, data consistency and model acquisition with laser stripes. *International Journal of Computer Integrated Manufacturing*, 11(4):293–310, 1998.

[56] R. Tsai. A versatile camera calibration technique for high-accuracy 3D machine vision metrology using off-the-shelf TV cameras and lenses. *IEEE Journal of Robotics and Automation*, 3(4):323–344, 1987.

[57] G. Turk and M. Levoy. Zippered polygon meshes from range images. In *SIGGRAPH'94*, pages 311–318, 1994.

[58] B. C. Vemuri and J. K. Aggarwal. 3D model construction from multiple views using range and intensity data. In *IEEE Conference on Computer Vision and Pattern Recognition*, pages 435–438, 1986.

[59] M. Wheeler, Y. Sato, and K. Ikeuchi. Consensus surface for modeling 3D objects from multiple range images. In *IEEE International Conference on Computer Vision*, pages 917–924, 1998.

[60] D. N. Wood, D. I. Azuma, K. Aldinger, B. Curless, T. Duchamp, D. H. Salesin, and W. Stuetzle. Surface light fields for 3D photography. In *SIGGRAPH'00*, pages 287–296, 2000.

[61] R. J. Woodham. Photometric method for determining surface orientation from multiple images. *Optical Engineering*, 19(1):139–144, 1980.

[62] Z. Zhang. Iterative point matching for registration of free-form curves and surfaces. *Internation Journal of Computer Vision*, 13(2):119–152, 1994.

7 Perception for Human Motion Understanding

Christopher R. Wren

Mitsubishi Electric Research Laboratories
Cambridge, Massachusetts, USA
wren@merl.com

The fact that people are embodied places powerful constraints on their motion. By leveraging these constraints, we can build systems to perceive human motion that are fast and robust. More importantly, by understanding how these constraint systems relate to one another, and to the perceptual process itself, we can make progress toward building systems that interpret, not just capture, human motion.

7.1 Overview

The laws of physics, the construction of the human skeleton, the layout of the musculature, the various levels of organization within the nervous system, the context of a task, and even forces of habits and culture all conspire to limit the possible configurations and trajectories of the human form. The kinematic constraints of the skeleton are instantaneous. They are always true, and serve to bound the domain of feasible estimates. The rest of these constraints exist to some degree in the temporal domain: given past observations, they tell us something about future observations.

These phenomena cover a wide range of the time scales. The laws of physics apply in a continuous, instantaneous fashion. The subtle limits of muscle action may play out on time scales of milliseconds. Temporal structure due to the nervous system may range from tenths of seconds to minutes. Depending on the definition of a task, the task context may change over fractions of a minute or fractions of an hour. The subtle influence of affect might change over hours or even days. Habits and cultural norms develop over a lifetime.

A truly complete model of human embodiment would encompass all of these things. Unfortunately most of these phenomena are beyond the scope of current modeling techniques. Neuroscience is only beginning to explain the impact of the structures of the peripheral nervous system on motion. Models of higher-level processes such as affect, task and culture are even farther away.

The things that we can model explicitly include the instantaneous geometric constraints (blobs, perspective, and kinematics) and the dynamic constraints of Newton's Laws. Blobs represent a visual constraint. We are composed of parts, and those parts appear in images as connected, visually

C.R. Wren: *Perception for Human Motion Understanding*, Studies in Computational Intelligence (SCI) **7**, 265–324 (2005)
www.springerlink.com

coherent regions. Perspective constraints model the relationship between multiple views of the human body caused by our 3-D nature and the perspective projection of the world onto a CCD by a lens inside a camera. Kinematic constraints are the skeletal or connective constraints between the parts of the body: the length of a limb, the mechanics of a joint, and so on. The instantaneous configuration of the body is the *pose*. The kinematic constraints define the space of valid poses.

Newton's Laws represent a set of dynamic constraints: constraints in time. The assumption of bounded forces in the system implies bounded accelerations. Bounded accelerations in turn imply smoothness of the pose trajectory in time. Since the articulated frame of the body is complex and involves revolute joints, this isn't simply a smoothness constraint. It is a shaping function that is related to the global mass matrix which is a non-linear, time-varying function of the pose.

The rest of the constraint layers (neuromuscular, contextual, and psychological) can currently only be modeled statistically through observation. Fortunately the recursive estimation framework discussed below offers a natural way to factor out these influences and treat them separately from the geometry and physics. Unfortunately, further factorization of the signal is a poorly understood problem. As a result, we will treat these separate influences as a single, unified influence. This is obviously a simplification, but it is currently a necessary simplification.

7.1.1 Recursive Filters

The geometric constraints discussed above are useful for regularizing pose estimation, but the dynamic constraints provide something even more important: since they represent constraints in time, they allow prediction into the future. This is important because for human motion observed at video rates, physics is a powerful predictor.

With a model of the observation process, predictions of 3-D body pose in the near future can be turned into predictions of observations. These predictions can be compared to actual observations when they are made. Measuring the discrepancy between prediction and observation provides useful information for updating the estimates of the pose. These differences are called innovations because they represent the aspects of the observations that were unpredicted by the model.

This link between model and observation is the powerful idea behind all recursive filters, including the well known Kalman filters. Kalman filters are the optimal recursive filter formulation for the class of problems with linear dynamics, linear mappings between state and observation, and white, Gaussian process noise. Extended Kalman filters generalize the basic formulation to include the case of analytically linearizable observation and dynamic models.

Recursive filters are able to cope with data in real time thanks to a Markovian assumption that the state of the system contains all the information needed to predict its behavior. For example, the state of a rigid physical object would include both the position and velocity. There is no need for the filter to simultaneously consider all the observations ever made of the subject to determine it's state. The update of the state estimate only requires combining the innovation with the dynamic, observation, and noise models.

The complete recursive loop includes measurement, comparison of predicted observation to actual observation, corresponding update of state estimate, prediction of future state estimate, and rendering of the next predicted observation. This is the basic flow of information in a Kalman filter, and applies equally well to recursive filters in general.

For the case of observing the human body, this general framework is complicated by the fact that the human body is a 3-D articulated system and the observation process is significantly non-trivial. Video images of the human body are high-dimensional signals and the mapping between body pose and image observation involves perspective projection. These unique challenges go beyond the original design goals of the Kalman and extended Kalman filters and they make the task of building systems to observe human motion quite difficult.

7.1.2 Feedback for Early Vision

Most computer vision systems are modularized to help reduce software complexity, manage bandwidth, and improve performance. Often, low-level modules, comprised of filter-based pattern recognition pipelines, provide features to mid-level modules that then use statistical or logical techniques to infer meaning. The mid-level processes are made tractable by the dimensionality reduction accomplished by the low-level modules, but these improvements can incur a cost in robustness. These systems are often brittle: they fail when the assumptions in a low-level filter are violated. Once a low-level module fails, the information is lost. Even in the case where the mid-level module can employ complex models to detect the failure, there is no way to recover the lost information if there is no downward flow of information that can be used to avert the failure. The system is forced to rely on complex heuristics to attempt approximate repair[43].

Dynamic constraints enable the prediction of observations in the near future. These predictions, with the proper representation, can be employed by low-level perceptual processes to resolve ambiguities. This results in a more robust system, by enabling complex, high-level models to inform the earliest stages of processing. It is possible to retain the advantages of modularity in a closed-loop system through carefully designed interfaces.

For example, the DYNA system[51] measures the 2-D locations of body parts in the image plane using an image-region tracker. The system then estimates 3-D body part locations from stereo pairs of 2-D observations. Finally the full body pose is estimated from these 3-D observations using a 3-D, non-linear model of the kinematics and dynamics of the human body. This system is well modularized and fast, but but would be very brittle if it relied on information only flowing from low-level processes to high-level interpretation. Instead, predictions from the dynamic model are incorporated as prior information into the probabilistic blob tracker. The tracker is the first process to be applied to the pixels, so given this feedback, there is no part of the system that is bottom-up. Even this lowest-level pixel classification process incorporates high-level model influence in the form of state predictions represented as prior probabilities for pixel classification.

This influence is more significant than simply modifying or bounding a search routine. Our classifier actually produces different results in the presence of feedback: results that reflect global classification decisions instead of locally optimal decisions that may be misleading or incomplete in the global context. This modification is made possible due to the statistical nature of our blob tracker. Prior information generated by the body model transforms the bottom-up, maximum likelihood blob tracker into a maximum *a posteriori* classifier. Thanks to the probabilistic nature of the blob tracker, it is possible to hide the details of the high-level processes from the low-level processes, and thereby retain the speed and simplicity of the pixel classification.

7.1.3 Expression

An appropriate model of embodiment allows a perceptual system to separate the necessary aspects of motion from the purposeful aspects of motion. The necessary aspects are a result of physics and are predictable. The purposeful aspects are the direct result of a person attempting to express themselves through the motion of their bodies. Understanding embodiment is the key to perceiving expressive motion.

Human-computer interfaces make measurements of a human and use those measurements to give them control over some abstract domain. The sophistication of these measurements range from the trivial keyclick to the most advanced perceptual interface system. Once the measurements are acquired the system usually attempts to extract some set of features as the first step in a pattern recognition system that will convert those measurements into whatever domain of control the application provides. Those features are usually chosen for mathematical convenience or to satisfy an *ad hoc* notion of invariance.

The innovations process discussed above is a fertile source of features that are directly related to the embodiment of the human. When neuromuscular, contextual or psychological influences affect the motion of the body, these effects will appear in the innovations process if they are not explicitly modeled. This provides direct access for learning mechanisms to these influences without compounding them with the effects of physics, kinematics, imaging, or any other process that can be explicitly modeled by the system. This tight coupling between appearance, motion and behavior is a powerful implication of this framework.

7.2 Theoretic Foundations

This section will expand on the ideas presented in Section 7.1, while linking them to their roots in stochastic estimation theory. We begin with a grounding in the basic theories, which can be explored in more details in Gelb[2]. Then we proceed to expand on those ideas to find inspiration.

The fundamental idea presented in Section 7.1 is that perception is improved when it is coupled with expectations about the process being observed: specifically a model with the ability to make qualified predictions into the future given past observations. A logical framework for creating and employing this kind of model in a perceptual system can be found in the control and estimation literature. Since the human body is a physical system, it shares many properties with the general class of dynamic systems. It is instructive to approach the task of understanding human motion in the same way that an engineer might approach the task of observing any dynamic system.

One possible simplified block diagram of a human is illustrated in Figure 7.1. The passive, physical reality of the human body is represented by the *Plant*. The propagation of the system forward in time is governed by the laws of physics and is influenced by signals, $\mathbf{u}$, from *Control*. On the right, noisy observations, $\mathbf{y}$, can be made of the Plant. On the left, high level goals, $\mathbf{v}$, are supplied to the Controller.

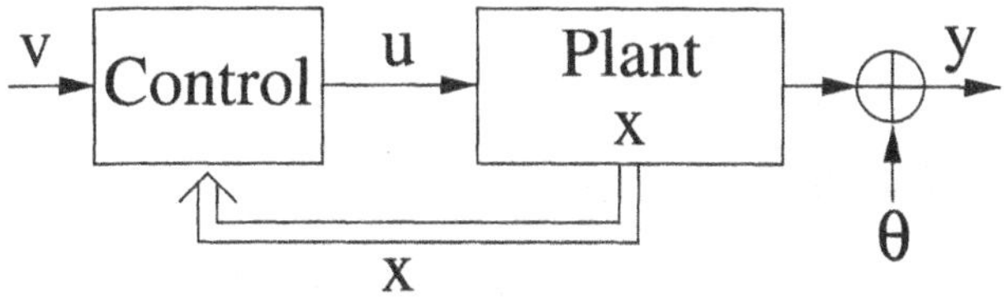

Fig. 7.1: A systems view of the human body

The observations are a function of the system state according to some measurement process, $h(\cdot)$. In our case this measurement process corresponds to the imaging process of a camera. As such, it is a non-linear, incomplete transform: cameras do not directly measure velocity, they are subject to occlusion, and they project 3-D information into 2-D observations:

$$\mathbf{y}_{t+1} = h(\mathbf{x}_{t+1}) + \boldsymbol{\theta}_t \tag{1}$$

The measurement process is also noisy. $\boldsymbol{\theta}_t$ represents additive noise in the observation. The $\boldsymbol{\theta}_t$ are assumed to be samples from a white, Gaussian, zero-mean process with covariance $\boldsymbol{\Theta}$:

$$\boldsymbol{\theta}_t \leftarrow \mathcal{N}(\mathbf{0}, \boldsymbol{\Theta}) \tag{2}$$

The state vector, $\mathbf{x}$, completely defines the configuration of the system in phase-space. The Plant propagates the state forward in time according to the system constraints. In the case of the human body this includes the non-linear constraints of kinematics as well as the constraints of dynamics. The Plant also reacts to the influences of the control signal. For the human body these influences come as muscle forces. It is assumed that the Plant can be represented by an unknown, non-linear function $f(\cdot, \cdot)$:

$$\mathbf{x}_{t+1} = f(\mathbf{x}_t, \mathbf{u}_t) \tag{3}$$

The control signals are physical signals, for example, muscle activations that result in forces being applied to the body. The Controller obviously represents a significant amount of complexity: muscle activity, the properties of motor nerves, and all the complex motor control structures from the spinal cord up into the cerebellum. The Controller has access to the state of the Plant, by the process of proprioception:

$$\mathbf{u}_t = c(\mathbf{v}_t, \mathbf{x}_t) \tag{4}$$

The high-level goals, $\mathbf{v}$, are very high-level processes. These signals represent the place where intentionality enters into the system. If we are building a system to interact with a human, then we get the observations, $\mathbf{y}$, and what we're really interested in is the intentionality encoded in $\mathbf{v}$. Everything else is just in the way.

7.2.1 A Classic Observer

A classic observer for such a system takes the form illustrated in Figure 7.2. This is the underlying structure of recursive estimators, including the well known Kalman and extended Kalman filters.

The *Observer* is an analytical model of the physical Plant:

$$\mathbf{x}_{t+1} = \boldsymbol{\Phi}_t \mathbf{x}_t + \mathbf{B}_t \mathbf{u}_t + \mathbf{L}_t \boldsymbol{\xi}_t \tag{5}$$

The unknown, non-linear update equation, $f(\cdot, \cdot)$ from Equation 3, is mod-

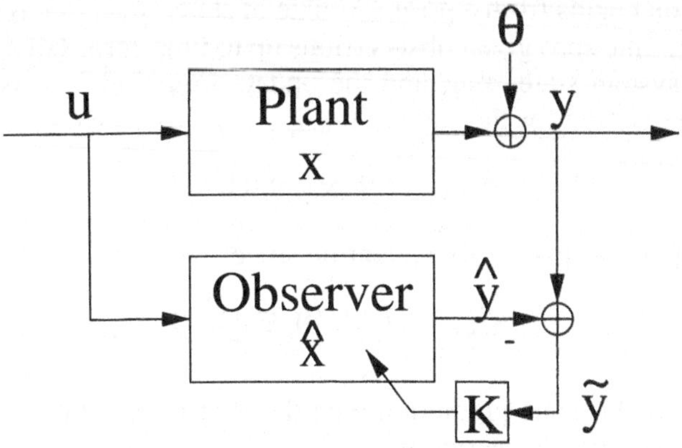

Fig. 7.2: Classic observer architecture

eled as the sum of two non-linear functions: $\Phi(\cdot)$ and $B(\cdot)$. $\Phi(\cdot)$ propagates the current state forward in time, and $B(\cdot)$ maps control signals into state influences. $\mathbf{\Phi}_t$ and $\mathbf{B}_t$ from Equation 5 are linearizations of $\Phi(\cdot)$ and $B(\cdot)$ respectively, at the current operating point. The right-hand term, $(\mathbf{L}_t\boldsymbol{\xi}_t)$, represents the effect of noise introduced by modeling errors on the state update. The $\boldsymbol{\xi}_t$ are assumed to be samples from a white, Gaussian, zero-mean process with covariance Ξ that is independent of the observation noise from Equation 2:

$$\boldsymbol{\xi}_t \leftarrow \mathcal{N}(\mathbf{0}, \Xi) \tag{6}$$

The model of the measurement process is also linearized. $\mathbf{H}_t$ is a linearization of the non-linear measurement function $h(\cdot)$:

$$\mathbf{y}_{t+1} = \mathbf{H}_t \mathbf{x}_{t+1} + \boldsymbol{\theta}_t \tag{7}$$

The matricies $\mathbf{\Phi}_t$ and $\mathbf{H}_t$, are performed by computing the Jacobian matrix. The Jacobian of a multivariate function of $\mathbf{x}$ such as $\mathbf{\Phi}(\cdot)$ is computed as the matrix of partial derivatives at the operating point $\mathbf{x}_t$ with respect to the components of $\mathbf{x}$:

$$\mathbf{\Phi}_t = \nabla_{\mathbf{x}_\mathbf{x}} \mathbf{\Phi}\big|_{\mathbf{x}=\mathbf{x}_t} = \begin{bmatrix} \frac{\partial \mathbf{\Phi}_1}{\partial \mathbf{x}_1}\Big|_{\mathbf{x}=\mathbf{x}_t} & \frac{\partial \mathbf{\Phi}_1}{\partial \mathbf{x}_2}\Big|_{\mathbf{x}=\mathbf{x}_t} & \cdots & \frac{\partial \mathbf{\Phi}_1}{\partial \mathbf{x}_n}\Big|_{\mathbf{x}=\mathbf{x}_t} \\ \frac{\partial \mathbf{\Phi}_2}{\partial \mathbf{x}_1}\Big|_{\mathbf{x}=\mathbf{x}_t} & \ddots & & \vdots \\ \vdots & & \ddots & \\ \frac{\partial \mathbf{\Phi}_m}{\partial \mathbf{x}_1}\Big|_{\mathbf{x}=\mathbf{x}_t} & \cdots & \cdots & \frac{\partial \mathbf{\Phi}_m}{\partial \mathbf{x}_n}\Big|_{\mathbf{x}=\mathbf{x}_t} \end{bmatrix}$$

This operation is often non-trivial.

Estimation begins from a prior estimate of state: $\hat{\mathbf{x}}_{0|0}$, that is the estimate of the state at time zero given observations up to time zero. Given the current estimate of system state, $\hat{\mathbf{x}}_{t|t}$, and the update Equation 5, it is possible to compute a prediction for the state at $t+1$:

$$\hat{\mathbf{x}}_{t+1|t} = \mathbf{\Phi}_t \hat{\mathbf{x}}_{t|t} + \mathbf{B}_t \mathbf{u}_t \tag{8}$$

Notice that $\boldsymbol{\xi}_t$ is not part of that equation since:

$$E[\boldsymbol{\xi}_t] = E\left[\mathcal{N}(\mathbf{0}, \mathbf{\Xi})\right] = \mathbf{0} \tag{9}$$

Combining this state prediction with the measurement model provides a prediction of the next measurement:

$$\hat{\mathbf{y}}_{t+1|t} = \mathbf{H}_t \hat{\mathbf{x}}_{t+1|t} \tag{10}$$

Again, $\boldsymbol{\theta}_t$ drops out since:

$$E[\boldsymbol{\theta}_t] = E\left[\mathcal{N}(\mathbf{0}, \mathbf{\Theta})\right] = \mathbf{0} \tag{11}$$

Given this prediction it is possible to compute the residual error between the prediction and the actual new observation $\mathbf{y}_{t+1}$:

$$\tilde{\mathbf{y}}_{t+1} = \boldsymbol{\nu}_{t+1} = \mathbf{y}_{t+1} - \hat{\mathbf{y}}_{t+1|t} \tag{12}$$

This residual, called the innovation, is the information about the actual state of the system that the filter was unable to predict, plus noise. A weighted version of this residual is used to revise the new state estimate for time $t+1$ to reflect the new information in the most recent observation:

$$\hat{\mathbf{x}}_{t+1|t+1} = \hat{\mathbf{x}}_{t+1|t} + \mathbf{K}_{t+1} \tilde{\mathbf{y}}_{t+1} \tag{13}$$

In the Kalman filter, the weighting matrix is the well-known Kalman gain matrix. It is computed from the estimated error covariance of the state prediction, the measurement models, and the measurement noise covariance, $\mathbf{\Theta}$:

$$\mathbf{K}_{t+1} = \mathbf{\Sigma}_{t+1|t} \mathbf{H}_t^T \left[\mathbf{H}_t \mathbf{\Sigma}_{t+1|t} \mathbf{H}_t^T + \mathbf{\Theta}_{t+1}\right]^{-1} \tag{14}$$

The estimated error covariance of the state prediction is initialized with the estimated error covariance of the prior state estimate, $\mathbf{\Sigma}_{0|0}$. As part of the state prediction process, the error covariance of the state prediction can be computed from the error covariance of the previous state estimate using the dynamic update rule from Equation 5:

$$\mathbf{\Sigma}_{t+1|t} = \mathbf{\Phi}_t \mathbf{\Sigma}_{t|t} \mathbf{\Phi}_t^T + \mathbf{L}_t \mathbf{\Xi}_t \mathbf{L}_t^T \tag{15}$$

Notice that, since $\mathbf{u}_t$ is assumed to be deterministic, it does not contribute to this equation.

Incorporating new information from measurements into the system reduces the error covariance of the state estimate: after a new observation, the state estimate should be closer to the true state:

$$\mathbf{\Sigma}_{t+1|t+1} = [\mathbf{I} - \mathbf{K}_{t+1}\mathbf{H}_t]\,\mathbf{\Sigma}_{t+1|t} \tag{16}$$

Notice, in Equation 5, that that classic Observer assumes access to the control signal $\mathbf{u}$. For people, remember that the control signals represent muscle activations that are unavailable to a non-invasive Observer. That means that an observer of the human body is in the slightly different situation illustrated in Figure 7.3.

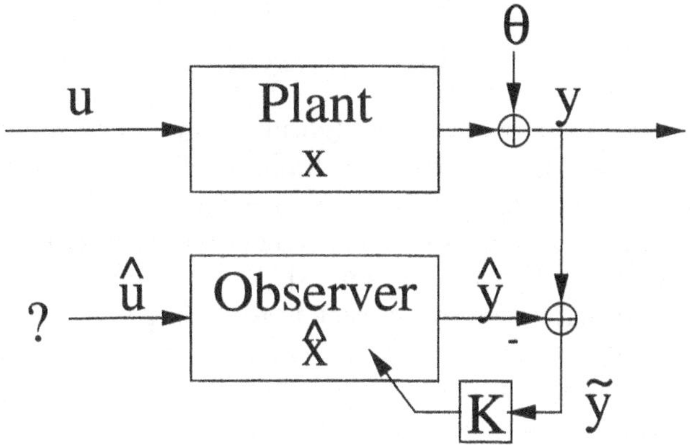

Fig. 7.3: An Observer of the human body can't access $\mathbf{u}$

7.2.2 A Lack of Control

Simply ignoring the $(\mathbf{B}_t\mathbf{u})$ term in Equation 5 results in poor estimation performance. Specifically, the update Equation 13 expands to:

$$\hat{\mathbf{x}}_{t+1|t+1} = \hat{\mathbf{x}}_{t+1|t} + \mathbf{K}_{t+1}(\mathbf{y}_{t+1} - \mathbf{H}_t(\mathbf{\Phi}_t\hat{\mathbf{x}}_{t|t} + \mathbf{B}_t\mathbf{u}_t)) \tag{17}$$

In the absence of access to the control signal $\mathbf{u}$, the update equation becomes:

$$\tilde{\hat{\mathbf{x}}}_{\mathbf{t+1|t+1}} = \hat{\mathbf{x}}_{\mathbf{t+1|t}} + \mathbf{K}_{\mathbf{t+1}}(\mathbf{y}_{\mathbf{t+1}} - \mathbf{H}_{\mathbf{t}}(\mathbf{\Phi}_{\mathbf{t}}\hat{\mathbf{x}}_{\mathbf{t|t}} + \mathbf{B}_{\mathbf{t}}\mathbf{0})) \tag{18}$$

The error ε between the ideal update and the update without access to the control signal is then:

$$\varepsilon = \left|\hat{\mathbf{x}}_{t+1|t+1} - \tilde{\hat{\mathbf{x}}}_{\mathbf{t+1|t+1}}\right| \tag{19}$$
$$= \mathbf{K}_{t+1}\mathbf{H}_t\mathbf{B}_t\mathbf{u}_t \tag{20}$$

Treating the control signal, $\mathbf{u}_t$, as a random variable, we compute the control mean and covariance matrix:

$$\bar{\mathbf{u}} = E[\mathbf{u}_t] \tag{21}$$
$$\mathbf{U} = E[(\mathbf{u}_t - \bar{\mathbf{u}})(\mathbf{u}_t - \bar{\mathbf{u}})^T] \tag{22}$$

If the control covariance matrix is small relative to the model and observation noise, by which we mean:

$$\|\mathbf{U}\| << \|\mathbf{\Xi}_t\| \tag{23}$$
$$\|\mathbf{U}\| << \|\mathbf{\Theta}_t\| \tag{24}$$

then the standard recursive filtering algorithms should be robust enough to generate good state and covariance estimates. However, as $\mathbf{U}$ grows, so will the error ε. For large enough $\mathbf{U}$ it will not be possible to hide these errors within the assumptions of white, Gaussian process noise, and filter performance will significantly degrade [3].

It should be obvious that we expect $\mathbf{U}$ to be large: if $\mathbf{u}$ had only negligible impact on the evolution of $\mathbf{x}$, then the human body wouldn't be very effective. The motion of the human body is influenced to a large degree by the actions of muscles and the control structures driving those muscles. This situation will be illustrated in Section 7.3.

7.2.3 Estimation of Control

It is not possible to measure $\mathbf{u}_t$ directly. It is inadvisable to ignore the effects of active control, as shown above. An alternative is to estimate $\mathbf{u}_{t+1|t}$. This alternative is illustrated in Figure 7.4: assuming that there is some amount of structure in $\mathbf{u}$, the function $g(\cdot,\cdot)$ uses $\hat{\mathbf{x}}$ and $\tilde{\mathbf{y}}$ to estimate $\hat{\mathbf{u}}$.

The measurement residual, $\tilde{\mathbf{y}}_{t+1}$ is a good place to find information about $\mathbf{u}_t$ for several reasons. Normally, in a steady-state observer, the measurement residual is expected to be zero-mean, white noise, so $E[\tilde{\mathbf{y}}_t] = 0$. From Equation 20 we see that without knowledge of $\mathbf{u}_t$, $\tilde{\mathbf{y}}_{t+1}$ will be biased:

$$E[\tilde{\mathbf{y}}_{t+1}] = \mathbf{H}_t\mathbf{B}_t\mathbf{u}_t \tag{25}$$

This bias is caused by the faulty state prediction resulting in a biased measurement prediction. Not only will $\tilde{\mathbf{y}}_{t+1}$ not be zero-mean, it will also not be

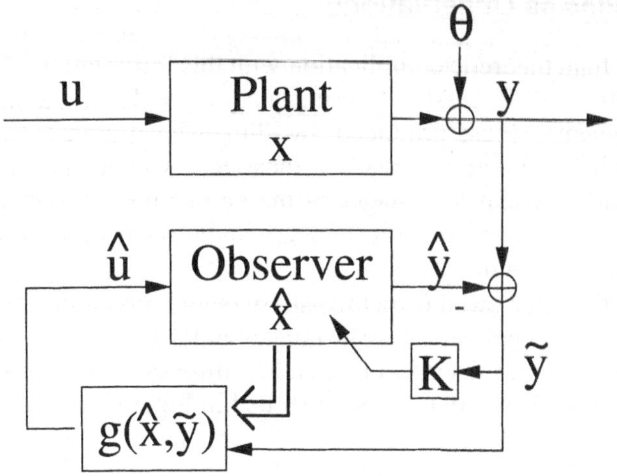

Fig. 7.4: An observer that estimates $\hat{\mathbf{u}}$ as well as $\hat{\mathbf{x}}$

white. Time correlation in the control signal will introduce time correlation in the residual signal due to the slow moving bias. Specific examples of such structure in the residuals will be shown in Section 7.3.

Learning the bias and temporal structure of the measurement residuals provides a mechanism for learning models of $\mathbf{u}$. Good estimates of $\mathbf{u}$ will lead to better estimates of $\mathbf{x}$ which are useful for a wide variety of applications including motion capture for animation, direct manipulation of virtual environments, video compositing, diagnosis of motor disorders, and others. However, if we remain focused on the intentionality represented by $\mathbf{v}$ on the far left of Figure 7.1, then this improved tracking data is only of tangential interest as a means to compute $\hat{\mathbf{v}}$.

The neuroscience literature[42] is our only source of good information about the control structures of the human body, and therefore the structure of $\mathbf{v}$. This literature seems to indicate that the body is controlled by the setting of goal states. The muscles change activation in response to these goals, and the limb passively evolves to the new equilibrium point. The time scale of these mechanisms seem to be on the scale of hundreds of milliseconds.

Given this apparent structure of $\mathbf{v}$, we expect that the internal structure of $g(\cdot,\cdot)$ should contain states that represent switches between control paradigms, and thus switches in the high-level intentionality encoded in $\mathbf{v}$. Section 7.3 discusses possible representations for $g(\cdot,\cdot)$ and Section 7.4 discusses results obtained in controlled contexts (where the richness of $\mathbf{v}$ is kept manageable by the introduction of a constrained context).

7.2.4 Images as Observations

There is one final theoretic complication with this formulation of an observer for human motion. Recursive filtering matured under the assumption that the measurement process produced low-dimensional signals under a measurement model that could be readily linearized: such as the case of a radar tracking a ballistic missile. Images of the human body taken from a video stream do not fit this assumption: they are high dimensional signals and the imaging process is complex.

One solution, borrowed from the pattern recognition literature, is to place a deterministic filter between the raw images and the Observer. The measurements available to the Observer are then low-dimensional features generated by this filter [46]. This situation is illustrated in Figure 7.5.

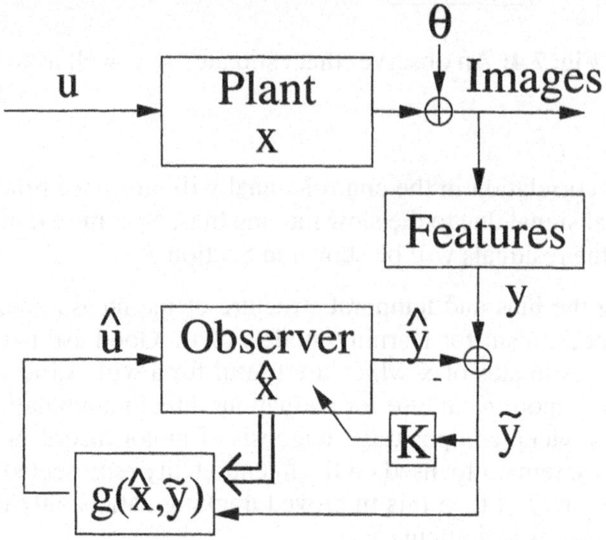

Fig. 7.5: Feature Extraction between image observations and the Observer

One fatal flaw in this framework is the assumption that it is possible to create a stationary filter process that is robust and able to provide all the relevant information from the image as a low dimensional signal for the Observer. This assumption essentially presumes a pre-existing solution to the perception problem. A sub-optimal filter will succumb to the problem of perceptual aliasing under a certain set of circumstances specific to that filter. In these situations the measurements supplied to the Observer will be flawed. The filter will have failed to capture critical information in the low-dimensional measurements. It is unlikely that catastrophic failures in feature extraction will produce errors that fit within the assumed white, Gaussian, zero-mean measurement noise model. Worse, the situation in Figure 7.5 provides no way for the predictions available in the Observer to avert these failures. This problem will be demonstrated in more detail in Section 7.5.

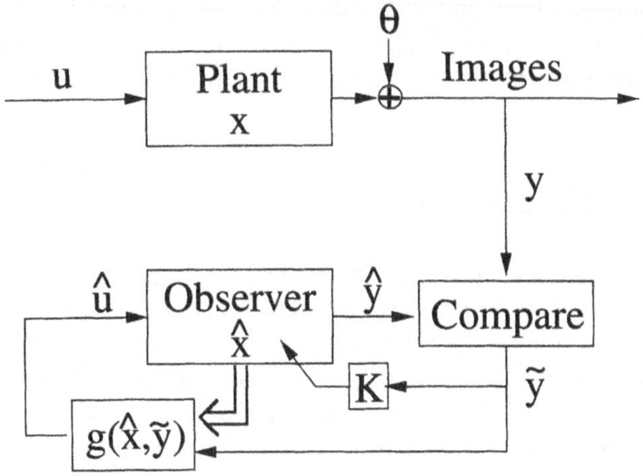

Fig. 7.6: The Observer driving a steerable feature extractor

A more robust solution is illustrated in Figure 7.6. A steerable feature extraction process takes advantage of observation predictions to resolve ambiguities. It is even possible to compute an estimate of the observation prediction error covariance, $(\mathbf{H}_t \boldsymbol{\Sigma}_{t+1|t} \mathbf{H}_t^T)$ and weight the influence of these predictions according to their certainty. Since this process takes advantage of the available predictions it does not suffer from the problems described above, because prior knowledge of ambiguities enables the filter to anticipate catastrophic failures. This allows the filter to more accurately identify failures and correctly propagate uncertainty, or even change modes to better handle the ambiguity. A fast, robust implementation of such a system is described in detail in Section 7.3.

7.2.5 Summary

So we see that exploring the task of observing the human from the vantage of classical control theory provides interesting insights. The powerful recursive link between model and observation will allow us to build robust and fast systems. Lack of access to control signals represent a major difference between observing built systems and observing biological systems. Finally that there is a possibility of leveraging the framework to help in the estimation of these unavailable but important signals.

For the case of observing the human body, this general framework is complicated by the fact that the human body is a 3-D articulated system and the observation process is significantly non-trivial. Video images of the human body are extremely high-dimensional signals and the mapping between body pose and image observation involves perspective projection. These unique challenges go beyond the original design goals of the Kalman and extended Kalman filters and they make the task of building systems to observe human motion quite difficult. The details involved in extending the basic framework to this more complex domain are the subject of the next section.

7.3 An Implementation

This section attempts to make the theoretical findings of the previous section more concrete by describing a real implementation. The DYNA architecture is a real-time, recursive, 3-D person tracking system. The system is driven by 2-D *blob features* observed in two or more cameras [4, 52]. These features are then probabilistically integrated into a dynamic 3-D skeletal model, which in turn drives the 2-D feature tracking process by setting appropriate prior probabilities.

The feedback between 3-D model and 2-D image features is in the form of a recursive filter, as described in the previous section. One important aspect of the DYNA architecture is that the filter directly couples raw pixel measurements with an articulated dynamic model of the human skeleton. In this aspect the system is similar to that of Dickmanns in automobile control [15], and results show that the system realizes similar efficiency and stability advantages in the human motion perception domain.

This framework can be applied beyond passive physics by incorporating various patterns of control (which we call 'behaviors') that are *learned* from observing humans while they perform various tasks. Behaviors are defined as those aspects of the motion that cannot be explained solely by passive physics or the process of image production. In the untrained tracker these manifest as significant structures in the innovations process (the sequence of prediction errors). Learned models of this structure can be used to recognize and predict this purposeful aspect of human motion.

The human body is a complex dynamic system, whose visual features are time-varying, noisy signals. Accurately tracking the state of such a system requires use of a recursive estimation framework, as illustrated in figure 7.7. The framework consists of several modules. Section 7.3.1 details the module labeled "2-D Vision". The module labeled "Projective Model" is described in [4] and is summarized. The formulation of our 3-D skeletal physics model, "Dynamics" in the diagram, is explained in Section 7.3.2, including an explanation of how to drive that model from the observed measurements. The generation of prior information for the "2-D Vision" module from the model state estimated in the "Dynamics" module is covered in Section 7.3.3. Sec-

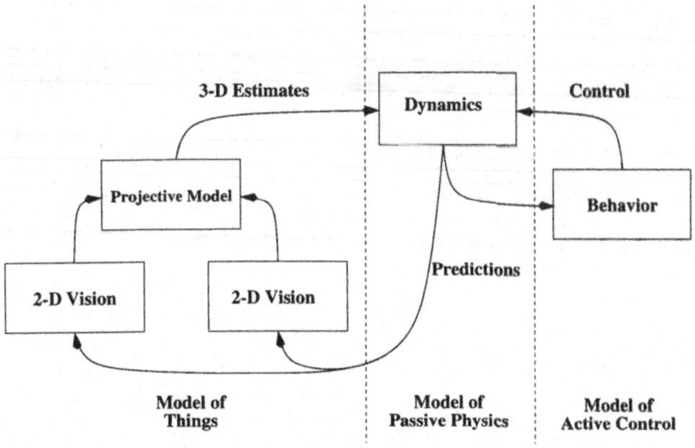

Fig. 7.7: The Recursive Filtering framework. Predictive feedback from the 3-D dynamic model becomes prior knowledge for the 2-D observations process. Predicted control allows for more accurate predictive feedback

tion 7.3.4 explains the behavior system and its intimate relationship with the physical model.

7.3.1 The Observation Model

Our system tracks regions that are visually similar appearance, and spatially coherent: we call these blobs. We can represent these 2-D regions by their low-order statistics. This compact model allows fast, robust classification of image regions.

Given a pair of calibrated cameras, pairs of 2-D blob parameters are used to estimate the parameters of 3-D blobs that exist behind these observations. Since the stereo estimation is occurring at the blob level instead of the pixel level, it is fast and robust.

This section describes these low-level observation and estimation processes in detail.

7.3.1.1 Blob Observations

If we describe pixels with spatial coordinates, (i, j), within an image, then we can describe clusters of pixels with 2-D spatial means and covariance matrices, which we shall denote $\boldsymbol{\mu}_s$ and $\boldsymbol{\Sigma}_s$. The blob spatial statistics are described in terms of these second-order properties. For computational convenience we will interpret this as a Gaussian model.

The visual appearance of the pixels, (y, u, v), that comprise a blob can also be modeled by second order statistics in color space: the 3-D mean, $\boldsymbol{\mu}_c$ and covariance, $\boldsymbol{\Sigma}_c$. As with the spatial statistics, these chromatic statistics

Fig. 7.8: A person interpreted as a set of blobs

are interpreted as the parameters of a Gaussian distribution in color space. We chose the YUV representation of color due to its ready availability from video digitization hardware and the fact that in the presence of white luminants it confines much of the effect of shadows to the single coordinate y [50].

Given these two sets of statistics describing the blob, the overall blob description becomes the concatenation $(ijyuv)$, where the overall mean is:

$$\boldsymbol{\mu} = \begin{bmatrix} \boldsymbol{\mu}_s \\ \boldsymbol{\mu}_c \end{bmatrix}$$

and the overall covariance is:

$$\boldsymbol{\Sigma} = \begin{bmatrix} \boldsymbol{\Sigma}_s & \boldsymbol{\Lambda}_{sc} \\ \boldsymbol{\Lambda}_{cs} & \boldsymbol{\Sigma}_c \end{bmatrix}$$

This framework allows for the concatenation of additional statistics that may be available from image analysis, such as texture or motion components. Figure 7.8 shows a person represented as a set of blobs. Spatial mean and covariance is represented by the iso-probability contour ellipse shape. The color mean is represented by the color of the blob. The color covariance is not represented in this illustration.

7.3.1.2 Frame Interpretation

To compute $p(\mathbf{O}|\boldsymbol{\mu}_k, \boldsymbol{\Sigma}_k)$, the likelihood that a given pixel observation, $\mathbf{O}$, is a member of a given blob, k, we employ the Gaussian assumption to arrive at the likelihood function:

$$p(\mathbf{O}|\mu_k, \Sigma_k) = \frac{\exp(-\frac{1}{2}(\mathbf{O} - \mu_k)^T \Sigma_k^{-1}(\mathbf{O} - \mu_k))}{(2\pi)^{\frac{m}{2}} |\Sigma_k|^{\frac{1}{2}}} \tag{26}$$

where $\mathbf{O}$ is the concatenation of the pixel spatial and chromatic characteristics.

$$\mathbf{O} = \begin{bmatrix} i \\ j \\ y \\ u \\ v \end{bmatrix} \tag{27}$$

Since the color and spatial statistics are assumed to be independent, the cross-covariance $\mathbf{\Lambda}_{sc}$ goes to zero, and the computation of the above value can proceed in a separable fashion [50].

For a frame of video data, the pixel at (i, j) can be classified with the Likelihood Ratio Test[46] by selecting, from the k blobs being tracked, that blob which best predicts the observed pixel:

$$\Gamma_{ij} = \arg\max_k \left[\Pr(\mathbf{O}_{ij}|\mu_k, \Sigma_k)\right] \tag{28}$$

where Γ_{ij} is the labeling of pixel (i, j). Due to the connected nature of people, it is possible to increase efficiency by growing out from initial position estimates to the outer edge of the figure. This allows the algorithm to only touch the pixels that represent the person and those nearby [50].

7.3.1.3 Model Update

Once the pixels are labeled by Γ, blob statistics can be re-estimated from the image data. For each class k, the pixels marked as members of the class are used to estimate the new model mean μ_k:

$$\hat{\mu}_k = E[O] \tag{29}$$

and the second-order statistics become the estimate of the model's covariance matrix Σ_k,

$$\hat{\Sigma}_k = E[(O - \mu_k)(O - \mu_k)^T] \tag{30}$$

7.3.1.4 A Compact Model

These updated blob statistics represent a low-dimensional, object-based description of the video frame. The position of the blob is specified by the two parameters of the distribution mean vector μ_s: i and j. The spatial extent of each blob is represented by the three free parameters in the covariance

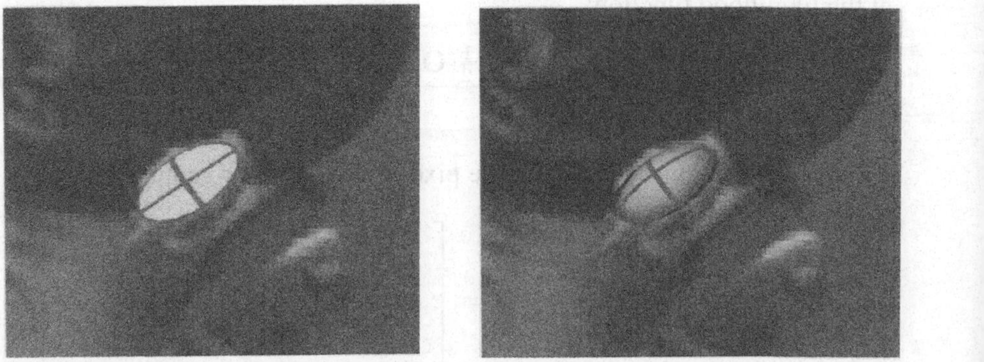

Fig. 7.9: Left: The hand as an iso-probability ellipse. Right: The hand as a 3-D blobject

matrix $\mathbf{\Sigma}_s$. A natural interpretation of these parameters can be obtained by performing the eigenvalue decomposition of $\mathbf{\Sigma}_s$:

$$\mathbf{\Sigma}_s \begin{bmatrix} | & | \\ L_1 & L_2 \\ | & | \end{bmatrix} = \begin{bmatrix} | & | \\ L_1 & L_2 \\ | & | \end{bmatrix} \begin{bmatrix} \lambda_1 & 0 \\ 0 & \lambda_2 \end{bmatrix} \tag{31}$$

Without loss of generality, $\lambda_1 \geq \lambda_2$, and $\|L_1\| = \|L_2\| = 1$. With those constraints, λ_1 and λ_2 represent the squared length of the semi-major and semi-minor axes of the iso-probability contour ellipse defined by $\mathbf{\Sigma}_s$. The vectors L_1 and L_2 specify the direction of these axes. Since they are perpendicular, they can be specified by a single parameter, say ω, the rotation of the semi-major axis away from the x axis. Thus we can represent the model $\{\boldsymbol{\mu}_s \mathbf{\Sigma}_s\}$ with the five parameters:

$$\{i, j, \lambda_1, \lambda_2, \omega\}$$

These parameters have the convenient physical interpretation of being related to the center, length, width, and orientation of an ellipse in the image plane, as shown in Figure 7.9.

Since the typical blob is supported by tens to hundreds of pixels, it is possible to robustly estimate these five parameters from the available data. The result is a stable, compact, object-level representation of the image region explained by the blob.

7.3.1.5 Recovery of a Three Dimensional Model

These 2-D features are the input to the 3-D blob estimation framework used by Azarbayejani and Pentland [4]. This framework relates the 2-D distribution of pixel values to a tracked object's 3-D position and orientation.

Inside the larger recursive framework, this estimation is carried out by an embedded extended Kalman filter. It is the structure from motion estimation framework developed by Azarbayejani to estimate 3-D geometry from images. As an extended Kalman filter, it is itself a recursive, nonlinear, probabilistic estimation framework. Estimation of 3-D parameters from calibrated sets of 2-D parameters is computationally very efficient, requiring only a small fraction of computational power as compared to the low-level segmentation algorithms[5]. The reader should not be confused by the embedding of one recursive framework inside another: for the larger context this module may be considered an opaque filter.

The same estimation machinery used to recover these 3-D blobs can also be used to quickly and automatically calibrate pairs of cameras from blob observation data[4].

After this estimation process, the 3-D blob has nine parameters:

$$\{x, y, z, l_1, l_2, l_3, w_1, w_2, w_3\}$$

As above, these parameters represent the position of the center of the blob, the length of the fundamental axes defined by an iso-probability contour, and the rotation of these axes away from some reference frame. Figure 7.9 shows the result of this estimation process.

7.3.2 Modeling Dynamics

There is a wide variety of ways to model physical systems. The model needs to include parameters that describe the *links* that compose the system, as well as information about the hard *constraints* that connect these links to one another. A model that only includes this information is called a *kinematic* model, and can only describe the static states of a system. The state vector of a kinematic model consists of the model state, $\mathbf{x}$, and the model parameters, $\mathbf{p}$, where The parameters, $\mathbf{p}$ are the unchanging qualities of the system. For example, the kinematics of a pendulum are described by the variable orientation of the hinge at the base, along with the parameters describing the mass of the weight and the length of the shaft.

A system in motion is more completely modeled when the *dynamics* of the system are modeled as well. A dynamic model describes the state evolution of the system over time. In a dynamic model the state vector includes velocity as well as position: $\mathbf{x}, \dot{\mathbf{x}}$, and the model parameters, $\mathbf{p}$. The state evolves according to Newton's First Law:

$$\ddot{\mathbf{x}} = \mathbf{W} \cdot \mathbf{X} \tag{32}$$

where $\mathbf{X}$ is the vector of external forces applied to the system, and $\mathbf{W}$ is the

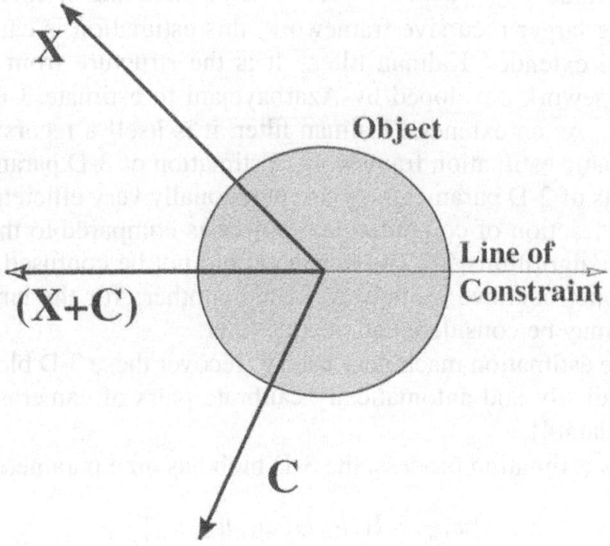

Fig. 7.10: The 2-D object is constrained to move on the indicated line. An external force, **X**, is applied to the object. The constraint force, **C**, keeps the object from accelerating away from the constraint

inverse of the system mass matrix. The mass matrix describes the distribution of mass in the system. In the pendulum example, the state vector would need to include not only the position of the hinge, but the rate of change as well. That information combined with the mass matrix, **W**, captures the momentum of the system. The presence of gravity acting on the system would be found in **X**.

7.3.2.1 Hard Constraints

Hard constraints represent absolute limitations imposed on the system. One example is the kinematic constraint of a skeletal joint. The model follows the *virtual work* formulation of Witkin[49]. The Witkin formulation has several advantages over reduced dimensionality solutions such as that described by Featherstone[16]: the constraints can be modified at run-time, and the modularity inherent in the mathematics drastically simplifies the implementation. The one significant disadvantage, which will be addressed below, is computational efficiency.

In a virtual work constraint formulation, all the links in a model have full range of unconstrained motion. Hard kinematic constraints on the system are enforced by a special set of forces $\mathbf{C}$:

$$\ddot{\mathbf{x}} = \mathbf{W} \cdot (\mathbf{X} + \mathbf{C}) \tag{33}$$

The constraints are specified as mathematical relationships between objects that are defined to be zero when the constraint is satisfied. The constraint forces C, are chosen to ensure that the constraints stay satisfied. In Figure 7.10 the constraint would be expressed as the distance between the line and the center of the object.

Using this formulation, an elbow would be represented by a two links with six degrees of freedom between them. Then a constraint would be written to project the motion of the joint down onto a one-dimensional manifold: no motion allowed in position, and only one degree of freedom allowed in orientation.

The constraints are functions of model state and time: $\mathbf{c}(\mathbf{x}, t)$. Constraints are defined as being satisfied when $\mathbf{c} = 0$. If a constraint is to remain satisfied, then the constraint velocity must remain zero, $\dot{\mathbf{c}} = 0$, and the constraint must not accelerate away from a valid state: $\ddot{\mathbf{c}} = 0$ or the system would soon be in an invalid state. The constraint forces from Equation 33 and Figure 7.10 keep the system from accelerating away from constraint satisfaction. The road to understanding these forces begins by differentiating $\mathbf{c}$:

$$\dot{\mathbf{c}} = \frac{\partial}{\partial \mathbf{x}} \mathbf{c}(\mathbf{x}, t) = \frac{\partial \mathbf{c}}{\partial \mathbf{x}} \dot{\mathbf{x}} + \frac{\partial \mathbf{c}}{\partial t} \tag{34}$$

and again:

$$\ddot{\mathbf{c}} = \frac{\partial \mathbf{c}}{\partial \mathbf{x}} \ddot{\mathbf{x}} + \frac{\partial \dot{\mathbf{c}}}{\partial \mathbf{x}} \dot{\mathbf{x}} + \frac{\partial^2 \mathbf{c}}{\partial t^2} \tag{35}$$

Combining with Equation 33 and setting $\ddot{\mathbf{c}} = 0$ yields a relationship between the model state, the applied external forces, and the constraint Jacobian, where the constraint restitution force $\mathbf{C}$ is the only unknown:

$$\frac{\partial \mathbf{c}}{\partial \mathbf{x}} \mathbf{W} \cdot (\mathbf{X} + \mathbf{C}) + \frac{\partial \dot{\mathbf{c}}}{\partial \mathbf{x}} \dot{\mathbf{x}} + \frac{\partial^2 \mathbf{c}}{\partial t^2} = 0 \tag{36}$$

The force in Figure 7.10 satisfies the relationship in Equation 36, as do many other possible vectors. Equation 36 is an under-determined system since the dimensionality of $\mathbf{c}$ will always be less than the dimensionality of $\mathbf{x}$, or the system would be fully constrained and wouldn't move at all. For example, in Figure 7.10, $\mathbf{c}$ is the distance between the center of the object and the line, a one dimensional value, while $\mathbf{x}$ is two dimensional, three if the object is allowed rotational freedom in the plane.

One problem with that choice of $\mathbf{C}$ in Figure 7.10 is that it will add energy to the system. Equation 36 only specifies that the force exactly counteract the component of $\mathbf{X}$ that is in violation of the constraints. Since con-

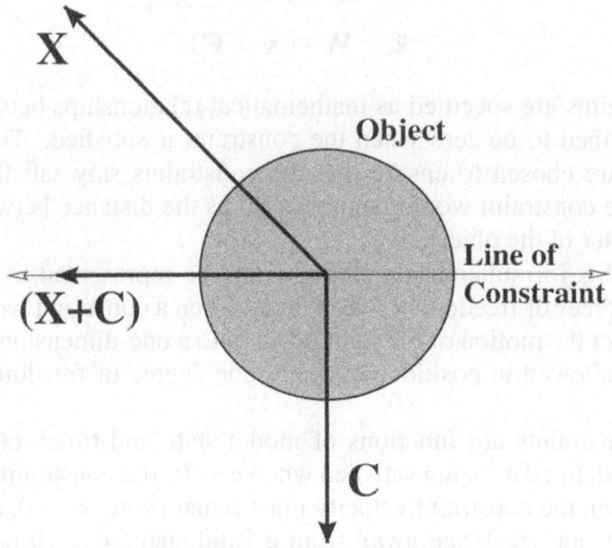

Fig. 7.11: No work is performed if the constraint force lies in the null space compliment of the constraint Jacobian

straints that add energy to the models will lead to instability, an additional requirement that the force **C** do no work on the system is employed, where work would be $\mathbf{C}\dot{\mathbf{x}}$. If **C** is always applied perpendicular to any direction of motion allowed by the constraint then this will be satisfied. In the example case of the object on the line, this means that **C** must be perpendicular to the line of constraint, as shown in Figure 7.11.

More generally, the valid displacements for the system are described by the null space of the constraint Jacobian:

$$\frac{\partial \mathbf{c}}{\partial \mathbf{x}} d\mathbf{x} = 0 \tag{37}$$

since valid displacements are required to leave $\mathbf{c} = 0$. The disallowed displacements are ones that do change **c**:

$$d\mathbf{x} = \boldsymbol{\lambda} \frac{\partial \mathbf{c}}{\partial \mathbf{x}} \tag{38}$$

So, to do no work, the constraint force **C** is required to lie in the same subspace:

$$\mathbf{C} = \boldsymbol{\lambda} \frac{\partial \mathbf{c}}{\partial \mathbf{x}} \tag{39}$$

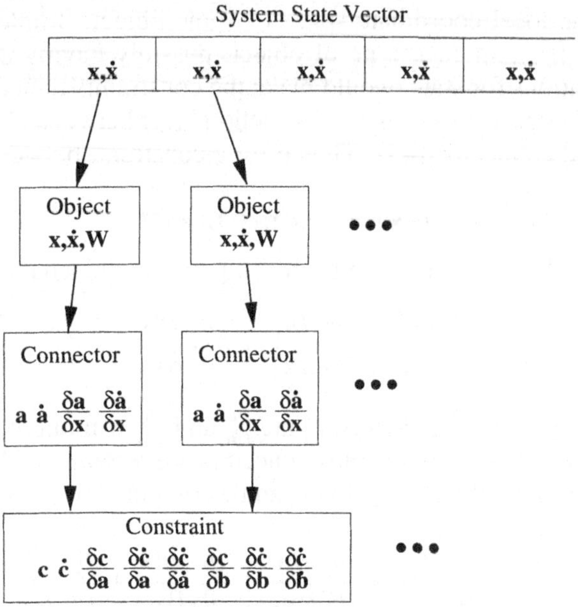

Fig. 7.12: The constraint system is made up of software modules that cooperate to construct Equation 40. Connectors are an abstraction that allows the Constraints to be more general, and hence more reusable

Combining that equation with Equation 36 results in a linear system of equations with only the one unknown, $\boldsymbol{\lambda}$:

$$-\left[\frac{\partial \mathbf{c}}{\partial \mathbf{x}}\mathbf{W}\frac{\partial \mathbf{c}}{\partial \mathbf{x}}^T\right]\boldsymbol{\lambda} = \frac{\partial \mathbf{c}}{\partial \mathbf{x}}\mathbf{W}\mathbf{X} + \frac{\partial \dot{\mathbf{c}}}{\partial \mathbf{x}}\dot{\mathbf{x}} + \frac{\partial^2 \mathbf{c}}{\partial t^2} \tag{40}$$

This equation can be rewritten to emphasize its linear nature. $\mathbf{J}$ is the constraint Jacobian, $\boldsymbol{\rho}$ is a known constant vector, and $\boldsymbol{\lambda}$ is the vector of unknown Lagrange multipliers:

$$-\mathbf{J}\mathbf{W}\mathbf{J}^T\boldsymbol{\lambda} = \boldsymbol{\rho} \tag{41}$$

To obtain $\mathbf{C}$, $\boldsymbol{\lambda}$ is substituted back into Equation 39. Many fast, stable methods exist for solving equations of this form.

All the components of Equation 40 that relate directly to the constraints are linearizations, so they must be recomputed at each integration step. This provides the opportunity to create and delete constraints at run-time simply by modifying the calculation of the constraint Jacobian.

Modularization Constraints are written as mathematical relationships between points in space. Often these points are located at some arbitrary location in the local coordinate space of some object. Implementing constraints to understand each type of object, possibly having different internal representations for state, would make the constraints unnecessarily complex. Witkin suggests inserting an abstraction layer between objects and constraints, called connectors[49]. Thus $\mathbf{c}$ for a constraint between two objects becomes:

$$\mathbf{c}(\mathbf{x}) = f(\mathbf{a}(\mathbf{x}_1), \mathbf{b}(\mathbf{x}_2)) \tag{42}$$

The constraint Jacobian can then be decomposed by the chain rule:

$$\frac{\partial \mathbf{c}}{\partial \mathbf{x}} = \frac{\partial \mathbf{c}}{\partial \mathbf{a}}\frac{\partial \mathbf{a}}{\partial \mathbf{x}_1} + \frac{\partial \mathbf{c}}{\partial \mathbf{b}}\frac{\partial \mathbf{b}}{\partial \mathbf{x}_2} \tag{43}$$

The constraint module can then compute $\frac{\partial \mathbf{c}}{\partial \mathbf{a}}$ and $\frac{\partial \mathbf{c}}{\partial \mathbf{b}}$ without regard to the underlying implementation while the connectors are responsible for calculating $\frac{\partial \mathbf{a}}{\partial \mathbf{x}}$. The constraint velocity Jacobian can be computed in the same way;

$$\frac{\partial \dot{\mathbf{c}}}{\partial \mathbf{x}} = \frac{\partial \dot{\mathbf{c}}}{\partial \mathbf{a}}\frac{\partial \mathbf{a}}{\partial \mathbf{x}_1} + \frac{\partial \dot{\mathbf{c}}}{\partial \dot{\mathbf{a}}}\frac{\partial \dot{\mathbf{a}}}{\partial \mathbf{x}_1} + \frac{\partial \dot{\mathbf{c}}}{\partial \mathbf{b}}\frac{\partial \mathbf{b}}{\partial \mathbf{x}_2} + \frac{\partial \dot{\mathbf{c}}}{\partial \dot{\mathbf{b}}}\frac{\partial \dot{\mathbf{b}}}{\partial \mathbf{x}_2} \tag{44}$$

Figure 7.12 shows how this information moves through the system. Each block represents a different software module. This abstraction is very powerful: in the system the same constraint code implements pin joints in 2-D models and ball joints in 3-D models. Because constraints involving rotational motion are somewhat more complex, the system differentiates between connectors without orientation, called Points, and connectors with orientation, called Handles.

Multiple Objects and Constraints Systems with only a single constraint are rather limiting. Multiple objects and constraints fit easily into the framework. For multiple objects, the state vector $\mathbf{x}$ becomes the concatenation of all the individual object state vectors. So in a 3-D model where every object has 6 degrees of freedom, with 5 objects the state vector would have dimensionality 30.

The mass matrix is similarly the concatenation of the individual mass matrices. Assuming static geometry for each object, the individual mass matrix is constant in the object local coordinate system. This mass matrix is transformed to global coordinates and added as a block to the global mass matrix. Since the global mass matrix is block diagonal, the inverse mass matrix is simply the concatenation of the individually inverted mass matrices, and so doesn't take an inordinate amount of time to compute.

Objects are enumerated with the order that they contribute to the global state vector. Constraints are similarly enumerated. A constraint between two objects contributes two blocks to the constraint Jacobian. The constraints ap-

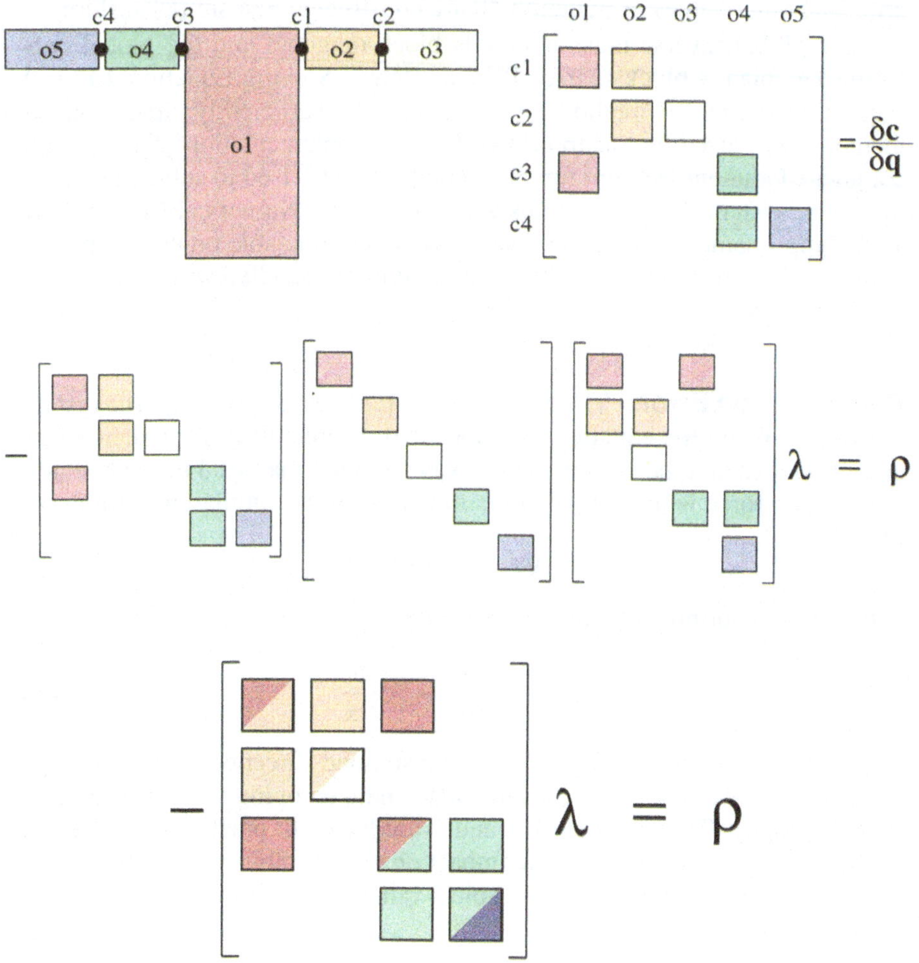

Fig. 7.13: Top: the individual constraint Jacobians each contribute one block per object that they affect to the global constraint Jacobian. **Middle:** each object also contributes to the block-diagonal inverse mass matrix from Equation 41. **Bottom:** Sparsely connected systems result in a block-sparse linear system

pear on the row according to the constraint's enumeration and the columns associated with the constrained objects. The structure of the constraint Jacobian is illustrated in Figure 7.13 for a model of the upper body with five links: torso, left upper arm, left lower arm, right upper arm, and right lower arm. The other values in Equation 40 are constructed in a similar fashion.

The global inverse mass matrix is block diagonal and the global constraint Jacobian is block sparse. Both are large. Solving Equation 40 for $\boldsymbol{\lambda}$ requires sparse matrix methods to be accomplished efficiently. Sparse matrix methods were used to construct $\mathbf{JWJ}^T$. An implementation of Linear Biconjugate Gradient Descent for sparse matrices was used to solve the resulting linear system. The algorithms were taken from Numerical Recipes[38]. These improvements made the constraint system tractable on contemporary hardware. The rest of the matrix manipulations are handled with a basic C++ matrix library.

Discretization Error The constraints of Equation 40 are only true instantaneously. When the equations are solved at discrete time steps then errors are introduced and the system drifts away from the manifold of valid states. A restoring force is used to keep the system from accumulating errors over time:

$$\ddot{\mathbf{x}} = \mathbf{W} \cdot (\mathbf{X} + \mathbf{C} + \mathbf{F}) \tag{45}$$

Where $\mathbf{F}$ is determined by the relationship:

$$F = \alpha \mathbf{c} \frac{\partial \mathbf{c}}{\partial \mathbf{x}} + \beta \dot{\mathbf{c}} \frac{\partial \mathbf{c}}{\partial \mathbf{x}} \tag{46}$$

This applies a restoring force in the constrained direction that brings the system back toward the nearest valid state and a damping force that reduces illegal velocity. The parameters α and β are fixed. In practice the selection of these parameters has very little impact on model stability since deviations from constraints remain small. A typical value for α is $1000\frac{N}{m}$ and a typical value for β is $4\frac{Ns}{m}$.

Distributed Integration Once the global forces are projected back into the allowable subspace and corrected for discretization error, all further computation is partitioned among the individual objects. This avoids computing the very large global version of Equation 33. This is possible since the inverse mass matrix $\mathbf{W}$ is block diagonal, so once the global value for $\mathbf{C}$ is determined, Equation 33 breaks down into a set of independent systems. This distributed force application and integration also provides the opportunity for objects to transform the applied forces to the local frame and to deal with forces and torques separately. This simplifies the implementation of the dynamics subsystem significantly, since each link is treated as a free six degree of freedom body.

7.3.2.2 Soft Constraints

Some constraints are probabilistic in nature. Noisy image measurements are a constraint of this sort, they influence the dynamic model but do not impose hard constraints on its behavior. As a result, the absolute constraint satisfaction described in the previous section is not appropriate.

Soft constraints are more appropriately expressed as a potential field acting on the dynamic system. The addition of a potential field function to model a probability density function pushes the model toward the most likely value. In general a soft constraint might be any function:

$$\mathbf{X}_{\text{soft}} = f_{\text{soft}}(\mathbf{S}, \mathbf{x}, \dot{\mathbf{x}}, \mathbf{p}) \tag{47}$$

where $\mathbf{S}$ is some parameterization over the family of potential fields specified by $f(\cdot)$.

The simplest function is the constant potential. Gravity is well-modeled by a constant field over the scales of the model. So the potential field is simply:

$$\mathbf{X}_{\text{soft}} = mg \tag{48}$$

where g is acceleration due to gravity, and m is the mass of the link affected by $\mathbf{X}_{\text{soft}}$.

A soft constraint that attracts a body part to a specific location is somewhat more complex:

$$\mathbf{X}_{\text{soft}} = k(\mathbf{x}_0 - \mathbf{x}) \tag{49}$$

where $\mathbf{x}_0$ is the desired position and k is a constant multiplier that affect the "softness" of the constraint. Care must be taken when choosing k to avoid introducing instabilities into the model. Values of k that are too large start to turn the soft constraint into something more like a hard constraint. In this case the constraint would be better modeled by the techniques described above.

It is also possible to construct anisotropic constraints

$$\mathbf{X}_{\text{soft}} = \frac{(\mathbf{x} - \mathbf{x}_0)}{\|(\mathbf{x} - \mathbf{x}_0)\|}(\mathbf{x} - \mathbf{x}_0)K^{-1}(\mathbf{x} - \mathbf{x}_0) \tag{50}$$

where K is a shaping matrix that determines the weighting of various directions. This allows soft constraints to have stronger influence in a particular direction. This is useful for modelling the influence of the blob observations discussed above, or any other regular, non-isotropic force field.

Note that functions may be arbitrarily complex. A good example is a controller of the form described in Section 7.3.4. Despite their complexity, the dynamics engine may represent them as a time-varying potential field. The forces applied by the controller simply become another force affecting the dynamic evolution of the model. The neuroscience literature supports this model[42].

7.3.2.3 Observation Influence

The 3-D observations described in Section 7.3.1 supply constraints on the underlying 3-D human model. Due to their statistical nature, observations are easily modeled as soft constraints. Observations are integrated into the dynamic evolution of the system by describing them with potential fields, as discussed in Section 7.3.2.2. These potential fields apply forces to the body model, causing it evolve over time toward the observations. The strength of these fields is related to the Kalman gain in a classic Kalman filter.

7.3.3 The Inverse Observation Model

In the open-loop system, the vision system uses a Maximum Likelihood framework to label individual pixels in the scene (Equation 28). To close the loop, we need to incorporate information from the 3-D model. This means generating 2-D statistical models from the 3-D body model that can be utilized by the vision system to improve its decisions.

The current state of the model $(\mathbf{x}_t, \dot{\mathbf{x}}_t)$ specifies the best estimate of the configuration of the body in model space given past observations. The first step in predicting future observations is to propagate the state forward according to the dynamic constraints described above. The external forces acting on the system can be assumed to be constant for the period of forward prediction (33ms in the case of video rate observations), or can be predicted forward in time by behavior models as described below in Section 7.3.4.

Once a best estimate of the future configuration of the system, $\mathbf{x}_{t+\Delta}$, has been computed, the next step is to generate a set of hypothetical 3-D observations that we would expect to see given that configuration. This involves generating distributions that represent 3-D ellipsoids that fit the observable portions of the model links. These distributions are transformed into the camera reference frame as described in Section 7.3.1.5. In this frame they are described as in Section 7.3.1 by either their second order statistics:

$$\{\mu_k^\circ, \Sigma_k^\circ\}$$

or, by their free parameters:

$$\{x, y, z, l_1, l_2, l_3, w_1, w_2, w_3\}$$

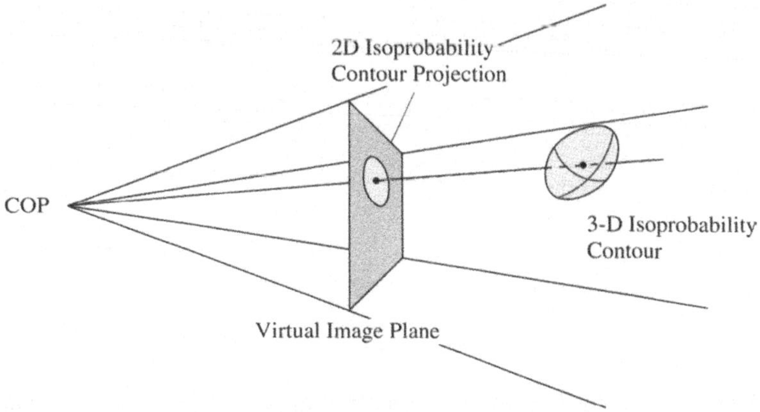

Fig. 7.14: Te true Projection of a 3-D Gaussian distribution onto a 2-D image plane is not a 2-D Gaussian distribution

The process of identifying the observable portions of these links is discussed in Section 7.3.2.3.

To be used as a prior for the classification decision in Equation 28, these 3-D distributions must be rendered into 2-D image coordinates using perspective projection. These projected distribution will be described by their second order statistics:

$$\{\boldsymbol{\mu}_k^\star, \boldsymbol{\Sigma}_k^\star\}$$

or alternatively by the free parameters:

$$\{i, j, \lambda_1, \lambda_2, \omega\}$$

The computation of $\boldsymbol{\mu}_k^\star$ from $\boldsymbol{\mu}_k^\circ$ is a straightforward application of the forward projective camera model. The parameters x, y, z map into the parameters i, j by perspective projection:

$$\boldsymbol{\mu}_k^\star = \begin{bmatrix} i \\ j \end{bmatrix} = \begin{bmatrix} x \\ y \end{bmatrix} \frac{1}{1+z} \tag{51}$$

The true perspective projection of a 3-D Gaussian distribution over $\Re^3$ is not a Gaussian distribution over the image coordinates $\Re^2$. It is necessary to employ an approximation to perspective projection that will yield a Gaussian distribution in image coordinates to obtain a value for $\boldsymbol{\Sigma}_k^\star$

Orthographic projection of a Gaussian distribution does result in a Gaussian distribution. This process involves integrating over the Gaussian in the direction of projection. Orthographic projection of the 3-D prior onto a XY plane passing through the mean is thus equivalent to taking the marginal of

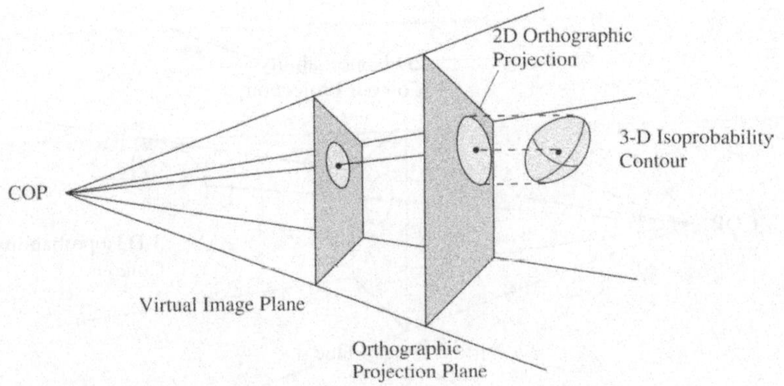

Fig. 7.15: Scaled-Orthographic projection approximation for 3-D Gaussian distribution onto a 2-D Gaussian distribution

a zero-mean Gaussian distribution:

$$\mathcal{N}\left(\mathbf{0}, \begin{bmatrix} \sigma_x & \lambda_{xy} \\ \lambda_{yx} & \sigma_y \end{bmatrix}\right) = \int_{-\infty}^{\infty} \mathcal{N}\left(\mathbf{0}, \begin{bmatrix} \sigma_x & \lambda_{xy} & \lambda_{xz} \\ \lambda_{yx} & \sigma_y & \lambda_{yz} \\ \lambda_{zx} & \lambda_{zy} & \sigma_z \end{bmatrix}\right) \partial z \quad (52)$$

Orthographic projection does not account for the scaling effect of perspective projection, so simply using orthographic projection would result in priors with significantly exaggerated covariances. A solution is to use the scaled orthographic approximation to perspective projection. Scaled orthographic projection uses perspective projection to map an intermediate 2-D orthographic projection into the virtual image plane. Since the plane of orthographic projection is parallel to the virtual image plane, this operation is equivalent to a scale. Scaling a Gaussian distribution retains the Gaussian nature, so we have the approximation we need. As illustrated in Figure 7.15, by placing the plane of orthographic projection at z, we can compute the 2-D blob covariance prior, $\mathbf{\Sigma}_k^{\star}$, from the 3-D covariance $\mathbf{\Sigma}_k^{\circ}$:

$$\mathbf{\Sigma}_k^{\star} = \begin{bmatrix} \sigma_i & \lambda_{ij} \\ \lambda_{ji} & \sigma_j \end{bmatrix} = \begin{bmatrix} \frac{1}{1+z} & 0 \\ 0 & \frac{1}{1+z} \end{bmatrix} \begin{bmatrix} \sigma_x & \lambda_{xy} \\ \lambda_{yx} & \sigma_y \end{bmatrix} \begin{bmatrix} \frac{1}{1+z} & 0 \\ 0 & \frac{1}{1+z} \end{bmatrix} \quad (53)$$

The result of this process is a prior distribution on image observations in the next frame:

$$p(\mathbf{O}_{ij} | \boldsymbol{\mu}_k^{\star}, \mathbf{\Sigma}_k^{\star}) \quad (54)$$

Integrating this information into the 2-D statistical decision framework of Equation 28 results in a Maximum *A Posteriori* decision rule for pixel clas-

Physics-only Model

Physics + Behavior Model

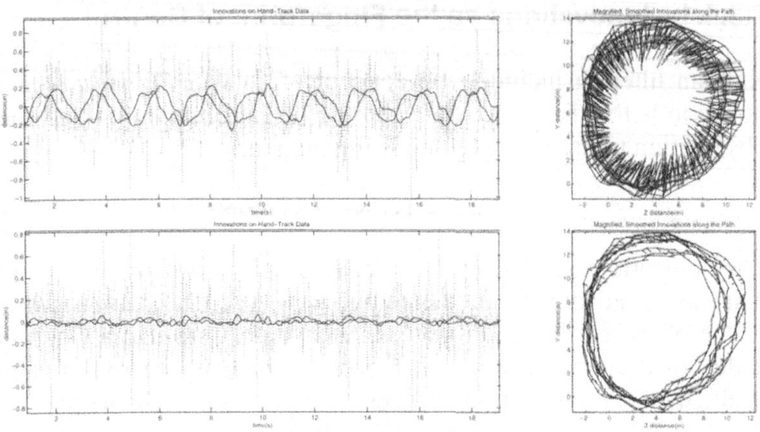

Fig. 7.16: Modeling tracking data of circular hand motion. Passive physics alone leaves significant structure in the innovations process. **Top Left:** Smoothing the innovations reveals unexplained structure. **Top Right:** Plotting the Innovations along the path makes the purposeful aspect of the action clear. **Bottom:** In this example, using a learned control model to improve predictions leaves only white process noise in the innovations process. The smoothed innovations stay near zero

sification:

$$\Gamma_{ij} = \arg\max_k \left[p(\mathbf{O}_{ij}|\boldsymbol{\mu}_k, \boldsymbol{\Sigma}_k)^\alpha \cdot p(\mathbf{O}_{ij}|\boldsymbol{\mu}_k^\star, \boldsymbol{\Sigma}_k^\star)^{1-\alpha} \right] \tag{55}$$

where α is the weighting parameter that indicates the importance of the prior information:

$$0 \leq \alpha \leq 1 \tag{56}$$

7.3.4 A Model for Control

Observations of the human body reveal an interplay between the passive evolution of a physical system (the human body) and the influences of an active, complex controller (the nervous system). Section 7.3.2 explains how, with a bit of work, it is possible to model the physical aspects of the system. However, it is *very* difficult to explicitly model the human nervous and muscular systems, so the approach of using observed data to estimate probability distributions over control space is very appealing.

7.3.4.1 Innovations as the Fingerprint of Control

Kalman filtering includes the concept of an *innovations process*. The innovation is the difference between the actual observation and the predicted observation transformed by the Kalman gain:

$$\boldsymbol{\nu}_t = \mathbf{K}_t(\mathbf{y}_t - \mathbf{H}_t\boldsymbol{\Phi}_t\hat{\mathbf{x}}_{t-1}) \tag{57}$$

The innovations process $\boldsymbol{\nu}$ is the sequence of information in the observations that was not adequately predicted by the model. If we have a sufficient model of the observed dynamic process, and white, zero-mean Gaussian noise is added to the system, either in observation or in the real dynamic system itself, then the innovations process will be white. Inadequate models will cause correlations in the innovations process.

Since purposeful human motion is not well modeled by passive physics, we should expect significant structure in the innovations process.

A simple example is helpful for illustrating this idea. If we track the hand moving in a circular motion, then we have a sequence of observations of hand position. This sequence is the result of a physical thing being measured by a noisy observation process. For this simple example we are making the assumption that the hand moves according to a linear, constant velocity dynamic model. Given that assumption, it is possible to estimate the true state of the hand, and predict future states and observations. If this model is sufficient, then the errors in the predictions should be solely due to the noise in the system.

The upper plots in Figure 7.16 show that model is not sufficient. Smoothing $\boldsymbol{\nu}$ reveals this significant structure (top left). Plotting the innovations along the path of observations makes the relationship between the observations and the innovations clear: there is some process acting to keep the hand moving in a circular motion that is not accounted for by the model (top right). This unanticipated process is the purposeful control signal that being applied to the hand by the muscles.

In this example, there is one active, cyclo-stationary control behavior, and its relationship to the state of the physical system is straightforward. There is a one-to-one mapping between the state and the phase offset into the cyclic control, and a one-to-one mapping between the offset and the control to be applied. If we use the smoothed innovations as our model and assume a linear control model of identity, then the linear prediction becomes:

$$\hat{\mathbf{x}}_t = \boldsymbol{\Phi}_t\hat{\mathbf{x}}_{t-1} + \mathbf{I}\mathbf{u}_{t-1} \tag{58}$$

where $\mathbf{u}_{t-1}$ is the control signal applied to the system. The lower plots in Figure 7.16 show the result of modeling the hand motion with a model of passive physics and a model of the active control. The smoothed innovations are basically zero: there is no part of the signal that deviates from our model except for the observation noise.

In this simple, linear example the system state, and thus the innovations, are represented the same coordinate system as the observations. With more complex dynamic and observations models, such as described in Section 7.3.2, they could be represented in any arbitrary system, including spaces related to observation space in non-linear ways, for example as joint angles.

The next section examines a more powerful form of model for control.

7.3.4.2 Multiple Behavior Models

Human behavior, in all but the simplest tasks, is not as simple as a single dynamic model. The next most complex model of human behavior is to have *several* alternative models of the person's dynamics, one for each class of response. Then at each instant we can make observations of the person's state, decide which model applies, and then use that model for estimation. This is known as the *multiple model* or *generalized likelihood* approach, and produces a generalized maximum likelihood estimate of the current and future values of the state variables [48]. Moreover, the cost of the Kalman filter calculations is sufficiently small to make the approach quite practical.

Intuitively, this solution breaks the person's overall behavior down into several "prototypical" behaviors. For instance, we might have dynamic models corresponding to a relaxed state, a very stiff state, and so forth. We then classify the behavior by determining which model best fits the observations. This is similar to the multiple model approach of Friedmann, and Isard[17, 23].

Since the innovations process is the part of the observation data that is unexplained by the dynamic model, the behavior model that explains the largest portion of the observations is, of course, the model most likely to be correct. Thus, at each time step, we calculate the probability $Pr^{(i)}$ of the m-dimensional observations $\mathbf{Y}_k$ given the ith model and choose the model with the largest probability. This model is then used to estimate the current value of the state variables, to predict their future values, and to choose among alternative responses.

7.3.4.3 Hidden Markov Models of Control

Since human motion evolves over time, in a complex way, it is advantageous to explicitly model temporal dependence and internal states in the control process. A Hidden Markov Model (HMM) is one way to do this, and has been shown to perform quite well recognizing human motion[45].

The probability that the model is in a certain state, S_j, given a sequence of observations, $\mathbf{O}_1, \mathbf{O}_2, \ldots, \mathbf{O}_N$, is defined recursively. For two observations, the density associated with the state after the second observation, $\mathbf{q}_2$, being S_j is:

$$\Pr(\mathbf{O}_1, \mathbf{O}_2, \mathbf{q}_2 = S_j) = \left[\sum_{i=1}^{N} \pi_i b_i(\mathbf{O}_1) \mathbf{a}_{ij} \right] b_j(\mathbf{O}_2) \tag{59}$$

where π_i is the prior probability of being in state i, and $b_i(\mathbf{O})$ is the probability of making the observation $\mathbf{O}$ while in state i. This is the Forward algorithm for HMM models.

Estimation of the control signal proceeds by identifying the most likely state given the current observation and the last state, and then using the observation density of that state as described above. If the models are trained relative to a passive-physics model, then likely it will be necessary to run a passive-physics tracker to supply the innovations that will be used by the models to select the control paradigm for a second tracker. We restrict the observation densities to be either a Gaussian or a mixture of Gaussians. For behaviors that are labeled there are well understood techniques for estimating the parameters of the HMM from data[39].

7.3.4.4 Behavior Alphabet Auto-Selection

Classic HMM techniques require the training data to be labeled prior to parameter estimation. Since we don't necessarily know how to choose a gesture alphabet *a priori*, we cannot perform this pre-segmentation. We would prefer to automatically discover the optimal alphabet for gestures from gesture data. The COGNO architecture performs this automatic clustering[12].

Unfortunately, the phrase "optimal" is ill-defined for this task. In the absence of a task to evaluate the performance of the model, there is an arbitrary trade-off between model complexity and generalization of the model to other data sets[47]. By choosing a task, such as discriminating styles of motion, we gain a well-defined metric for performance.

One of our goals is to observe a user who is interacting with a system and be able to automatically find patterns in their behavior. Interesting questions include:

- Is this (a)typical behavior for the user?
- Is this (a)typical behavior for anyone?
- When is the user transitioning from one behavior/strategy to another behavior/strategy?
- Can we do filtering or prediction using models of the user's behavior?

We must find the behavior alphabets that pick out the salient movements relevant to the above questions. There probably will not be one canonical alphabet for all tasks but rather many alphabets each suited to a group of tasks. Therefore we need an algorithm for automatically generating and selecting effective behavior alphabets. The goal of finding an alphabet that is suitable for a machine learning task can be mapped to the concept of feature selection.

The examples in Section 7.4 employ the COGNO algorithm[12] to perform unsupervised clustering of the passive-physics innovations sequences. Unsupervised clustering of temporal sequences generated by human behavior is a very active topic in the literature[44, 1, 31, 37].

7.3.5 Summary

This section presents a framework for human motion understanding, defined as estimation of the physical state of the body combined with interpretation of that part of the motion that cannot be predicted by passive physics alone. The behavior system operates in conjunction with a real-time, fully-dynamic, 3-D person tracking system that provides a mathematically concise formulation for incorporating a wide variety of physical constraints and probabilistic influences. The framework takes the form of a non-linear recursive filter that enables pixel-level processes to take advantage of the contextual knowledge encoded in the higher-level models.

The intimate integration of the behavior system and the dynamic model also provides the opportunity for a richer sort of motion understanding. The innovations are one step closer to the original intent, so the statistical models don't have to disentangle the message from the means of expression.

Some of the benefits of this approach including increase in 3-D tracking accuracy, insensitivity to temporary occlusion, and the ability to handle multiple people will be demonstrated in the next section.

7.4 Results

This section will provide data to illustrate the benefits of the DYNA framework. The first part will report on the state of the model within DYNA and the quantitative effects of tracking improvements. The rest will detail qualitative improvements in human-computer interface performance in the context of several benchmark applications.

7.4.1 Tracking Results

The dynamic skeleton model currently includes the upper body and arms. The full dynamic system loop, including forward integration and constraint satisfaction, iterates on a 500MHz Alpha 21264 at 600Hz. Observations come in from the vision system at video rate, 30Hz, so this is sufficiently fast

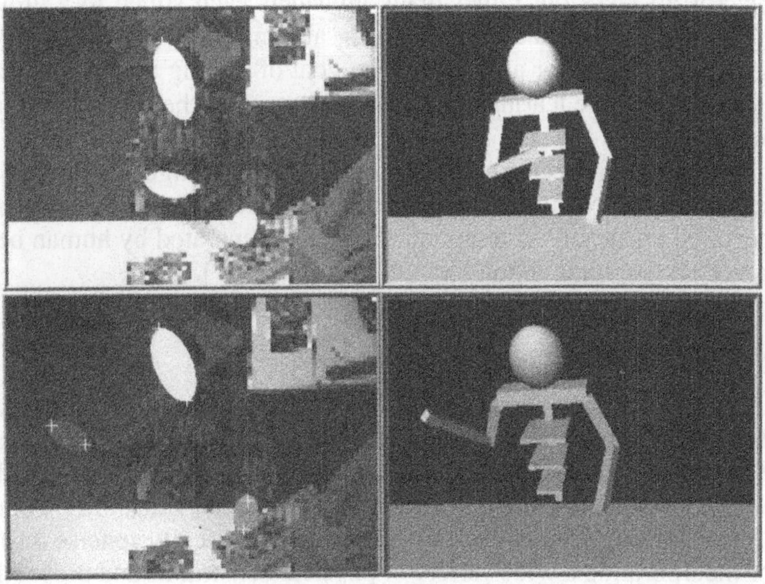

Fig. 7.17: Left: video and 2-D blobs from one camera in the stereo pair. **Right:** corresponding configurations of the dynamic model

for real-time operation. Figure 7.17 shows the real-time response to various target postures. The model interpolates those portions of the body state that are not measured directly, such as the upper body and elbow orientation, by use of the model's intrinsic dynamics, the kinematic constraints of the skeleton, and and the behavior (control) model.

The model also rejects noise that is inconsistent with the dynamic model. This process isn't equivalent to a simple isometric smoothing, since the mass matrix of the body is anisotropic and time-varying. When combined with an active control model, tracking error can be further reduced through the elimination of overshoot and other effects. Table 7.18 compares noise in the physics+behavior tracker with the physics-only tracker noise. It can be seen that there is a significant increase in performance.

The plot in Figure 7.19 shows the observed and predicted X position of the hand and the corresponding innovation trace before, during and after the motion is altered by a constraint that is modeled by the system: arm kinematics. When the arm reaches full-extension, the motion is arrested. The system is able to predict this even and the near-zero innovations after the event reflect this. Non-zero innovations before the event represent the controlled acceleration of the arm in the negative X direction. Compare to the case of a collision between a hand and the table illustrated in Figure 7.20. The table is not included in the system's model, so the collision goes unpredicted. This

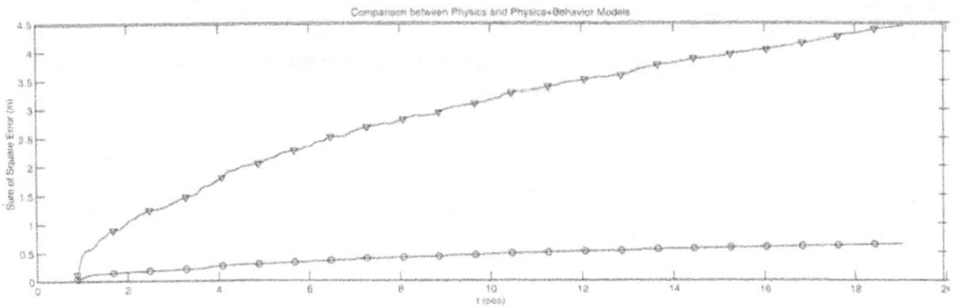

Fig. 7.18: Sum Square Error of a Physics-only tracker (triangles) vs. error from a Physics+Behavior Tracker

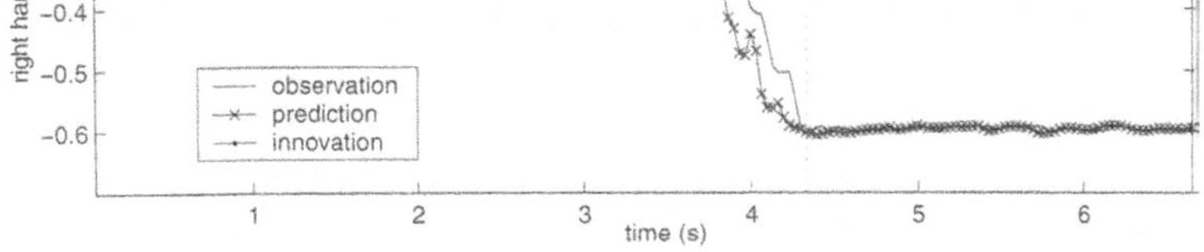

Fig. 7.19: Observed and predicted X position of the hand and the corresponding innovation trace before, during and after expression of a modeled constraint: arm kinematics

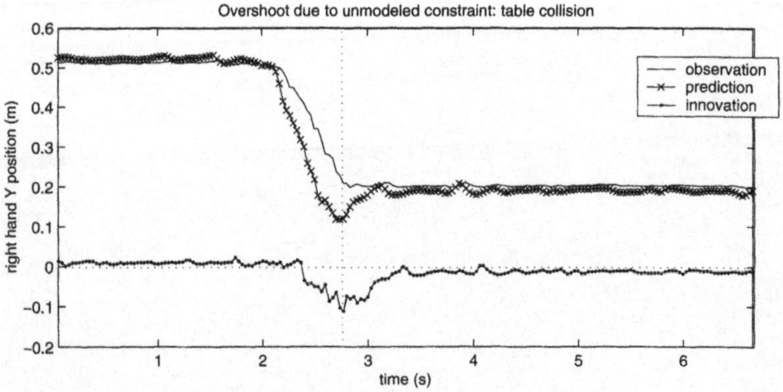

Fig. 7.20: Observed and predicted Y position of the hand and the corresponding innovation trace before, during and after expression of a un-modeled constraint: collision with a table

results in overshoot, and a corresponding signal in the innovations process after the event.

Figure 7.21 illustrates one of the most significant advantages to tracking of feedback from higher-level models to the low-level vision system. The illustrated sequence is difficult to track due to the presence of periodic, binocular, flesh-flesh occlusions. That is, one hand is occluded by the other from both camera viewpoints in a periodic fashion: in this example at approximately 1Hz. The binocular nature of the occlusion events doesn't allow for view selection to aid tracking: there is no unambiguous viewpoint available to the system. Flesh-flesh occlusions are particularly difficult for tracking systems since it's easier to get distracted by an object with similar appearance (like another hand) than it is to be distracted by an object with a very different appearance (like a green shirt sleeve). The periodic nature of the occlusions means that the system only has a limited number of unambiguous observations to gather data before another occlusion again disrupts tracker stability.

Without feedback, the 2-D tracker fails if there is even partial self-occlusion, or occlusion of an object with similar appearance (such as another person), from a single camera's perspective. In the even more demanding situation of periodic, binocular, flesh-flesh occlusions, the tracker fails horribly. The middle pair of plots in Figure 7.21 show the results. The plots from a cross-eyed stereo pair. The low-level trackers fail at every occlusion causing the instantaneous jumps in apparent hand position reported by the system. Time is along the X axis, from left to right. The other two axes represent Y and Z position of the two hands. The circular motion was performed in the Y-Z plane, so X motion was negligible. It is not shown in the plot.

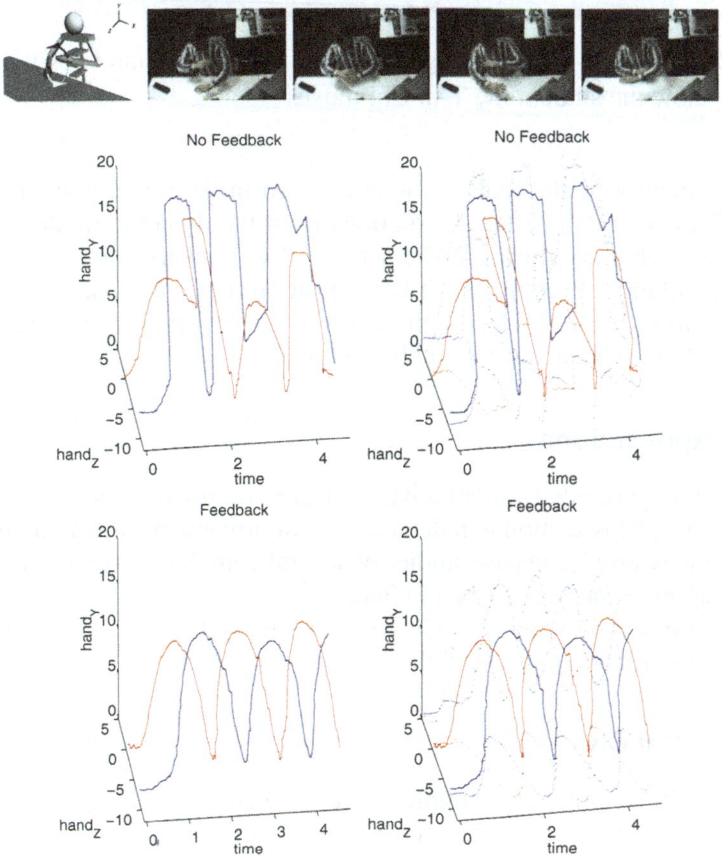

Fig. 7.21: Tracking performance on a sequence with significant occlusion. **Top:** A diagram of the sequence and a single camera's view of **Middle:** A graph of tracking results without feedback (cross-eyed stereo pair). **Bottom:** Correct tracking when feedback is enabled (cross-eyed stereo pair)

Fig. 7.22: The T'ai Chi sensei gives verbal instruction and uses it's virtual body to show the student the T'ai Chi moves

The situation with feedback, as illustrated in the lower pair of plots in Figure 7.21, is much better. Predictions from the dynamic model are used to resolve ambiguity during 2-D tracking. The trackers survive all the occlusions and the 3-D estimates of hand position reveal a clean helix through time (left to right), forming rough circles in the Y-Z plane. With models of behavior, longer occlusions could be tolerated.

7.4.2 Applications

Section 7.4.1 provided quantitative measures of improvement in tracking performance. This section will demonstrate improvements in human-computer interaction by providing case studies of several complete systems that use the perceptual machinery described in Section 7.3.

The three cases are the T'ai Chi instructional system, the Whack-a-Wuggle virtual manipulation game, and the strategy game Netrek.

7.4.2.1 T'ai Chi

The T'ai Chi sensei is an example of an application that is significantly enhanced by the recursive framework for motion understanding simply by benefiting from the improved tracking stability. The sensei is an interactive instructional system that teaches the human a selection of upper-body T'ai Chi gestures[7].

The sensei is embodied in a virtual character. That character is used to demonstrate gestures, to provide instant feedback by mirroring the student actions, and to replay the student motions with annotation. Figure 7.4.2.1 shows some frames from the interaction: the sensei welcoming the student on the left, and demonstrating one of the gestures on the right. The interaction is accompanied by an audio track that introduces the interaction verbally and marks salient events with musical cues.

There are several kinds of feedback that the sensei can provide to the student. The first is the instant gratification associated with seeing the sen-

Fig. 7.23: Visual and Auditory cues are used to give the student feedback. The sensei mimics the student motions and indicates problems to be worked on

sei mirror their motions. This allows the student to know that the sensei is attending to their motions and gives immediate feedback regarding their perceived posture relative to the ideal gestures they were just shown. When the sensei decides that feedback is appropriate this mirroring stops: indicating to the student that the interaction paradigm has changed. In this feedback mode the sensei can either critique individual gestures or remind the user of the global order of the sequence. The left and center images in Figure 7.23 show an example of a critique of a specific gesture. Visual and musical cues indicate the temporal location of the error and the sensei's gaze direction indicates the errant hand. The right image in Figure 7.23 is an example of feedback regarding the overall structure of the T'ai Chi sequence.

There are several technologies at work making these interactions possible. The mirroring is accomplished simply by piping the tracking data through the inverse-kinematics engine inside the sensei's body. This is technically simple, but it is imperative that it be robust. It is a meaningful event when the sensei stops mirroring: it signals to the student that they should stop and prepare to receive feedback. Tracking failures can cause the sensei to pause for an indeterminate length of time. Even worse, tracking failures can cause the sensei to spontaneously contort into one of many unpleasant configurations. Either event will obviously detract from the student's learning experience.

The technology behind the identification and interpretation of T'ai Chi gestures is somewhat more complex. To summarize: the sensei learns T'ai Chi gestures by watching a human perform the motions. The system builds Hidden Markov Model (HMM) of each gesture. The HMMs are comprised of a sequence of states with Markov temporal dynamics and Gaussian output probabilities. In effect they are capturing a mean path through parameter space that represents the gesture, and a covariance that specifies an envelope around the mean path that represents the observed variability. The gestures are recognized in the usual way [10]. Once a gesture is recognized, the lattice is examined to find the point at which the observed gesture differ the most

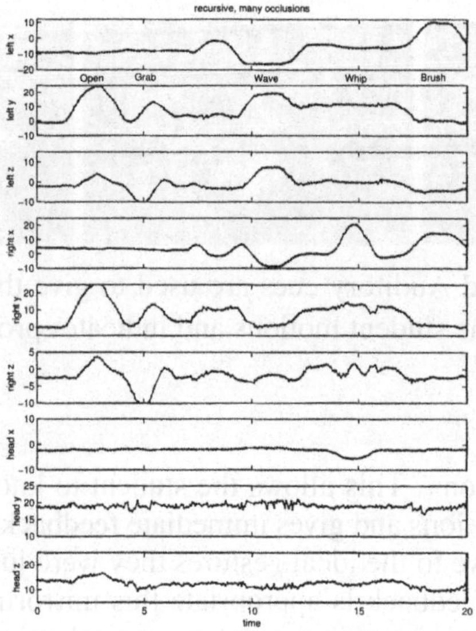

Fig. 7.24: This example is performed carefully to avoid self-occlusions

from the learned ideal, weighted by the allowable variation [7]. Tracking errors can have a large impact on this process. If a critical chunk of a gesture is missed due to tracking error then it may be unrecognizable, or worse, the system may attempt to correct illusory motion errors and confuse the student.

The individual T'ai Chi gestures are relatively complex. Each gesture takes several second to perform. Critiques often involve playback of observed and ideal gestures, and as a result a single critique may last several tens of seconds. The result is that the frequency of interaction is relatively low. Tracking errors that result in misclassified gestures or illusory motion errors can thus lead to a significant break in the experience. Failures thus lead to frustration. Frustration is antithetical to the underlying goal of T'ai Chi: relaxation and focus.

The plot in Figure 7.24 shows a flawless execution of a five gesture sequence as tracked by a bottom-up 3-D blob tracker. The individual gestures are hand-labeled at the top of the plot. The plots show, from top to bottom, the X, Y and Z position of the left hand, right hand, and head. Y is positive up. X is positive to the student's right. Z is positive away from the user. The following discussion will focus on two salient features. The first feature is the initial bump in the Y axis of both hands. This corresponds to the up and down motion of "Opening Form". The second feature is the double bump in the right hand X position at the end of the sequence. This corresponds to the right hand motion in "Single Whip" and the motion of both hands to the

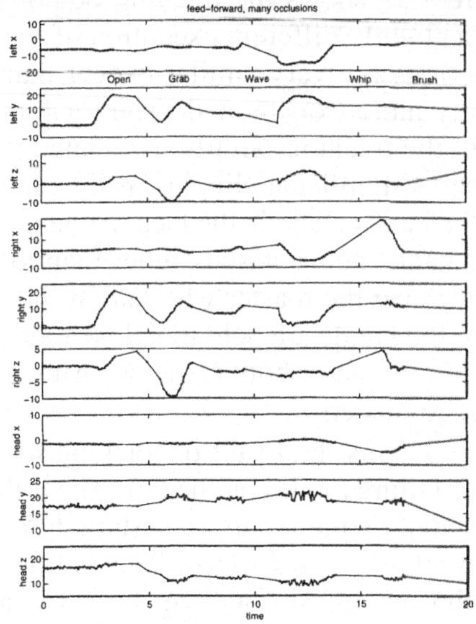

Fig. 7.25: A pessimal example that demonstrates all possible failure modes

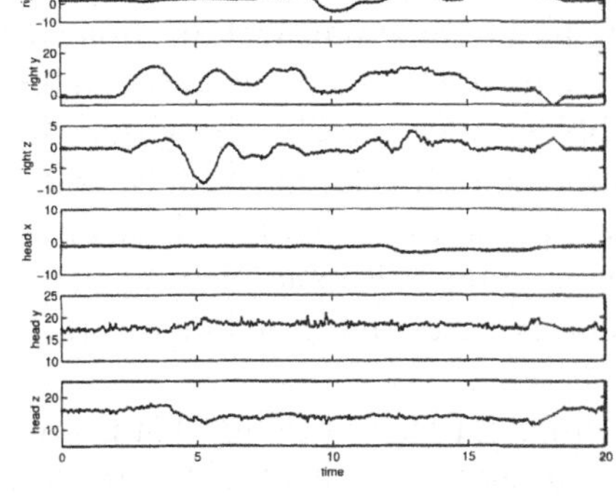

Fig. 7.26: The recursive tracker is is able to resolve the ambiguities

right in "Brush Knee".

The plot in Figure 7.25 shows the tracking output from the bottom-up 3-D blob tracker for a slightly different execution of the sequence. To a human observer the motions are very similar except that the motions in the failure case are slightly more exaggerated. The tracking data looks significantly different due to the tracking failures associated with occlusions. The first salient feature shows significant clipping as the top of the gesture is lost when one or both of the hands occlude the face in one or both of the cameras. This is caused by the hands being raised higher than ideal (or, alternatively if the student is shorter than the master who taught the sensei). The second feature is lost because the hands are held too close together. This causes the 2-D blob trackers to fail in one or both cameras. Almost no tracking data is acquired after 14s due to this error.

The plot in Figure 7.26 is the result of tracking motion very similar to that represented by the failure modes on the left. The difference is that in this case the motion is tracked by the recursive system described in Section 7.3. Both features are present and well-formed. This is true despite the fact that the hands were up high enough in opening form to cause occlusion with the face and close enough together to cause 2-D tracking failures at the end of the sequence. The recursive structure of the tracker allows the system to integrate information at all levels of modeling to produce good tracks in these ambiguous cases, and sensei is better able to evaluate the student's true departures from the ideal and provide more appropriate feedback.

7.4.2.2 Whack-a-Wuggle

Whacka is a virtual manipulation game. A virtual actor mirrors the motions of the human's upper body. The goal is for the human to touch objects in the virtual world vicariously through the virtual actor. The world contains static objects (Wuggles) and moving objects (Bubbles). Wuggles can always be whacked by returning to the same physical location since they do not move in the virtual environment, and the mapping between the physical and virtual worlds is fixed. Bubbles move through the virtual space, so they require hand-eye coordination to pop. In addition each Bubble is at a different depth away from the player in the virtual world. Due to poor depth perception in the virtual display, popping a Bubble requires the human to reach up to the approximate location of the bubble and then perform a limited search in depth.

There are a very small number of things that people do with their hands when playing Wacka. Hands are usually either resting, whacking, or popping. Whacking quickly becomes a ballistic action once the human learns where the Wuggles map into the physical world. Popping always involves performing the more complex visual-motor feedback loop required to find

Fig. 7.27: In Whack-a-Wuggle a clown mirrors the player's body motions. By guiding the clown's arms, the player can Whack Wuggles (the objects with eyes) and pop Bubbles (the objects floating in the air)

the Bubbles because they have random and only partially observable depth. The pace of the game is fast enough to discourage motivated players from wasting time performing extraneous gesticulations.

This context was used to test the method of alphabet selection described in Section 7.3.4.4. The player's motions were tracked and hand labeled for three types of behavior: *Whacking*, *Popping*, or *Tracker Failure*. Task 1 was to recognize these three classes of behavior. Task 2 was to distinguish the playing styles of different people.

The best alphabet for distinguishing the three players consisted of 3 elements, with 10 states each. Figure 7.28 shows alphabet traces for the three players over approximately one minute of play. These traces are the features used for player identification. Player identification performance was 75% for three players. Gesture recognition performance was also over 70%.

These results demonstrate the feasibility of recognizing gestures from model innovations. Recent work on temporal sequence clustering [44, 1, 31, 37] provides the clustering tools required for similarly attacking more complex domains.

7.5 Background

The implementation described above combines a rich 3-D model of the human body with probabilistic image features and a recursive formulation to realize robust tracking of the human form. DYNA also explicitly incorporates learned patterns of control into the body model. This level of modeling

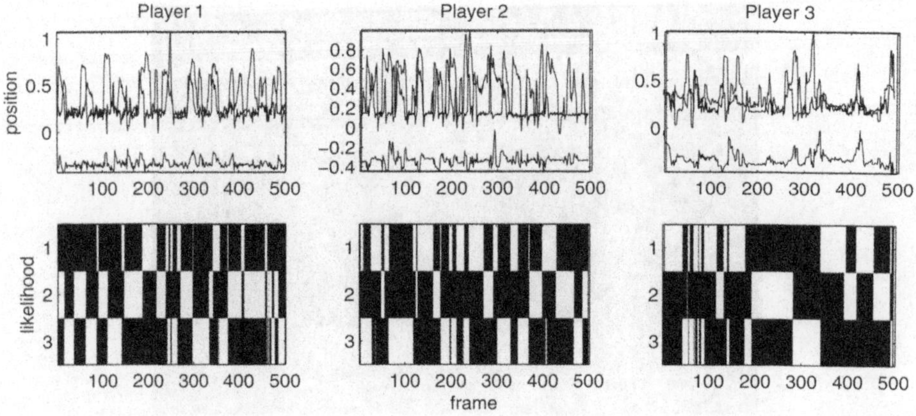

Fig. 7.28: Top: Tracking data. **Bottom:** Corresponding likelihood trace of the identification alphabet

detail and careful feedback design greatly improves robustness and opens the door to very subtle analysis of human motion. However, not all applications require this level of detail. There is now a rich literature on human motion perception, and it is full of alternative formulations. This section will introduce you to some of the key papers and place them within the context of the DYNA framework.

Perhaps the first efforts at body tracking were by Badler and O'Rourke in 1980[33], followed by Hogg in 1988 [32]. These early efforts used edge information to drive a kinematic model of the human body. These systems require precise hand initialization, and can not handle the full range of common body motion.

Following this early work using kinematic models, some researchers began using dynamic constraints to track the human body. Pentland and Horowitz employed non-rigid finite element models driven by optical flow in 1991[35], and Metaxas and Terzopolous' 1993 system used deformable superquadrics [26, 30] driven by 3-D point and 2-D edge measurements.

Authors have applied variations on the basic kinematic analysis-synthesis approach to the body tracking problem[41, 6, 20]. Gavrila and Davis[18] and Rehg and Kanade[40] have demonstrated that this approach has the potential to deal with limited occlusions, and thus to handle a greater range of body motions.

The DYNA framework is most closely related to the behavior modeling work of Pentland and Liu in 1995[36], Blake[23] and Bregler[8]. Those papers introduce the idea of modeling human behavior as the interaction between low-level observations and dynamic models in a recursive framework.

7.5.1 Particle Filtering

It is useful to compare the approach advocated here with that taken by particle filtering approaches such as the CONDENSATION algorithm proposed by Isard and Blake[24]. Both approaches have their roots in recursive filtering and control theory as described in Section 7.2.

The fundamental difference between recursive methods and particle filtering is in the representation of the probability density from which the state estimate is drawn. Recursive filtering methods, of which the Kalman filter is an example, represent the density parametrically. The CONDENSATION algorithm is a framework for representing the density as a set of samples drawn from the underlying distribution and is therefore not parametric. The Multiple Hypothesis Testing (MHT) framework from the control and estimation community[3] provides an extension to unimodal recursive filtering frameworks that allows multi-modal, parametric representations. Work by Rehg[11] shows how unimodal trackers such as the one described here may be embedded in the MHT framework.

All recursive filtering frameworks require a mechanism for propagation forward in time of the estimated density. In Kalman filtering this is accomplished in parameter space using a unimodal model of state dynamics to update both the mean, $\hat{\mathbf{x}}_t$, and variance, $\mathbf{\Sigma}_t$, of the density. Kalman filters assume a linear model of the state dynamics, $\mathbf{\Phi}$. The mean and variance updates are accomplished with the well known closed-form equations[2].

In the CONDENSATION algorithm, the distribution is represented non-parametrically, so it must employ a different mechanism for state update. Specifically the distribution is represented as a set of samples, S_t, in model state space with an associated set of weights Π_t. The system dynamics are represented as a density conditional on the current model state: $p(X_{t+1}|X_t)$. Model update begins by selecting a sample, $s_t^{(i)}$, from the set S_t with probability determined by the set of weights Π_t. A new sample is then drawn from the corresponding density $p(X_{t+1}|X_t) = s_t^{(i)}$, and becomes the sample $s_{t+1}^{(i)}$ in the updated representation. The new weight $\pi_{t+1}^{(i)}$ is determined by comparing the sample $s_{t+1}^{(i)}$ to the image evidence.

A tracker using CONDENSATION must maintain a large number of samples to adequately represent the underlying distribution. The literature calls for approximately 1500 samples to track a bouncing ball with a single degree of freedom, and as many as 15,0000 samples to track the hand during a drawing task [25]. The advantage is that the distribution is not required to match a pre-defined parametric form. Since the Kalman filter chooses a parametric representation that is unimodal, it is possible for ambiguous observations to move that single mode away from the truth. Dynamic state updates will likely cause the location of the mode to diverge further from the truth over time. The sampled representation of the density in CONDENSATION means that it is possible for a portion of the probability mass to be allocated to divergent hypotheses. As long as there are a sufficient number of samples available to adequately represent the distribution, alternate hypotheses may be propagated forward as long as there are observations to support them.

One issue with Kalman-derived filtering techniques is that it is difficult to formulate closed-form update equations for arbitrary parametric density representations as is done for the unimodal Gaussian case. MHT is a method for maintaining a set of N unimodal Gaussian hypotheses in the Kalman filtering framework without the need to reformulate the update equations. The MHT method is essentially a bank of Kalman filters. Each Kalman filter makes predictions according to its own, possibly unique, state estimate and dynamic model. The resulting innovations sequence is a measure of how well each filter is explaining the observed data. Filters with large innovations are poor predictors and are candidates for reinitialization [3]. Due to the fact that each filter accounts parametrically for some component of the underlying density, only a handful of hypotheses are required, as compared to the thousands of point hypotheses required by CONDENSATION. The literature shows that MHT can track a human figure, even with relatively weak underlying unimodal trackers, with as few as 10 hypotheses[11].

These methods therefore represent extremes of a continuum where computational cost can be traded off against modeling cost. For systems where the cost of model-building is very high compared to the cost of computation, CONDENSATION is a good, general method for tracking. For situations where the dynamics is reasonably well understood and the cost of computation is very high (as in the case of real-time human motion understanding), then parametric recursive filtering, possibly in conjunction with an MHT framework, is a better choice.

7.5.1.1 Recent Work on CONDENSATION

In the time since the first description of CONDENSATION in 1996 [23] there have been several documenting modifications to CONDENSATION that help avoid the curse of dimensionality involved in tracking systems with many degrees of freedom or with complex dynamic models. Mostly the improvements in computational efficiency of the algorithm are achieved by incorporating more sophisticated domain models. The pure form of CONDENSATION was the general tracker that discovered the dynamics from the observation at high computational cost. These extensions involve replacing flocks of hypotheses with domain-specific models that are better at capturing the true dynamics of the systems. These extensions essentially fill in the spectrum of trackers defined by pure CONDENSATION on one end, and the MHT framework on the other.

MacCormick and Blake [29] propose an extension that culls samples that represent inconsistent model states. This requires the CONDENSATION implementation to have more domain knowledge. For the task of tracking single degree of freedom motion of human bodies, this improvement allows tracking 2-D movement of head and shoulders of two people with 2000 samples. That is compared with tracking three people, two stationary and one moving with a single degree of freedom with 1000 samples.

Duetscher, North, Bascle, and Blake [14] adds the notion of kinematic singularities and end-stops. By adding significant domain knowledge they bring CONDENSATION much closer to MHT, requiring 50 samples to track arm motion with two degrees of freedom (elbow flexion and and shoulder rotation). They also track walking motion that is similar to, but less energetic than the dancing motions tracked by the MHT system in [11] with only 10 hypotheses.

7.5.2 Analysis-Synthesis

The analysis-synthesis approach is a widely used approximation to the general framework of recursive filtering. There are two major differences between the recursive framework advocated here and the analysis-synthesis (AS) framework: AS includes a significant feed-forward component, and lacks a model of system dynamics. Each of these issues are discussed in detail below.

7.5.2.1 Overview of Analysis-Synthesis

An AS tracker is composed of an Analysis module, a Synthesis module and a model of the system to be tracked. The Analysis module is composed of a set of image processing routines that extracts features from image observations. The Synthesis module renders the model state as a synthetic image composed of features congruent to those extracted by the image processing routines. The overall goal is to directly compare the model state to the observed image. An iterative process updates the model state until the best match between image and model is found.

Due to the complexity of real world image production, direct comparison is very difficult. The Synthesis module would have to generate photorealistic images of each hypothesized configuration. Further, direct image comparison would likely be dominated by aspects of the scene that may be uninteresting: the direction of illumination for example. As a result, the comparison is done in an abstract feature space. Image processing routines are used to extract these features from the image. In practice these features are often edge features, but in principle could be any low-level feature. The model state is rendered into this feature space by the Synthesis model. The set of features produced by the Synthesis module is then compared to the set of features extracted by the Analysis module. Discrepancies between analysis features and synthesis features represent a residual to be minimized by the system. The minimization of this residual is solved using iterative gradient descent in model space. This requires a model of how variation in feature space maps into variation in model space. Since Synthesis likely involves complex non-linear components from revolute model variables, perspective projection, and other processes, this model is usually a local, linear approximation. This places constraints on the gradient descent step size and leads to the search having a high computational cost.

In the case of tracking articulated bodies or body-parts, the model usually consists of a kinematic linkage. This model can be linearized around the current estimated operating point to yield a set of Jacobian matrices relating variation in feature space to model state perturbance. These Jacobians are used to transform the measured residuals into model adjustments. The iteration process may be weighted by temperature coefficients, or filtered by other, more sophisticated, gradient decent approaches.

7.5.2.2 The Feed-Forward Nature of Analysis

The analysis module is feed-forward, meaning that the image processing routines (often deterministic filters) operate without access to any context that may be available in the rest of the system. This one failing has a serious impact on the stability of the system. The Analysis module does not have access to the high-level constraints coded in the system model or Synthesis module that might help resolve low-level ambiguities. There is no one place where images and high-level knowledge are brought together. The two domains are separated by the feature abstraction. This barrier makes AS systems more modular. The benefits of breaking this barrier are discussed in detail in Section 7.4.1.

It is important to note that merely providing prior information to an Analysis module is not enough. The implementation must have enough mathematical flexibility to incorporate the information in a meaningful way. As an example, convolution trackers often use simple Kalman predictors to improve tracking performance, but is this a truly recursive system?

Consider the normal operation of a convolution tracker: an image patch is copied from an image, possibly matted, and then convolved with an image (or a region of an image). The maximum resulting value is accepted as the new location of the feature represented by the image patch. The easiest, and most common, way to incorporate this prior knowledge into the tracker is to constrain the search to the likely area. This does not result in a better answer. It merely results in a more efficient answer since less area needs to be searched[22]. If the "right" answer is within the oval it will be found faster, but there will be no difference between the result of this search and the result of the slower search. A truly recursive system would yield a different maximum.

This raises the question of feature selection. Probabilistic features such as the blobs presented in Section 7.3 provide a mathematically natural method of integrating high-level knowledge. Features such as convolution patches, edges, and other deterministic, filter-based features, do not provide easy integration of context.

7.5.2.3 Lack of System Dynamics

The second major difference between the two techniques is the form of the model update. In AS, a distance metric is induced in the chosen feature space. These gradients are then transformed into model space to determine the appropriate system update. The system update is completely driven by observations. The Recursive Filter update is shaped not only by observation models, but also the system dynamics, as described in Section 7.3. This enables the Recursive Filter to take advantage of the constraints of system dynamics when determining the appropriate system update. Since real dynamic systems are heavily constrained by the laws of physics and the structure of their control mechanisms, the system dynamics is an important source of context for the observer.

The lack of these constraints on the update mechanism in AS means that past model configurations do not inform the current choice of state estimate. Ambiguities or errors in feature extraction may result in the iteration process finding an explanation that, while plausible in isolation, is impossible due to the physical and practical limitations on the actual system being observed. One example of this sort of failure is the instantaneous flip between two solutions that are ambiguous in the absence of dynamic constraints.

7.5.2.4 How Real Systems Address These Shortcomings

When building an AS system, it is necessary to find ways to overcome these shortcomings in the framework. This section will discuss several systems, their contribution to AS work, and how they relate to true recursive filtering.

Gavrilla and Davis [18, 19] employ a kinematic model of the human body in their AS framework. The kinematic skeleton is augmented with a skin composed of tapered superquadrics. The subjects wear tight clothing, so the chains of tapered superquadrics are a good approximation to the external silhouette of the body. Edge energy is used as the common feature space: it is extracted from the image by a filter and rendered from the model. This, combined with the model parameter search for each image, makes the system computationally very expensive. This is an example of classic AS applied to the tracking of the human body. Later improvements included avoiding the inversion of Jacobian matrices to improve the performance of the search algorithm.

Kakadiaris and Metaxas [27] also use contour information combined with a detailed volumetric model of the human body. Unlike the above system, this system uses 3-D deformable models to track features in the image sequence. These low-level trackers represent a further dimensionality reduction, and thus undoubtedly improve the computational performance of the system. However, the trackers do not receive assistance from the body model and are subject to failures. The system partially deals with these failures by predicting occlusions. These predictions are used to select fortuitous views from the collection of available cameras in an attempt to avoid using data from ambiguous 2-D tracking results. Dynamics are used to make observation predictions, but these predictions are only used to initialize the feature search, as discussed above.

Delamarre and Faugeras [13] use the AS framework to track the human body. The silhouette is the common feature space. Silhouettes are tracked using active contours that are sensitive to optic flow and image intensity. The 3-D body model is a kinematic chain composed of spheres, parallelpipeds, and truncated cones. This model is projected into the image plane and the model is adjusted to match the observed silhouette. Since active contours are used, it is possible that prior information could help drive the contour tracker. This is a distinct advantage over AS implementations that use edge energy, or other deterministic features.

7.5.3 Recursive Systems

Some recursive systems have relied on optical flow for tracking. Similar to regularization work, they attempt to induce a low dimensional model as a minimum mean squared error explanation of noisy flow data. They modify the regularization by utilizing the benefits of Kalman filtering: prediction of observations and corresponding revision of the state estimate given new observations. This ideally eliminates the search inherent in the regularization framework.

Pentland and Horowitz [35] demonstrate the power of this framework for tracking 3-D and $2\frac{1}{2}$-D motions on deformable and articulated objects in video. They use modal analysis to project the flow field into relatively low-dimensional translation and deformation components. In this way they take advantage of the regularization approach to reduce the dimensionality of the optic flow data, but avoid the usual search in model parameter space. These low-dimensional signals are then used by an extended Kalman filter framework to estimate the new best estimate of pose.

The mapping of optic flow data to model influence requires the ability to predict model state and project that state into the image plane to determine proximity. This requires careful initialization. Although the framework should support prediction and resolution of ambiguities such as occlusion, those issues are not addressed in the the literature.

Bregler and Malik [8] present a very interesting recursive framework for tracking of the human body similar to DYNA that includes probabilistic feature extraction influenced by dynamic models, and the possibility for tightly coupled behavioral models. The system relies on optic flow data, but tracks blobs in that data. Unfortunately the early work on tracking focused entirely on 2-D features and 2-D models. As a result the motions tracked were all 2-D motions parallel to the camera. The dynamic models were all learned from observation. This is interesting, but also limits the system because it learns models in 2-D that are not good approximations of the real system dynamics due to the unobservability of truly 3-D motions in projective 2-D imagery.

In later work [9], the system was extended to 3-D but the extension came at the expense of the interesting recursive aspects of the earlier system. The 3-D system involved a return to more classical regularization work with the addition of clever representations for rotation that made the system tractable. Like most of work covered in this chapter the focus was on off-line motion capture, not real-time interface.

7.5.4 Bayesian Networks

Several systems have attempted to substitute explicit models at various levels of the tracking process with Bayesian networks.

Sherrah and Gong [43] use mean-shift trackers and face detectors in 2-D imagery to recover low-level features. In a process called *tracking by inference*, the tracked features are then labeled by a hand-crafted Bayesian Network. The simple trackers will fail during periods of ambiguity, but the network repairs the resulting mislabeling. This is accomplished by computing the most likely labeling given the observations and the network, which encodes a combination of distilled knowledge about dynamics and human habit.

Kwatra, Bobick and Johnson [28] present a similar system except that the low-level features are generated by a sophisticated static analysis method called GHOST[21] and the temporal integration modeling is concentrated into a Bayesian network. The probabilities in the network are maintained using a particle filtering method. This division is advantageous because the framework can accommodate any sort of static feature estimation. The statistical model of dynamics is weak compared to explicit kinematics and dynamics, but has the advantage of being light-weight.

Pavlović, Rehg, Cham and Murphy [34] present a system that tracks walking motion parallel to the image plane. This system dispenses with precise volumetric and kinematic models. Image correlation is used to track parts of the body. The trackers are initialized by a generalized form of MHT that consists of many Kalman filters embedded in a Bayes network. The Kalman filters employ learned dynamic models to make predictions, and the innovations process is used to drive model switching.

These systems have the advantage that the probabilistic models of dynamics and habit can be estimated from data and therefore can more easily accommodate idiosyncrasies of those datasets, but the models are soft compared to explicit kinematic and dynamic models, so that convenience can come at the cost of system fidelity.

7.5.5 Summary

Systems that visually track human motion fall into three basics categories: analysis-synthesis, recursive systems, and the more recent wave of fully statistical methods including particle filtering and Bayesian networks. Each of these methods has it's uses. Due to their modularity, analysis-synthesis systems are simple to build and can employ a wide range of features, but they rely on expensive searches for the synthesis component and allow features failures to propagate though the entire system. Particle filtering are an easy way to get many of the advantages of a truly recursive system with minimal modeling effort, but their lack of strong models requires massive Monte-Carlo simulations to adequately represent the underlying probability density. Fully recursive systems, such as the DYNA implementation presented above, require significant modeling effort, but this effort is rewarded with robustness, efficiency, and subtlety of analysis. Other recursive systems can allow the implementor to realize some of these benefits with less modeling effort when the application is less demanding.

7.6 Conclusion

Sophisticated perceptual mechanisms allow the development of rich, full-body human-computer interface systems. These systems are applicable to desk- and room-scaled systems, theatrical devices, physical therapy and diagnosis, as well as robotic systems: any interface that needs to interpret the human form in it's entirety. In this chapter you have seen the details of a computer vision system called DYNA that employs a three-dimensional, physics-based model of the human body and a completely recursive architecture with no bottom-up processes. This system is complex, but illustrated how careful modeling can improve robustness and open the door to very subtle analysis of human motion. Not all interface systems require this level of subtlety, but the key elements of the DYNA architecture can be tuned to the application.

Every level of processing in the DYNA framework takes advantage of the constraints implied by the embodiment of the observed human. Higher level processes take advantage of these constraints explicitly while lower level processes gain the advantage of the distilled body knowledge in the form of predicted probability densities. These predictions enable more robust classification decisions. Systems which omit a model of embodiment entirely, or try to hallucinate physical constraints in the image plane will fail to capture important aspects of human motion and the overall system performance will suffer. Understanding embodiment is crucial to perceiving human motion.

Systems that rely on any completely bottom-up processing will incur performance penalties to recover from inevitable low-level errors, if it is possible to recover at all. Recursive frameworks are also crucial to perceiving human motion.

As perceptual technologies continue to improve, they will enable an environment rich with interactivity. Computers will cease to be boxes that necessitate unnatural, and sometimes painful or damaging interaction. Computers disappear into the environment, and the things we do naturally will become the primary interface. Our most important experiences are interactions with other people, and as machine perception advances, computation will finally begin to engage in that conversation.

7.7 Acknowledgments

The author would like to thank the readers that improved this document with their comments and suggestions: Trevor Darrell, Eric Grimson, Yuri Ivanov, and Alex Pentland.

References

[1] J. Alon, S. Sclaroff, G. Kollios, and V. Pavlovic. Discovering clusters in motion time-series data. In *Proceedings of the IEEE Conference on Computer Vision and Pattern Recognition*, pages 375–381, 2003.

[2] The Analytical Sciences Corporation. *Applied Optimal Estimation*, 1996.

[3] M. Athans and C. B. Chang. Adaptive estimation and parameter identification using multiple model estimation algorithm. Technical Report 1976-28, Massachusetts Institute of Technology Lincoln Laboratory, Lexington, Massachusetts, USA, June 1976. Group 32.

[4] Ali Azarbayejani and Alex Pentland. Real-time self-calibrating stereo person tracking using 3-D shape estimation from blob features. In *Proceedings of 13th ICPR*, Vienna, Austria, August 1996. IEEE Computer Society Press.

[5] Ali Jerome Azarbayejani. *Nonlinear Probabilistic Estimation of 3-D Geometry from Images*. PhD thesis, Massachusetts Institute of Technology, February 1997. Media Arts and Sciences.

[6] A. Baumberg and D. Hogg. An efficient method for contour tracking using active shape models. In *Proceeding of the Workshop on Motion of Nonrigid and Articulated Objects*. IEEE Computer Society, 1994.

[7] David A. Becker. Sensei: A real-time recognition, feedback, and training system for t'ai chi gestures. Master's thesis, Massachusetts Institute of Technology Media Laboratory, 1997. also MIT Media Lab Perceptual Computing TR426.

[8] Christoph Bregler. Learning and recognizing human dynamics in video sequences. In *Proc. IEEE Conf. on Computer Vision and Pattern Recognition*, June 1997.

[9] Christoph Bregler and Jitendra Malik. Video motion capture. Technical Report UCB/CSD-97-973, University of California, Berkeley, 1997.

[10] Lee W. Campbell, David A. Becker, Ali Azarbayejani, Aaron Bobick, and Alex Pentland. Invariant features for 3-d gesture recognition. In *Second International Conference on Face and Gesture Recognition*, pages 157–62, Killington, VT, USA, 1996.

[11] Tat-Jen Cham and James M. Rehg. A multiple hypothesis approach to figure tracking. In *Workshop on Perceptual User Interfaces*, San Francisco, Calif., November 1998.

[12] Brian P. Clarkson and Alex Pentland. Unsupervised clustering of ambulatory audio and video. In *Proceedings of the International Conference of Acoustics Speech and Signal Processing*, Phoenix, Arizona, 1999.

[13] Quentin Delamarre and Olivier Faugeras. 3d articulated models and multi-view tracking with silhouettes. In *Proceedings of the Seventh International Conference on Computer Vision*. IEEE, 1999.

[14] J. Deutscher, B. North, B. Bascle, and A. Bake. Tracking through singularities and discontinuities by random sampling. In *Proceedings of the Seventh International Conference on Computer Vision*. IEEE, 1999.

[15] Ernst D. Dickmanns and Birger D. Mysliwetz. Recursive 3-d road and relative ego-state recognition. *IEEE Trans. Pattern Analysis and Machine Intelligence*, 14(2):199–213, February 1992.

[16] Roy Featherstone. *Coordinate Systems and Efficiency*, chapter 8, pages 129–152. Kluwer Academic Publishers, 1984.

[17] Martin Friedmann, Thad Starner, and Alex Pentland. Device synchronization using an optimal linear filter. In H. Jones, editor, *Virtual Reality Systems*. Academic Press, 1993.

[18] D. M. Gavrila and L. S. Davis. Towards 3-d model-based tracking and recognition of human movement: a multi-view approach. In *International Workshop on Automatic Face- and Gesture-Recognition*. IEEE Computer Society, 1995. Zurich.

[19] D. M. Gavrila and L. S. Davis. 3-d model-based tracking of humans in action: a multi-view approach. In *CVPR96*. IEEE Computer Society, 1996.

[20] Luis Goncalves, Enrico Di Bernardo, Enrico Ursella, and Pietro Perona. Monocular tracking of the human arm in 3d. In *International Conference on Computer Vision*, Cambridge, MA, June 1995.

[21] I. Haritaoglu, D. Harwood, and L. Davis. Ghost: A human body part labeling system using silhouettes. In *Fourteenth International Conference on Pattern Recognition*, pages 77–82, 1998.

[22] Thanarat Horprasert, Ismail Haritaoglu, David Harwood, Larry S. Davis, Christopher R. Wren, and Alex P. Pentland. Real-time 3d motion capture. In *Workshop on Perceptual User Interfaces*, San Francisco, Calif., November 1998.

[23] Michael Isard and Andrew Blake. Contour tracking by stochastic propagation of conditional density. In *Proc. European Conference on Computer Vision*, pages 343–356, Cambridge, UK, 1996.

[24] Michael Isard and Andrew Blake. Condensation - conditional density propagation for visual tracking. *Int. J. Computer Vision*, 1998. in press.

[25] Michael Isard and Andrew Blake. A mixed-state condensation tracker with automatic model-switching. In *Proc 6th Int. Conf. Computer Vision*, 1998.

[26] I. Kakadiaris, D. Metaxas, and R. Bajcsy. Active part-decomposition, shape and motion estimation of articulated objects: A physics-based approach. In *CVPR94*, pages 980–984, 1994.

[27] Ioannis Kakadiaris and Dimitris Metaxas. Vision-based animation of digital humans. In *Computer Animation*, pages 144–152. IEEE Computer Society Press, 1998.

[28] Vivek Kwatra, Aaron F. Bobick, and Amos Y. Johnson. Temporal integration of multiple silhouette-based body-part hypotheses. In *IEEE Computer Vision and Pattern Recognition*, December 2001.

[29] John MacCormick and Andrew Blake. A probabilistic exclusion principle for tracking multiple objects. In *Proceedings of the Seventh International Conference on Computer Vision*. IEEE, 1999.

[30] Dimitris Metaxas and Dimitris Terzopoulos. Shape and non-rigid motion estimation through physics-based synthesis. *IEEE Trans. Pattern Analysis and Machine Intelligence*, 15(6):580–591, 1993.

[31] David C. Minnen and Christopher R. Wren. Finding temporal patterns by data decomposition. In *Proceedings of the 6th International Conference on Automatic Face and Gesture Recognition*, 2004.

[32] K. Oatley, G. D. Sullivan, and D. Hogg. Drawing visual conclusions from analogy: preprocessing, cues and schemata in the perception of three dimensional objects. *Journal of Intelligent Systems*, 1(2):97–133, 1988.

[33] J. O'Rourke and N.I. Badler. Model-based image analysis of human motion using constraint propagation. *IEEE Trans. Pattern Analysis and Machine Intelligence*, 2(6):522–536, November 1980.

[34] Vladimir Pavlović, James M. Rehg, Tat-Jen Cham, and Kevin P. Murphy. A dynamic bayesian network approach to figure tracking using learned dynamic models. In *Proceedings of the Seventh International Conference on Computer Vision*. IEEE, 1999.

[35] A. Pentland and B. Horowitz. Recovery of nonrigid motion and structure. *IEEE Trans. Pattern Analysis and Machine Intelligence*, 13(7):730–742, July 1991.

[36] Alex Pentland and Andrew Liu. Modeling and predicition of human behavior. In *IEEE Intelligent Vehicles 95*, September 1995.

[37] Fatih Porikli and Tetsuji Haga. Event detection by eigenvector decomposition using object and frame features. In *PID*, 2004.

[38] William H. Press, Saul A. Teukolsky, William T. Vetterling, and Brian P. Flannery. *Numerical Recipes in C: the art of scientific computing*. Cambridge University Press, Cambridge, U.K., second edition, 1992.

[39] Lawrence R. Rabiner. A tutorial on hidden markov models and selected applications in speech recognition. *Proceedings of IEEE*, 77(2):257–285, 1989.

[40] J.M. Rehg and T. Kanade. Visual tracking of high dof articulated structures: An application to human hand tracking. In *European Conference on Computer Vision*, pages B:35–46, 1994.

[41] K. Rohr. Cvgip: Image understanding. *"Towards Model-Based Recognition of Human Movements in Image Sequences*, 1(59):94–115, 1994.

[42] R. Shadmehr, F. A. Mussa-Ivaldi, and E. Bizzi. Postural force fields of the human arm and their role in generating multi-joint movements. *Journal of Neuroscience*, 13(1):45–62, 1993.

[43] Jamie Sherrah and Shaogang Gong. Tracking discontinuous motion using bayesian inference. In *ECCV (2)*, pages 150–166, 2000.

[44] Padhraic Smyth. Clustering sequences with hidden markov models. In Michael C. Mozer, Michael I. Jordan, and Thomas Petsche, editors, *Advances in Neural Information Processing Systems*, volume 9, page 648. The MIT Press, 1997.

[45] Thad Starner and Alex Pentland. Real-time american sign language recognition from video using hidden markov models. In *Proceedings of International Symposium on Computer Vision*, Coral Gables, FL, USA, 1995. IEEE Computer Society Press.

[46] Charles W. Therrien. *Decision, Estimation, and Classification*. John Wiley and Sons, Inc., 1989.

[47] Vladimir N. Vapnik. *Statistical Learning Theory*. John Wiley & Sons, 1998.

[48] A. S. Willsky. Detection of abrupt changes in dynamic systems. In M. Basseville and A. Benveniste, editors, *Detection of Abrupt Changes in Signals and Dynamical Systems*, number 77 in Lecture Notes in Control and Information Sciences, pages 27–49. Springer-Verlag, 1986.

[49] Andrew Witkin, Michael Gleicher, and William Welch. Interactive dynamics. In *ACM SIGGraph, Computer Graphics*, volume 24:2, pages 11–21. ACM SIGgraph, March 1990.

[50] Christopher Wren, Ali Azarbayejani, Trevor Darrell, and Alex Pentland. Pfinder: Real-time tracking of the human body. *IEEE Trans. Pattern Analysis and Machine Intelligence*, 19(7):780–785, July 1997.

[51] Christopher R. Wren. *Understanding Expressive Action*. PhD thesis, Massachusetts Institute of Technology, March 2000. Electrical Engineering and Computer Science.

[52] Christopher R. Wren and Alex P. Pentland. Dynamic models of human motion. In *Proceedings of FG'98*, Nara, Japan, April 1998. IEEE.

8 Cognitive User Modeling Computed by a Proposed Dialogue Strategy Based on an Inductive Game Theory

Hirotaka Asai[1], Takamasa Koshizen[2], Masataka Watanabe[1], Hiroshi Tsujin[2], Kazuyuki Aihara[3]

1. Department of Quantum Engineering and Systems Science, Graduate School of Engineering, University of Tokyo, {asai,watanabe}@sk.q.t.u-tokyo.ac.jp
2. Honda Research Institute Japan Co. Ltd., {koshiz,tsujino}@jp.honda-ri.com
3. Department of Information a Science, Institute of Industrial Science, University of Tokyo

Abstract

This paper advocates the concept of user modeling (UM), which involves dialogue strategies. We focus on human-machine collaboration, which is endowed with human-like capabilities and in this regard, UM could be related to cognitive modeling, which deals with issues of perception, behavioral decision and selective attention by humans. In our UM, approximating a pay-off matrix or function will be the method employed in order to estimate user's pay-offs, which is basically calculated by user's action. Our proposed computation method allows dialogue strategies to be determined by maximizing mutual expectations of the pay-off matrix. We validated the proposed computation using a social game called ``Iterative Prisoner's Dilemma (IPD)" that is usually used for modeling social relationships based on reciprocal altruism. Furthermore, we also allowed the pay-off matrix to be used with a probability distribution function. That is, we assumed that a person's pay-off could fluctuate over time, but that the fluctuation could be utilized in order to avoid dead reckoning in a true pay-off matrix. Accordingly, the computational structure is reminiscent of the regularization implicated by the machine learning theory. In a way, we are convinced that the crucial role of dialogue strategies is to enable user models to be smoother by approximating probabilistic pay-off functions. That is, their user models can be more accurate or more precise since the

H. Asai et al.: *Cognitive User Modeling Computed by a Proposed Dialogue Strategy Based on an Inductive Game Theory*, Studies in Computational Intelligence (SCI) **7**, 325–351 (2005)
www.springerlink.com

dialogue strategies induce the *on-line* maintenance of models. Consequently, our improved computation allowing the pay-off matrix to be treated as a probabilistic density function has led to better performance, Because the probabilistic pay-off function can be shifted in order to minimize error between approximated and true pay-offs in others. Moreover, our results suggest that in principle the proposed dialogue strategy should be implemented to achieve maximum mutual expectation and uncertainty reduction regarding pay-offs for others. Our work also involves analogous correspondences on the study of pattern regression and user modeling in accordance with machine learning theory.

Key words: User modeling, Dialogue strategy, Inductive Game theory, Pay-off function, Mutual cooperation

8.1 Introduction

In recent years effective studies of *User Modeling* (UM) have attracted a renewed interest from researchers in the field of machine learning, cognitive science, and robotics. One of the fundamental objective of human - machine (including robot) interaction research is to design systems to be more usable, more useful, and to provide users with experiences fitting their specific background knowledge and objective. UM tackles the new essential challenges that have arisen to improve the cognitive way in which people interact with computational machines to do work, think, communicate, learn, observe, decide and so on. In a way, we are convinced that UM can cope with these challenges. The major characteristic of UM is its focus on the human emulation approach, which is based on the metaphor that to improve human-computer collaboration is to endow computers with *human-like capabilities*. Therefore, recently, UM seemed to be more related to cognitive modeling (CM) research which deals with issues of perception, how input is processed and understood, how output is produced, developed theories of the cognitive process related to human brain components that havebeen dedicated to brain science (Newell, 1983). However, it is still too complicated to model human cognition using knowledge from brain science, *e.g.,* Human Information Processor (HIP). Using psychological studies would be appropriate since they basically refer to human behaviors, and they have been used to analyze and model, in order to represent pay-offs of humans. In these studies, pay-offs can be treated as a sort of hidden or tangible or latent variable. In practice, UM aims at building a manifestation of humans based on their behavioral analyses, which is usually supported by psychological evidence. In fact, the UM study has already been engaged in deductive approaches in which psychology labeled each pay-offs of humans.

Strictly speaking, it is obvious that UM and CM have different perspectives and different purposes though these perspectives and purposes somehow overlap. Therefore, in our context, we take into account UM by integrating CM effectively with respect to user's pay-offs and characteristics, though the basic idea seems to be originated from the HIP (Newell, 1983).

Some of user modeling were derived from the need and desire to provide better support for human-computer collaboration (Fischer, 2001). User modeling, a 'collaborative' learning approach was used whenever one could assume that a user behaves in a similar way to other users (Basu, 1998 and Gervasio, 1998). In this approach, a model is built using data from a group of users, and it is then used to make predictions about an individual user. Practically, it reduces the data collection burden for individual users, though this prevents modeling the behavior of different types of users. In contrast, human emulation or content-based learning approach is built based on the metaphor that improves human-computer collaboration by endowing computers with *human-like capabilities*, as already described above. That is, human-like capabilities are expected to ensure long-lasting interaction by increasing the population of collaborative behaviors. After all, machines can recognize characteristics of a sole user. Basically, the content-based learning approach is *inductive* when a user's past behavior is a reliable indicator of his/her future behavior. In this way, user's data from his/her past experience is taken into account when building a predictive model. The predictive model is alternatively defined as a statistical model because statistical analysis is employed to generate predictive user models, simply called probabilistic generative models. However, this approach requires a system to collect fairly large amounts of data from each user, in order to enable the formulation of the statistical model.

In this paper, we attempt to deal with user modeling, mediated by our dialogic behavioral strategy. The proposed dialogue strategy can also be derived from a game theory (Nash, 1951). However, we utilize a particularly *inductive game theory* (Kaneko, 1999) where the individual player does not have any prior knowledge of the structure of the game. Instead, he/she accumulates experiences induced by occasional random trials in repeated play. This theory implies, in the end, maximizing each player's pay-off matrix or function by determining his/her behaviors. Our dialogic behavioral planning scheme is inspired by this inductive game theory. Players must consider each pay-off induced by their behaviors depending onthe surrounding situation. The inductive game theory aims at the formulation and emergence of individual views about society from experiences. Indeed, it allows game players to let only each payoff's expectations be maximized, and the relationship can eventually be cooperation rather than anti-cooperation. This is because such a game theory, proposed by (Kaneko, 1999) can be assumed to mediate the implications on relevant sociological, economical and even psychological literature. Generally, it is expected that a person should develop mutual strategies of dialogic behavior during the development of his or her life, in order to be able to communicate with others. As a consequence, our dialogic behavioral planning will allow players to generate models based on experiences, which are obtained from playing the social game in a recurrent situation. In the first paragraph, we pointed out the importance of user modeling. That is, we assumed that such a repeated social cooperative game could let players continually communicate by approximating other payoffs, according to the probabilistic generative models. To sustain such a communication, they must believe that longer will eventually be more profitable (*e.g.*, pay-off to each other) than only maximizing a their individual player's pay-off in the short-term. As a result, we expect that the pay-off expectation of both players will be maximized in the long-term. Thus, this kind of social cooperative game can be regarded as human studies with psychological and neuroscience literatures. For example, there is a well-known repeated game, called *iterative prisoner's dilemma (IPD)*. The IPD game has been used by investigators from a wide range of disciplines to model social relationships based on reciprocal altruism (Axelrod and Hamilton, 1981;Axelrod, 1984;Boyd, 1988;Nesse, 1990;Trivers, 1971). Interestingly, a result of the game can be to opt for immediate gratification attaining the maximum pay-off for that round. It may overlook or fail to consider the future consequences of defection. That means that players who resist the temptation to defect for short-term gain and instead persist in mutual cooperation may be better guided by the future consequences of their decisions.

The proposed computation will be implemented and validated using the IPD game. That is, we allow the IPD to cope with the approximation of a true pay-off matrix by estimating each type of players, *pay-off estimation* as well as by providing a dialogue strategy. The updated version of the proposed computation will be described by introducing a probability distribution function in the pay-off matrix, to deal with a *dead reckoning* problem regarding the true pay-off in others. The probabilistic form of our algorithm will improve our original computation with respect to the pay-off approximation. Overall, the dialogue strategy portion of the proposed computation could play the role of smoothing (probabilistic) generative models, which are used for estimating each player's pay-off. Since the dialogue strategy allows players to pose self-control, the reciprocal expectation of their payoffs will be maximized.

Additionally, the parametric form of probabilistic generative models could be more suitable to come up with the pay-off approximation. In a conclusive manner, our UM suggests to utilize the dialogue strategy that is obtained by approximating a *probabilistic* pay-off function. The proposed dialogue strategy must also take into account the following points:

–Maximum mutual expectation

–Uncertainty reduction

This paper will describe a new scheme of UM, which is combined with CM. In Section 8.2, we will show how the UM has been explored so far using machine learning theory. In Section 8.3, we will explain the link between social psychology and game theory. The major concept of our proposition - user modeling by a long-lasting dialogue strategy is described in Section 8.4. In Section 8.5, the proposed algorithm, and computation results will be presented with respect to the UM utilizing a long-lasting dialogue strategy, a concept is derived from the social game theory. Finally, we will conclude the presentation of our proposed computation and comment on future work.

8.2 Machine Learning and User Modeling

User modeling presents a number of challenges for machine learning that has hindered its application in user modeling, including: the need for large data sets; the need for labeled data; conflict drift; and computational complexity (Webb, 2001). Many applications of machine learning in user modeling focused on developing models of cognitive processes, usually called cognitive modeling (CM). The true purpose of integrating UM and CM includes discovering users' characteristics, which are on the cognitive process that underlie users' behavior. However, user modeling presents a number of very significant challenges for machine learning applications. In most problems, it is natural that learning algorithms require many training examples to be accurate (Valiant, 1984). In *predictive statistical models for user modeling*, this parameter represents an aspect of a user's future behavior based on the outcomes of possible behavior analysis. This often provides a major drawback as updating the user models based on the historical behavioral outputs is difficult, since the learning scheme is entirely *off-line*, and it requires significantly large amounts of training data to parameterize the aspect of users. As a consequence, their learning problems fail to its *ill-posed problem* of training outcome many times. As a result, the burden of collecting data in many cases must be seriously considered to allow the learning problem to catch up in real world competence.

Additionally, off-line learning prevents updating user models, when new types of information, are incorporated in the user models. We expect that our dialogue strategy will bring the learning of user modeling into be *smoother* (*i.e.*, more precise). This brings *on-line* maintenance of user modeling in order to more accurately estimate pay-offs in others. This also brings the question of the dialogue strategy allowing a machine's action to be done in collaboration with humans. In order to attain those objectives, the dialogue strategy ought to take into account a long-lasting interaction between machines and humans. In order to evaluate such a smoothing operation the long-lasting dialogue strategy will ensure satisfaction levels of humans to machine's actions. Nevertheless, machine-learning theory has only provided a mathematical criterion to evaluate trained models (usually called generative models) with respect to its generalization. Thus, the issue is to estimate a user's pay-off, and the dialogue strategy can be undertaken by having machines to generate *self-control* actions. Computationally, a mutual expectation between man and machine will lead to a maximum mutual expectation, which could approximately correspond to a user's satisfaction. In short, our proposed dialogue strategy suggests not only to consider the traditional computational effect but also to regard the psychological effect because the computation has to deal with pay-offs of humans. Additionally, the probabilistic pay-off's function given by the computation can be suitable to manifest uncertainty of the psychological aspect involved in man-machine interactions.

This point will be discussed later.

8.3 Social Psychology and Game Theory

In the previous section, we described the importance of social psychology, which is incorporated in the computational aspect of our dialogue strategy. Therefore, we present the relationship between social psychology and game theory that is the base of our proposed computation.

The scientific discipline attempts to understand and explain how the thought, feeling, and behavior of individuals are influenced by the actual, imagined, or implied presence of others. A fundamental perspective in social psychology emphasizes the combined effects of both the individual and the situation on human behavior. Interestingly, recent studies report researcher attempts to quantitatively model phenomena, which have occurred in social psychology using game theory.

What economists call game theory, psychologists call theory of social situation. Although game theory is related to 'parlor' games, most of the studies in game theory focus on how groups of people interact. In principle, there are two main branches of the game theory, cooperative and non-cooperative (defection). In classical game theory, Nash is also initiated a kind of noncooperative game theory (Nash, 1951). In principle, it is assumed that players are rational in the sense that they have high abilities of logical reasoning and knowledge of the structure of the game. Based on their abilities and prior knowledge, the individual player could make a decision precisely. We also call this theory *deductive game theory* because deduction is appropriate for the study of societies where players are well informed, such as small games played by experts. On the other hand, the inductive game theory assumes that players' may learn some parameters of the game and strategies of others as well as the payoffs from their own behavior. In a sense, the payoffs will need to be approximated by the parameters that model their own behavior.

One way to describe a game is the players (or individuals) participating in this game, and for each player, listing the alternative choices (called actions or strategies) available to that player. Usually, alternative choices can be taken depend on expected utility whose concept enters economic analysis of an individual's preferences over alternative bundles of consumption goods (Debreu, 1964). In the case of a *two*-player game, the actions of the first player form the rows of a pay-off matrix, and the actions of the second player the columns. In the game theory, the entries in the matrix are two numbers representing the *utility* to the first and second player respectively. In our context, the utility is approximately identified with the pay-off in others. The most representative game is *prisoner's dilemma (PD)*. The game allows players to choose *confess* or *non-confess* to the crime. The game can be represented by the pay-off matrix or function. It is noted that the pay-off matrix can be taken into account if the matrix is changed over time.

In short, the PD game has several characteristics. For example, no matter what value the matrix has, the partner is always best to confess (same action). In contrast, the other feature of the game is that it changes in a significant way if the game is repeated, or if the players interact with each other again in the future. In this case in the first round the suspects may reason that they should not confess because if they do not their partner will not confess in the second round of the game. The IPD game illustrates the theoretical question of analyzing the possibility of being rewarded or punished in the future for current behavior. The conclusion is that players ought to choose a cooperative behavior rather than a non-cooperative one.

In order to do that, players must approximate the pay-off matrix by estimating pay-offs in others.

The Iterated PD (IPD), sometimes called repeated PD (RPD), is simply an extension of the PD. The point of IPD is that when the game is played repeatedly, the probability of reciprocation determines whether defection or cooperation will maximize reinforcement in the long-term (Baker and Rachlin, 2001). The IPD implies that the pay-off's expectations of players ought to satisfy the two conditions:

– Players must interact repeatedly with social partners over the course of a lifetime

– Players must be able to discriminate against those who do not reciprocate altruism (Axelrod and Hamilton, 1981; Trivers, 1971).

Therefore, cooperation allows players to reflect self-control in order to maximize the high expectation of each pay-off by knowing the others as well as knowing themselves.

In fact, the IPD with self-control has been considered by Baker (Baker and Rachlin, 2001). They used the IPD game calculating the probability of reciprocation between two players to determine whether defection or cooperation would maximize reinforcement in the long-term. In a sense, cooperation reflects the fact that man uses self-control, which allows players to attain a probability of reciprocation of 1.0 in the long-term.

8.4 User Modeling by a Long-Lasting Dialogue Strategy

As mentioned above, user modeling may be built using an approach that consolidates both features of collaborative learning and content-based learning. We want our long-lasting dialogue strategy to follow this approach. There are previous studies related to user modeling, which take dialogue strategy into account (Litman, 2000). In practice, they use a spoken dialogue system, though multimodal dialogue has been, to date, combined with the spoken dialogue system (Andre, 1998)(Noma, 2000). Essentially, the most serious problem is *poor* speech recognition so they propose multimodal information that is dedicated from psychological experiences between infant and adult. Legerstee et al., has studied about the social expectancies between infants and adults (Legerstee, 2001). The social expectancies are defined as infants' expectancy for affective sharing. They investigated the role of maternal affect mirroring on the development of prosocial behaviors and social expectancies in 3-month-old infants. Prosocial behavior was characterized as infants' positive behavior and increased attention toward their mothers. These findings indicate that there is a relation between affects mirroring and social expectation of infants.

Psychologically, our dialogue strategy aims constitute the relationship between infant and adult (*mother*). That is, a machine is a sort of learner who needs to train the maternal affect mirroring with respect to the development of prosocial behavior and social expectancies. In a sense, the mother corresponds to users, and the machine attempts to share affection by estimating the pay-off of users. Our dialogue strategy permits the machinery development for attaining prosocial behaviors and social expectancies in accordance with user models. It allows machines to acquire user models, which may be difficult to learn in a short-term interaction with users only. Hence, the dialogue strategy must ensure a long-lasting interaction, which is able to gain a maximum user pay-off. In order to do so, we will provide a dialogue strategy, which maintains co-operative (friendly) behavior rather than increasing the population of unfriendly behavior by repeated interaction between players. Each player then will need to know the pay-off in others. Our proposed dialogue strategy aims at using multimodal information for specifying user models by maximizing user pay-offs in long-lasting interactions between machine and user. In real-world competence, the dialogue is psychologically expected to gain user's satisfaction by machines allowing users to induce behavioral plans related to social co-operative behaviors. The past user modeling research has assumed *persistency of interest* (Lieberman, 1995), whereby users maintain their behavior or interests over time. Interests are one of the fundamental pay-off of a user. That is, the induction of maintaining or modulating user's interests, endowed by the dialogue strategy, could lead to reduce the amount of training data. Therefore, this also can bring about a significant reduction of computational complexity across user modeling.

Furthermore, such user modeling cascaded with dialogue can be considered as psychologically effective so that reciprocal communication between machine and human is most likely feasible. For instance, the dialogue is expected to increase user's satisfaction by eliminating - behaviors that can be most suitable for a specific user's according to his or her pay-off's estimation. Gaining the satisfaction of users is really crucial for realizing a long-lasting interaction between a user and a machine. User modeling will still be able to estimate the interests, which vary with time.

8.5 Our Dialogue Strategy and Computations

In this section, we first provide a proposed algorithm with respect to our dialogue strategy. In order to show the computation of the proposed algorithm, we use the finite non zero-sum two-person game called IPD. We will first provide a brief description of IPD. In the game, two players independently choose to either cooperate with each other, or not, and each action is rewarded based on the maximization of the pay-off. Giving the reward, during the game, entirely depends on the interaction of both players' choices in that round. In the game (theory), players presumably play a role of decision makers, who makes the final choice among the alternatives.

The pay-off matrix represents known pay-offs to players (individuals) in a strategic situation given choices made by other players in the same situation. In practice, there are 'four' outcomes resulting from the two possible actions: cooperation (C) and defection (D); player A and player B cooperate (CC), player A cooperates and player B defects (CD), player A defects and player B cooperates (DC), or player A and player B defect (DD). The payoff of player A for the outcomes are arranged as follows: DC ≥ CC ≥ DD ≥ CD, and CC ≥ (CD+DC) ≥ 2. Each value of the pay-off matrix corresponds to a different outcome of social interaction.

Fig.1 describes the proposed algorithm with our dialogue strategy, which can be aimed to estimate the pay-off in others. In this way, many interactions between the two players must be undertaken. It is assumed that players initially do not know about each pay-off. Thus, probabilistic (generative) models to estimate the pay-off in others are set to *uniform*. The probabilistic models can be more accurate after repeated interactions are taken by our dialogue strategy, which considers both maximum mutual expectation and uncertainty reduction. The algorithm also calculates a probabilistic error of true and approximating pay-off matrix. After all, we can expect to obtain a desired probabilistic model.

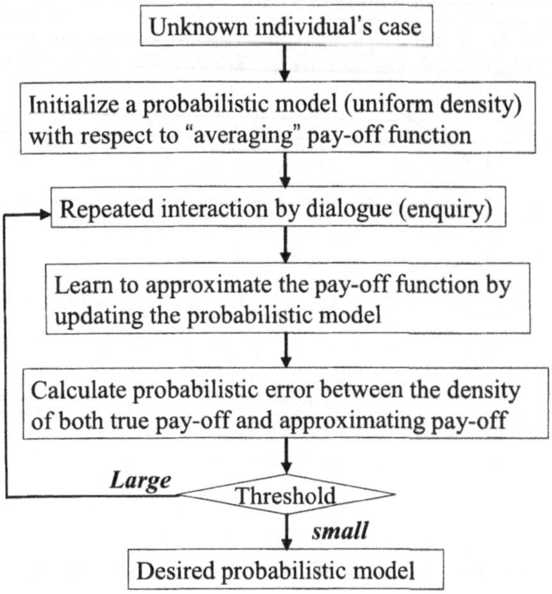

Fig. 8.1. The principal flow of our proposed algorithm

The primary objective of the proposed algorithm is to allow each player to find out the elements of each pay-off matrix, which is used for estimating pay-offs in others. To simplify the simulation, several conditions were assumed as follows,

– A true pay-off matrix is unchangeable.

– The pay-off matrix is unknown between two players.

– Each player's behavior can be used for modeling to estimate the pay-off state of players. The model generated by learning is called 'a (probabilistic) generative model'. The (probabilistic) generative models are used for approximating the pay-off matrix.

– An effective dialogue strategy could precisely approximate the pay-off matrix, (particularly when players interact with each others).

– During the dialogue interaction between the two players, they attempt to follow collaborative behavioral patterns, which lead to a maximum expectation with respect to the pay-off matrix.

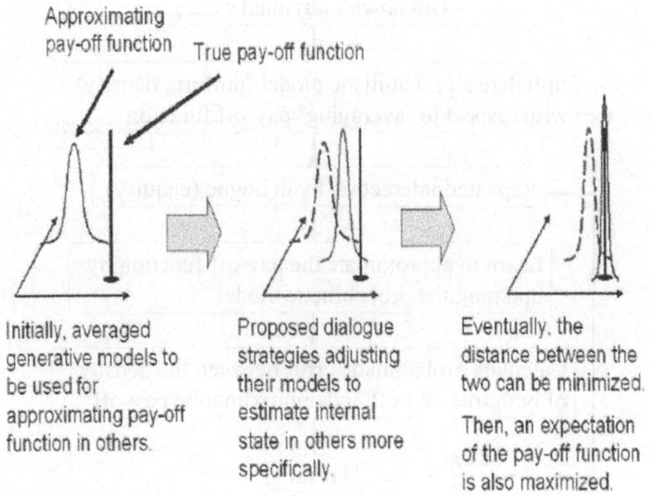

Fig.8.2. Computational view of our proposed algorithm shown in Fig.8.1

The computation of the proposed algorithm allows each player explicitly to inquire about his/her pay-off. As a result, each player is able to compare the true value and the estimated value of his/her pay-off, though the estimated values were previously predicted by the pay-off's approximation. That is, the probabilistic model, alternatively called the user model, which was obtained by machine learning, can calculate the estimated value of the pay-off. Importantly, a mutual expected error can be partially calculated from the estimated and true value of the pay-off. If the mutual error is greater than the given threshold, the interaction between the two players is reiterated. In practice, the IPD game constrains allowing players to be reciprocated by minimizing the error of the mutual expectation.

Figure .8.2 describes a prospective computation of our proposed algorithm. In the figure, the three main steps are represented by the initialization of the probabilistic models, the middle stage of our dialogue strategy for estimating pay-offs in others and the final stage to minimize the distance between an approximated and a true pay-off given by δ-function.

Next, we will elaborate two computations of our proposed algorithm because our dialogue strategy assumes that generative models could be suitable to approximate pay-off functions in others. Moreover, we assume that another major role of the dialogue strategy is to make the models precise and thus the probabilistic property would also be compatible with this assumption.

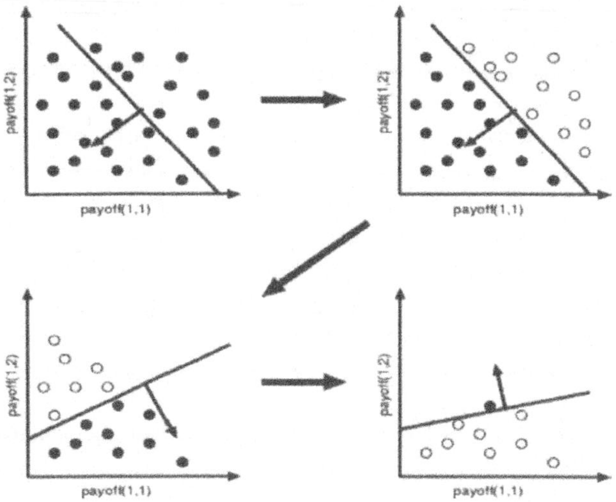

Fig. 8.3. Computational schema of proposed algorithm. An optimal hyperplane is used to distinguish possible dialogue actions in terms of maximizing mutual expectation. The hyperplane, which is built by its normal vector, is used for specifying possible (dialogue) actions at the end. O denotes irrelevant dialogue actions, whereas • denotes relevant dialogue actions to maximize mutual expectation

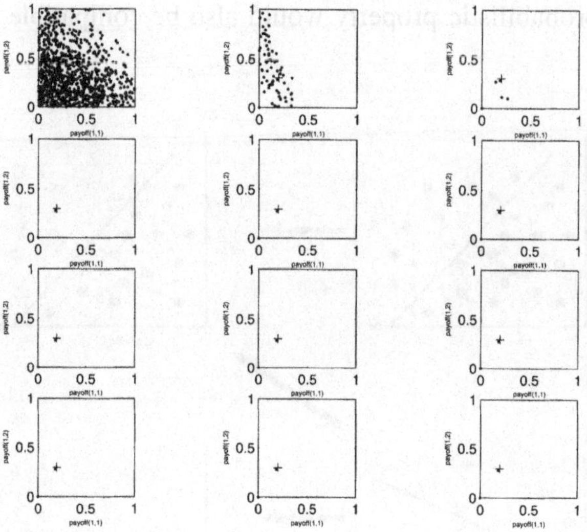

Fig. 8.4. A simulation result based on our proposed dialogue scheme. All plotted data was normalized

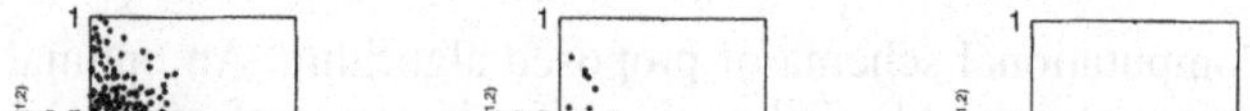

Fig. 8.5. A simulation result based on our proposed dialogue scheme with *type 1*. All plotted data was normalized. The initial variance is relatively smaller that of Fig.4, before the dialogue strategy (type1) is undertaken

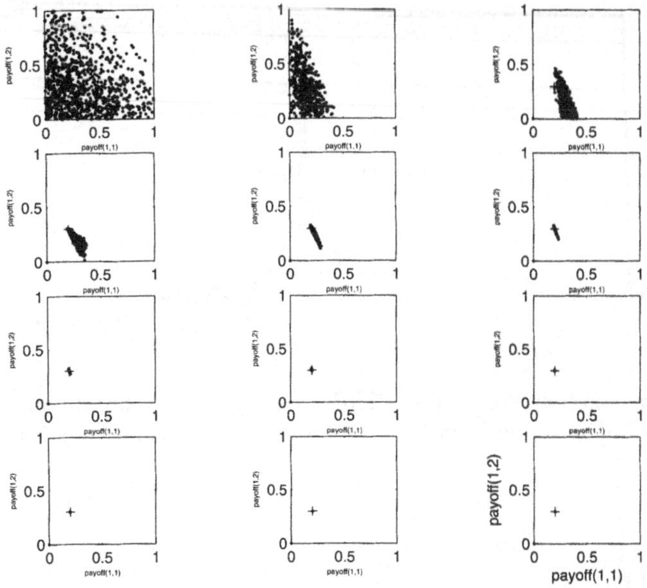

Fig. 8.6. A simulation result based on our proposed dialogue scheme with *type* 2

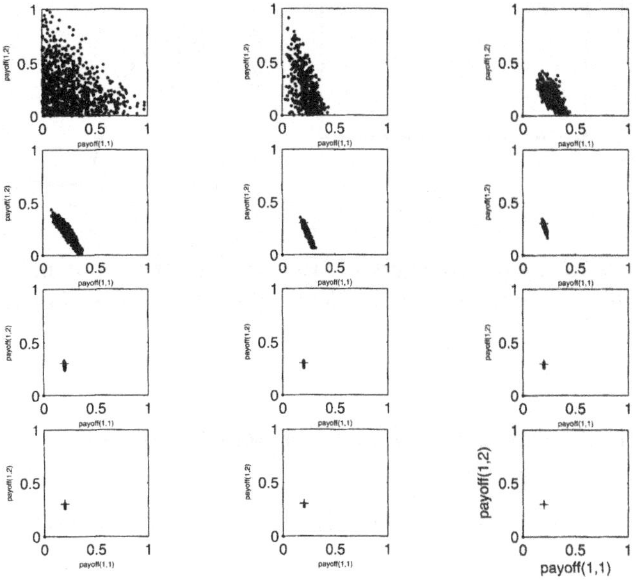

Fig. 8.7. A simulation result based on our proposed dialogue scheme with *type* 2. All plotted data was normalized. The initial variance is relatively smaller that of Fig.6, before the dialogue strategy (type2) is undertaken

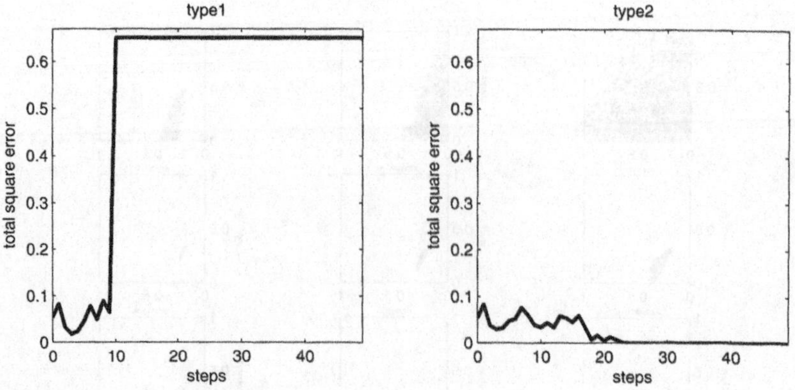

Fig. 8.8. The total squared errors (TSEs) for pay-off's approximation are calculated with dialogue actions type1-2. The results were calculated by Eq.16

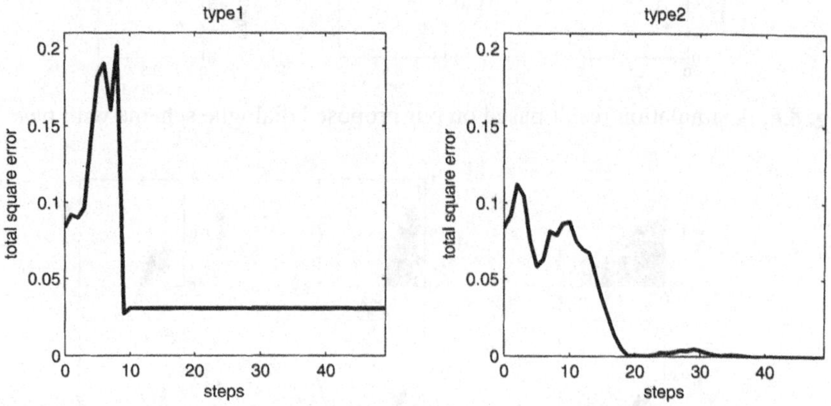

Fig. 8.9. The total squared errors (TSEs) for pay-off's approximation are calculated with dialogue actions type1-2. The results were calculated by Eq.17

8.5.1 Proposed Dialogue Strategy (Type 1)

Assuming that each action set Π_1, Π_2 of players is described as follows,

$$\begin{aligned} \Pi_1 &= \{k \mid k = 1,2,\cdots,m\} \\ \Pi_2 &= \{l \mid l = 1,2,\cdots,n\} \end{aligned} \quad (1)$$

Now, let p and q are the frequency of dialogue actions taken by player1 and player2 respectively. That is, p_i and q_j denotes the frequency at which player 1 chooses i-th dialogue action, whereas player 2 chooses j-th dialogue action.

$$p = (p_1 \cdots p_i \cdots p_m) \quad i \in \Pi_1$$
$$q = (q_1 \cdots q_j \cdots q_n) \quad j \in \Pi_2 \tag{2}$$

In addition, m and n denote the number of dialogue actions. Thus, the pay-off matrix is defined as follows:

$$M = \begin{pmatrix} a_{1,1} & \cdots & a_{1,n} \\ \vdots & \ddots & \vdots \\ a_{m,1} & \cdots & a_{m,n} \end{pmatrix} \tag{3}$$

where $a_{i,j}$ corresponds to the pay-off's value of both players. By approximating the pay-off matrix (Eq.(3)), a player will be able to predict future dialogue strategies of the other player. If player1 and player2 select the i-th and j-th actions, respectively, player1 obtains the profit of $a_{i,j}$.

So far, we have described the conceptual computation of our proposed algorithm. However, the algorithm must be simplified, in order to implement it.

The expected value of the pay-off is given by the following equation,

$$E(p,q) = pMq^T \tag{4}$$

The player's strategy is also obtained based on the statistical 'frequency' of possible dialogue actions, in which players have their dialogue interactions during the IPD game.

$$\hat{E}_{mutual}(p,q) = pAq^T + q\hat{B}p^T \tag{5}$$

where $\hat{E}_{mutual}$ corresponds to the summation between the actual pay-off matrix of player1 A and the estimated pay-off matrix of player2 $\hat{B}$. Note that the symbol ^ means simply that certain variables or vectors are estimated.

Let $\hat{b}_1, \hat{b}_2, \hat{b}_3, \hat{b}_4$ be the estimated components of player2's pay-off matrix $\hat{B}$. Then, this matrix is written as follow:

$$\hat{B}^i = \begin{pmatrix} \hat{b}_1^i & \hat{b}_2^i \\ \hat{b}_3^i & \hat{b}_4^i \end{pmatrix} \tag{6}$$

$$E^{\hat{B}}(q,p) = q\hat{\bar{B}}p^T = \frac{1}{N}\sum_{i=1}^{N} q\hat{B}^i p^T \tag{7}$$

where, $E^{\hat{B}}$ is identical to the right term of the expectation shown in Eq.(7). N denotes the number of possible dialogue strategies, which are undertaken in the pay-off matrix. The symbol $\bar{}$ represents the mean of the pay-off matrix $\hat{B}^i$.

It is assumed that the dialogue undertaken by player1 want to satisfy the mathematical relationship, defined by player2's expected payoff matrix $\hat{\bar{B}}$. In a sense, the estimated player2's payoff matrix can be the criterion set before player1's dialogue is undertaken.

Behavioral selection is made by our dialogue strategy based on the criterion given by Eq.(8).

In principle, the dialogue strategies that are taken by player1 are selectively obtained based on the criterion in Eq.(8).

Obviously, each player chooses the two dialogue types (namely, type 1 and type 2) of the pay-off matrix B for player 2. Importantly, the dialogue strategy must also be undertaken in terms of maximizing each pay-off. In order to compute the proposed dialogue strategy, we provide the following equation that player1 uses to determine specific strategies. These strategies can be selected with respect to maximizing the mutual expectation in Eq. (5).

Now, we assume that the estimated pay-off matrix $\hat{\bar{B}}$ is represented by $\hat{\bar{B}}^{\dagger\zeta}$, when † denotes $\hat{\bar{B}}$ with a maximized mutual expectation in Eq.(5).

By being given a dialogue strategy, the possible consequence of mutual expectation can result in either $\hat{E}^{\hat{\bar{B}}^{\dagger\delta}}_{mutual}(q,p) \geq \hat{E}_{mutual}(q,p)$ or $\hat{E}^{\hat{\bar{B}}^{\dagger\delta}}_{mutual}(q,p) \leq \hat{E}_{mutual}(q,p)$.

In other words, player2's estimated pay-off matrix $\hat{\bar{B}}$ can be updated by selecting a new possible dialogue action, which will be able to satisfy the equation's condition shown in Eq.(8).

Figure. 8.3 shows that an optimal hyperplane was used to distinguish possible actions either relevant or irrelevant in terms of maximizing mutual expectation in Eq.(5).

We defined the hyperplane that was built by its normal vector $\left(\frac{1}{2},\frac{1}{2},\frac{1}{2},\frac{1}{2}\right)$. Applying the hyperplane is to determine a specific dialogue strategy approximating pay-offs in others. The figure describes a number of interactions between players that can be effective to obtain an appropriate action to estimate pay-offs of player2. It must be noticed that dialogue strategies take place by inferring the hyperplane.

More specifically, we will now explain the computational aspect corresponding to the issue described in Fig. 8.3.

Our computation enables a predictive hyperplane to be manifested by an inequality. The proposed dialogue strategy forces possible actions to satisfy the following inequality,

$$E^{B}(q,p) > E^{\hat{\bar{B}}}(q,p) \tag{8}$$

Therefore, Eq.(8) can be described as follows:

$$qBp^{T} > q\hat{\bar{B}}p^{T} \tag{9}$$

and, Eq.(9) can be written as follows:

$$x_1b_1 + x_2b_2 + x_3b_3 + x_4b_4 > E^{\hat{\bar{B}}}(q,p) \tag{10}$$

where, x_1, x_2, x_3 and x_4 denotes possible dialogue strategies of each player.

Here, we assume the following mathematical relationships:

$$qp = x_1, q(1-p) = x_2, (1-q)p = x_3, (1-q)(1-p) = x_4 \tag{11}$$

Hence, Eq.(10) becomes

$$x_1b_1 + x_2b_2 + x_3b_3 + x_4b_4 > x_1\hat{\bar{b}}_1^{i} + x_2\hat{\bar{b}}_2^{i} + x_3\hat{\bar{b}}_3^{i} + x_4\hat{\bar{b}}_4^{i} \tag{12}$$

$E^{B}(q,p)$ is written as follow:

$$E^{B}(q,p) = (q,1-q)\begin{pmatrix} b_1 & b_2 \\ b_3 & b_4 \end{pmatrix}\begin{pmatrix} p \\ 1-p \end{pmatrix} \tag{13}$$

Thus, we obtain the following mathematical relationships.

$$\left.\begin{array}{ll} q\hat{b}_1 + (1-q)\hat{\bar{b}}_3 > q\hat{\bar{b}}_2 + (1-q)\hat{\bar{b}}_4 & then \quad p=1 \\ q\hat{\bar{b}}_2 + (1-q)\hat{\bar{b}}_4 > q\hat{\bar{b}}_1 + (1-q)\hat{\bar{b}}_3 & then \quad p=0 \end{array}\right\} \tag{14}$$

The predictive dialogue action taken by player1, which maximizes the expectation of player2's pay-off, can be obtained by Eq.(14).

In this way, the expectation $E^A(p,q)$, determines the predictive dialogue action taken by player2.

Fig. 4-5 describes the computational aspect, resulting from several simulations considering the proposed dialogue strategy. In those figures, each graph particularly represents (1,1) and (1,2) component in the matrix, which projects to a Cartesian coordinated space. More precisely, a true pay-off matrix B was constituted by (0.2,0.45,0,0.35). Initially, each possible dialogue action belonging to an estimated pay-off matrix $\hat{\bar{B}}$ was generated by a normal distribution function, with a mean (gravitation) vector of (0.25,0.25,0.25,0,25). Determining possible dialogue actions, capable of giving well defined estimations of the true pay-off matrix B, was done with our proposed computation schematically given by Figures 8.4-8.7. By increasing the number of dialogue interactions, as shown upper left to lower right, the variance of estimated components can be gradually shrank. + also denotes the true value of the (1,1) and (1,2) component.

In addition, 'type 1' implicates when players choose their dialogue actions determinately, subject to the approximation of true pay-off matrix.

8.5.2 The Improved Dialogue Strategy (Type 2)

However, when we have dialogue actions in daily life it is hard to perceive the other's pay-off without uncertainty. The uncertainty causes 'dead reckoning', which makes it difficult to approximate a true pay-off matrix in others. This is a major drawback of the proposed dialogue strategy described above. Thus, the proposed dialogue strategy considers a pay-off matrix, which is modeled by a probability distribution (density) function, in order to deal with the uncertainty of a true pay-off in others.

The following equation allows our dialogue strategy to take into account the uncertainty, which could be modeled by a gaussian noise N(0, σ). The noise N(0,σ) represents a *zero* mean with a variance σ.

Therefore, the prospective dialogue strategy taken by player1 is computed by taking into account the following probabilistic noise as follows:

$$\hat{b}^i = \hat{b}^k + N(0,\delta) \tag{15}$$

where, $\hat{b}^i$ can be obtained by shifting the original pay-off matrix in Eq.(6). We suggest that the addictive noise effect to the pay-off may play a crucial role in the stability of dialogue strategies, and prevents in particular from having to use dead reckoning. This could also be reminiscent of a regularization effect in machine learning theory. In practice, the regularization has been applied in wide range of sensory technology. In our case, the proposed dialogue strategy incorporating Eq.(15) in our dialogue strategy is capable of having real world competence. This may be true for intelligent sensory technology, for instance, proposed by (Koshizen, 2002). Such a technology learns cross-correlation among different sensors - selecting sensors that can be the best ones for minimizing a predictive localization error of a robot, by modeling the uncertainty (Koshizen, 2001). That means probabilistic models, computed by sonar and infrared sensors, were employed to estimate each robot's location.

Figures 8.6–8.7 describes the computational aspect, resulting from several simulations considering the proposed dialogue strategy 'type 2' implicates when player choose their dialogue actions statistically, subject to the approximation of true pay-off matrix.

Altogether, the players interacted 5000 times. Interactions were made of 100 sets, and each set consisted of 50 steps. The initial value of possible numbers of the pay-off matrix was 1000 points. All components of the pay-off matrix were *normalized.* The plotted points represent dialogue actions, which were taken by player1 during their interactions. The rule of their interactions was assumed to follow the context of the modified IPD game. As a result, the actual pay-off matrix of player2 was cooperative, so a pay-off matrix was approximated by inhibiting anticooperative actions during the interactions.

Figures 8.8–8.9 illustrates the Total Squared Error (TSE), which corresponds to the Euclid distance between a true pay-off matrix and an approximated pay-off matrix. The TSE was calculated during the IPD game between the two players. In our computation, the TSE is given by the following equation.

$$TSE = \sum_{i=1}^{4} (b^i - \hat{\bar{b}}^i)^2 \tag{16}$$

Furthermore, let us penalize the TSE shown in Eq.(16). That is,

$$TSE^* = \sum_{i=1}^{4}(b^i - \hat{b}^i)^2 + \lambda\Omega(f) \tag{17}$$

where, $\Omega(f)$ denotes the smoothness function, normally called *regularization term* or *regularizer* (Tikonov, 1963). In machine learning theory, it is known that the *regularizer* $\Omega(f)$ represents the complexity term. It can also express either the smoothness of the approximated function f given by •. The regularizer has been applied into real world application (Vauhkonen, 1998)(Koshizen and Rosseel, 2001).

A crucial difference when Fig.8 compares Fig.9 is, the size of variance before the dialogue strategies (type1 or/and type2) are undertaken.

Furthermore, the second term of Eq.(17) corresponds to the smoothness of (probabilistic) generative models, which are obtained by a learning scheme. The models can be used for selective purposes in order to acquire the *unique* model that fits the 'true' density function best. Therefore, the result from the learning scheme can be further minimized. Generally, the process is called *model selection*. Theoretically, results brought by Eq.(17) are closely related to Eq.(16)(*ex.*, Hamza, 2002). In our case, TSE^* is not calculated by the regularizer $\Omega(f)$ explicitly, though it is implicitly brought by the actual calculation according to Eq.(17).

We attempt to enhance our proposed computation by taking into account Eq.(17). That is, the dialogue strategy must be able to reduce the uncertainty of other's pay-offs. In practice, players inquire about their own pay-offs explicitly. The inquiry reducing the uncertainty of player2's pay-off $\hat{B}$ corresponds to $\Omega(f)$, which considers the past experiences of their inquiries. Additionally, λ may provide self-control of interactive dialogue to the inquiries. In fact, many inquiries would sometimes be regarded as troublesome to others. Therefore, self-control will be needed.

During dialogue interaction, the inquiry by each player is performed in our dialogue strategy, in order to reduce the uncertainty of a true pay-off of player2. The uncertainty is also modeled by probabilistic density functions. Thus, we expect that our proposed computation shown in Eq.(17) is capable of providing a minimization better than *TSE* in Eq.(15), *i.e.*, $TSE^* < TSE$. Fig. 4 to 9 represents several computational results, which were obtained by the original model and the improved model. The biggest difference between the original and the improved models was that the approximatedpay-offfunction*f*involved a probabilistic property. It certainly affects the dialogue strategy, which is capable of making generative models smooth by reducing the uncertainty. In order to be effective, a long-lasting interaction between the two players must be ensured, as described before. The probabilistic property can cope with fluctuations of a pay-off in others. This can often resolve a problem where there is no longer a unique best strategy such as in the IPD game. The initial variances created by each action, are relatively large (Figs. 8.4 and 8.6) whereas in Fig. 8.5 and 8.7 they are *smaller*. In these figures, + denotes a true pay-off's value and (•) denotes an approximated value calculated by a pay-off function. Since Figs. 8.6 and 8.7 were obtained by computational results from the probabilistic pay-off function, the approximated values could be close to the true value. The inverse can also be true as shown in Fig.8.4 and 8.5. Additionally, Figs .8.8 and 8.9 illustrate TSE for each case represented in Figs 8.4, 8.5 and Figs 8.6, 8.7. The final TSE is 0.650999 (Fig.8.8: left), 0.011754 (Fig. 8.8: right), 0.0000161 (Fig. 8.9: left) and 0.000141 (Fig. 8.9: right) respectively. From all the results, one can see that the computation involving a probabilistic pay-off function showed better performances with respect to TSE because it avoided the dead-reckoning problem across the pay-off in others.

Pattern Regression	User Modeling
Pattern Classification	User Classification
Discriminant Function	Pay-off Function
Mean Squared Error for Discriminant Function Approximation (Standard Expectation Maximization)	Total Squared Error for Pay-off Function Approximation (Mutual Expectation Maximization)
Regularization to parameterize Degree of generalization	Regularization to parameterize Degree of Satisfaction

Fig. 8.10. Analogous correspondences between pattern regression and user modeling

Figure. 8.10 shows analogous correspondences between pattern regression and user modeling. From the figure, we can clearly see a lot of structural similarities for each element such as classification, functional approxima-tion, and regularization. We can also see cross-correlations between pattern regression and user modeling.

8.6 Conclusion and Future Work

In this paper, we theoretically presented a computational method for user modeling (UM) that can be used for estimating pay-offs of a user. Our proposed system allows a pay-off matrix in others to be approximated based on inductive game theory. This means that behavioral examples in others need to be employed with the pay-off approximation. The inductive game theory involves social cooperative issues, which take into account a dialogue strategy in terms of maximizing the pay-off function. We reminded that the dialogue strategy had to bring into play long-lasting interactions with each other, so the approximating pay-off matrix could be used for estimating pay-offs in others. This makes a substructure of inducing social cooperative behavior, which leads to the maximum reciprocal expectation thereof.

In our work, we provided a computation model of the social cooperation mechanism, using inductive game theory. In the theory, predictive dialogue strategies were assumed by implementation based on behavioral decisions taken by others. Additionally, induction is taken as a general principle for the cognitive process of each individual.

The first simulation was carried out using the IPD game to minimize a total squared error (TSE), which was calculated by both a true and an approximated pay-off matrix. It is noted that minimizing the TSE can essentially be identical to maximizing expectation of a pay-off matrix. In the simulation, inquiring about pay-offs in others was considered as a computational aspect of the dialogue strategy. Then, the second simulation, in which a pay-off matrix can be approximated by a probability distribution function (PDF), was undertaken. Since we assumed that the pay-off matrix could fluctuate over time, a probabilistic form of pay-off matrix would be suitable to deal with the uncertainty. Consequently, the result, obtained by the second simulation (Section 8.5.2) provided better performances because of escaping from the dead-reckoning problem of a pay-off matrix.

Moreover, we also pointed out the significance to introduce how the probabilistic pay-off function could cope with a real world competence when the behavioral analysis was used to model pay-offs in others. In principle, the behavioral analysis can be calculated by sensory technology based on vision and audition. Additionally, there would be no sense the pay-off matrix to change in daily communication. This means that sensing technology has to come up with a way to reduce the uncertainty of someone's pay-offs. Consequently, this could lead to approximating pay-off distribution function accurately. Furthermore, in the second simulation, we pointed out that the proposed dialogue strategy could play a role in refining the estimated pay-off function. This is reminiscent of model selection problem in machine learning theory. In the theory, (probabilistic) generative models are selectively eliminated in terms of generalization performance. Our dialogue strategy brings into play the on-line maintenance of user models. That is, the dialogue strategy, leads to a long-lasting interaction, which allowed user models to be selected, in terms of approximating a true pay-off's density function. More specifically, the dialogue strategy would allow inquiry to reduce the uncertainty of pay-offs in others. The timing and content quality when inquiring to others, should also be noted as being a human-like dialogue strategy involving cognitive capabilities. Our study has shown that inductive game theory effectively provides a theoretical motivation to elicit the proposed dialogue strategy, which is feasible with maximum mutual expectation and uncertainty reduction. Nevertheless, substantial studies will still require establishing our algorithm by considering with the inductive game theory.

Another future extension of this work could be applied to our proposed computation with humanoid robot applications, allowing humanoid robots to be able to carry out reciprocal interactions with humans. Our computation of UM suggests that users who resist the temptation to defect for short-term gain and instead persist in mutual cooperation between robots and humans. A long-lasting interaction will thus require other's pay-off's estimations. Importantly, the long-lasting interaction could also be used to evaluate how much the robots gain the satisfaction of humans. We are convinced that this could be one of the most faithful aspects particularly when humanoid robots are considered with man-machine interactions. Consequently, our work provides a new scheme of man-machine interaction, which is computed by maximizing a mutual expectation of pay-off functions in others.

References

1 Newell, A. Unified theories of cognition, Cambridge, MA:Harvard University Press, 1983.
2 Newell, A. Unified theories of cognition, Cambridge, MA:Harvard University Press, 1983.
3 Fischer, G. User modeling in Human-Computer Interaction. *User modeling and User-Adapted Interaction*, 11:65--86, 2001.
4 Basu, C., Hirsh, H. and Cohen, W. Recommendation as Classification: Using and Content-Based Information in Recommendation. In: AAAI98-Proceedings of the Fifteenth National Conference on Artificial Intelligence, Madison, Wisconsin, 714-720, 1998.
5 Gervasio, M., Iba, W. and Langley, P. Learning to Predict User Operations for Adaptive Scheduling. In : AAAi98-Proceedings of the Fifteenth National Conference on Artificial Intelligence, Madison, Wisconsin, 721-726, 2001.
6 Nash, J.F., *Annals of Mathematics*, 54:286-295, 1951.
7 Kaneko, M and Matsui, A., Inductive Game Theory: Discrimination and Prejudices. *Journal of Public Economic Theory*, Blackwell Publishers, Inc., 1(1):101-137, 1999.
8 Axelrod, R. and Hamilton, W.D., The evolution of cooperation. Science, 211, 1390-1396, 1981.
9 Axelrod, R.M., The Evolution of Cooperation, New York: Basic Books, 1984.
10 Boyd, R., Is the repeated Prisoner's dilemma a good model of reciprocal altruism ?. Ethol. Sociobiol, Vol.9, 211-222, 1988.
11 Nesse, R.M., Evolutionary Explanations of Emotions, Human Nature, 1, 261-289, 1990.
12 Trivers R., Social Evolution, Menlo Part, CA: Cummings, 1985.
13 Webb, G.I., Pazzani, M.J. and Billsus, D., Machine Learning for User Modeling, User Modeling and User-Adapted Interaction, 11(1-2), 19-29.
14 Valiant L.G., A Theory of The Learnable Communications of the ACM, 27, 1134-1142, 1984.
15 Debreu, G., A Continuity properties of paretian utility. *International Economic Review*, Vol.5, pp.285-293, 1964.
16 Baker, F. and Rachlin, H. Probability of reciprocation in repeated prisoner's dilemma games. Journal of Behavioral Decision Making, 14, 51-67, John Wiley & Sons, Ltd.

17 Andre, E. Rist, T., Mueller, J., Integrating Reactive and Scripted Behaviors in a Life-Llike Presentation Agent. Proc. Int. Conf. Autonomous Agents, 261-268.
18 Noma, T., Zhao, L. and Badler, N.I. Design of a Virtual Human Presenter, IEEE Computer Graphics and Applications, 20(4), 79-85, 2000.
19 Legerstee, M., Barna, J. and DiAdamo, C., Precursors to the development at intention of 6 months: Understanding people and their actions, Developmental Psychology, 36(5), 627-634.
20 Lieberman, H., Letizia:An Agent that Assists Web Browsing. In:IJCAI95-Proceedings of the Fourteenth International Joint Conference on Artificial Intelligence, Montreal, Canada 924-929.
21 Dempster, A.P., Laird, N.M. and Rubin, D.B., Maximum Likelihood from Incomplete Data via the EM Algorithm, J.Roy.Statist.Soc.Ser, B39, 1-38, 1977.
22 Koshizen,T., Improved Sensor Selection Technique by Integrating, J. of Intel. and Robo. Syst., 2001.
23 Koshizen, T., Yamada, S. and Tsujino, H., Semantic Rewiring Mechanism of Neural Cross-supramodal Integration based on Spatial and Temporal Properties of Attention. Neurocomputing, 52-54, 643-648, 2003.
24 Tikhonov, A.N., On Solving ill-posed problem and method of regularization. Doklady Akademii Nawk USSR, 153, 501-504, 1963.
25 Vauhkonen, M., Vadasz, D. and Kaipio, J.P., Tiknonov Regularization and Prior Information in Electrical Impedance Tomography. IEEE Transaction on Madical Imaging, 17, 2, 285-293,2001.
26 Koshizen, T. and Rosseel, Y. A New EM Algorithm using the Tikonov Regularizer and the GMB-REM Robot's Position Estimation System. Int. J. of Knowle. Intel. Engi. Syst., 5, 2-14, 2001.
27 Hamza, A.B., Krim, H. and Unal G.B., Unifying probabilistic and variational estimation. IEEE Signal Processing Magazine, 37-47,2002.

The manufacturer's authorised representative in the EU is Springer Nature Customer Service Centre GmbH, Europaplatz 3, 69115 Heidelberg, Germany. If you have any concerns regarding our products, please contact ProductSafety@springernature.com

Printed and bound by CPI Group (UK) Ltd, Croydon, CR0 4YY
30/09/2026
02225573-0010